Hearing with the Mind

OXFORD STUDIES IN MUSIC THEORY

Series Editor
Steven Rings

Studies in Music with Text,
David Lewin

Audacious Euphony: Chromatic Harmony and the Triad's Second Nature,
Richard Cohn

Music as Discourse: Semiotic Adventures in Romantic Music,
Kofi Agawu

Beating Time and Measuring Music in the Early Modern Era,
Roger Mathew Grant

Mahler's Symphonic Sonatas, Seth Monahan

Pieces of Tradition: An Analysis of Contemporary Tonal Music,
Daniel Harrison

Music at Hand: Instruments, Bodies, and Cognition,
Jonathan De Souza

Foundations of Musical Grammar,
Lawrence M. Zbikowski

Organized Time: Rhythm, Tonality, and Form,
Jason Yust

Flow: The Rhythmic Voice in Rap Music,
Mitchell Ohriner

Performing Knowledge: Twentieth-Century Music in Analysis and Performance,
Daphne Leong

Enacting Musical Time: The Bodily Experience of New Music,
Mariusz Kozak

Hearing Homophony: Tonal Expectation at the Turn of the Seventeenth Century,
Megan Kaes Long

Form as Harmony in Rock Music,
Drew Nobile

Desire in Chromatic Harmony: A Psychodynamic Exploration of Fin de Siècle
Tonality, Kenneth M. Smith

A Blaze of Light in Every Word: Analyzing the Popular Singing Voice,
Victoria Malawey

*Sweet Thing: The History and Musical Structure of a Shared American Vernacular
Form,* Nicholas Stoia

*Hypermetric Manipulations in Haydn and Mozart:
Chamber Music for Strings, 1787–1791,*
Danuta Mirka

How Sonata Forms: A Bottom-Up Approach to Musical Form,
Yoel Greenberg

Exploring Musical Spaces: A Synthesis of Mathematical Approaches,
Julian Hook

The Musical Language of Italian Opera, 1813– 1859,
William Rothstein

Times A-Changin': Flexible Meter as Self-Expression in Singer-Songwriter Music,
Nancy Murphy

Swinglines: Rhythm, Timing, and Polymeter in Musical Phrasing,
Fernando Benadon

Tonality: An Owner's Manual,
Dmitri Tymoczko

Sounds as They Are: The Unwritten Music in Classical Recordings,
Richard Beaudoin

Theorizing Music Evolution: Darwin, Spencer, and the Limits of the Human,
Miriam Piilonen

Swinglines: Rhythm, Timing, and Polymeter in Musical Phrasing,
Fernando Benadon

Embodied Expression in Popular Music: A Theory of Musical Gesture and Agency,
Timothy Koozin

Seeing Voices: Analyzing Sign Language Music,
Anabel Maler

Hearing with the Mind: Proto-Cognitive Music Theory in the Scottish Enlightenment,
Carmel Raz

Hearing with the Mind

*Proto-Cognitive Music Theory in the
Scottish Enlightenment*

CARMEL RAZ

Oxford University Press is a department of the University of Oxford.
It furthers the University's objective of excellence in research, scholarship,
and education by publishing worldwide. Oxford is a registered trade mark of
Oxford University Press in the UK and in certain other countries.

Published in the United States of America by Oxford University Press
198 Madison Avenue, New York, NY 10016, United States of America.

© Oxford University Press 2025

This is an open access publication, available online and distributed under the terms of a Creative
Commons Attribution-Non Commercial-No Derivatives 4.0 International licence (CC BY-NC-ND 4.0),
a copy of which is available at https://creativecommons.org/licenses/by-nc-nd/4.0/.
Subject to this license, all rights are reserved.

Inquiries concerning reproduction outside the scope of the above should be sent
to the Rights Department, Oxford University Press, at the address above.

You must not circulate this work in any other form and you must impose this
same condition on any acquirer

CIP data is on file at the Library of Congress

ISBN 9780197786178

DOI: 10.1093/9780197786208.001.0001

Printed by Integrated Books International, United States of America

to the memory of my father, Tzvi Raz

Contents

Acknowledgments	xi
List of Figures	xv
Introduction	1
1. Potter, Musician, Merchant, Scribe: The Many Lives of John Holden (1729–1772)	18
2. An Eighteenth-Century Theory of Musical Cognition? John Holden's *Essay towards a Rational System of Music* (1770)	42
3. "Our Nurses' Tunes": John Holden and Scottish Psalmody	92
4. To "Fill Up, Completely, the Whole Capacity of the Mind": Listening with Attention in Late Eighteenth-Century Scotland	125
5. Rhythm as a Universal "Science of Man": Walter Young's "Essay on Rhythmical Measures" (1790)	156
6. What's in a Game? Rediscovering the Music Theory of Anne Young	201
Afterword	249
Appendix	257
Bibliography	259
Index	277

Acknowledgments

I first encountered John Holden's *Essay towards a Rational System of Music* (1770) while a graduate student writing a dissertation on the intertwined histories of music and the nerves. I was immediately fascinated by the treatise's emphasis on attention and its role in music perception. As the topic had little to do with my research at the time, I filed it away, although I enjoyed encouraging conversations about Holden with a number of my mentors at Yale University. I would like to thank Pat McCreless for encouraging my interest in Scottish music theory, Rick Cohn for his enthusiasm right from the start, and Brian Kane for many stimulating conversations on the topic of attention. I began working more seriously on Scottish music theory while at the Society of Fellows of Columbia University, and there I would like to thank my fellow fellows Ben Breen, Maggie Cao, David J. Gutkin, Chris Florio, María González Pendás, Heidi Hausse, Arden Hegele, Hidetaka Hirota, Whitney Laemmli, Max Mishler, Dan-el Padilla Peralta, Joelle Abi Rached, Rebecca Woods, and Grant Wythoff, as well as Eileen Gillooly, Emily Bloom, Elaine Sisman, Reinhold Martin, and Christopher L. Brown for all their support. Since taking up the position of group leader at the Max Planck Institute for Empirical Aesthetics in Frankfurt am Main, I have been extremely fortunate to enjoy the steadfast encouragement of the institute's directors Melanie Wald-Fuhrmann, Fredrik Ullén, and David Poeppel. My sincere thanks are also due to the librarians at the MPIEA, and in particular to Kerstin Schoof, Nancy Schön, Carl Then, and Stefan Strein. Also in Frankfurt, I am deeply indebted to Diana Gleiss for demolishing countless bureaucratic obstacles, to Maja Fricke for her administrative help, and to Felix Bernoully and Lorna Bittner for their expert work on the images in this book.

This book includes more developed versions of work presented in article form in the *Journal of Music Theory*, *Music Theory Spectrum*, and *SMT-V*; I am greatly indebted to the editors and staff of these journals, in particular to Poundie Burstein, Laura Emmery, Noah Kahrs, Julie Pedneault-Deslauriers, and Peter H. Smith, as well as to Roger Grant and the journals' anonymous peer reviewers, for many helpful suggestions. I would also like to thank Duke University Press for granting me permission to reproduce material that

xii ACKNOWLEDGMENTS

directly overlaps with the 2018 article in this book. Chapters of this book were presented as talks at the Max Planck Institute for Empirical Aesthetics and at Yale, Columbia, Glasgow, and Cornell universities; I am grateful to all audience members for the responses offered during these events, and in particular to Annelies Andries, Eamonn Bell, Joe Dubiel, Dan Harrison, Andrew Hicks, Alexandra Kieffer, Marius Kozak, Gundula Kreuzer, Ellen Lockhart, Nathan Martin, Nicholas Mathew, Roger Moseley, Roger Parker, Judith Peraino, Jessica Peritz, Ben Piekut, Ian Quinn, Annette Richards, Ellen Rosand, Braxton Shelley, Ben Steege, David Yearsley, and Anna Zayaruznaya for their insightful questions.

Many people generously shared their research and their sources with me over the years. I owe special thanks to George Kennaway for letting me read the manuscript of *John Gunn, Musician Scholar in Enlightenment Britain* before it went to press, to Jean Mackenzie of the Scottish Pottery Society for her help in tracking down various pottery and bell-ringing leads, to Margaret Doris for material on John Gunn's library, to Jeanice Brooks and the Duke of Buccleuch and Queensberry for archival findings pertaining to Anne Young, to Dean Cooke for the photo of the frontispiece of Margaret Bryan's *Compendious System of Astronomy*, and to Michael Foulds for sharing his photo of the inside wall of Glasgow Tolbooth Steeple. I would also like to thank Philip Woodworth and Lawrence Holden for their interest in the Holden family, and Ellie Hisama for encouraging my initial interest in Anne Young. Important insights were also provided by Will Deringer regarding the commercial background of Holden's mathematics, by Michael Kassler on British printing culture, by Danny K. S. Walden with regard to Holden's temperament, by Leonie Krempien with regard to nineteenth-century German university archives, and by Courtney Weiss Smith on Walter Young's theoretical approach to prosody. Benjamin Vernot and Elena Essel patiently guided me into the amazing world of DNA analysis in our shared study of Anne Young's *Musical Games*. Special thanks are due to Tom Service for including Anne Young and John Holden on his "serious fun" episode of "Music Matters." My heartfelt thanks also extend to Elias Mazzucco of the British Library for tracking down the image of Schultze's "Musical Balloon," and to Adrienne Rusinko of the Princeton University Library Special Collections for kindly digitizing a great many historical games. I owe a great debt to my colleagues in Scotland, and in particular Andrew Bull for his indefatigable research assistance, Brianna Robertson Kirkland for her excellent advice on all things Scottish, and David McGuiness and Lavinia Cooper for their musical skill and insight. My thanks also go to Joseph Reid, David Scott, and Graeme S. Millen for information about the archival materials of Glasgow Kilwinning Lodge No. 4. I am greatly indebted to the fabulous folks at Oxford University Press, and in particular to Norm Hirschy, Rada Radojicic, and Gopinath T. A., as well as to the series editor, Steve Rings, whose advice on conceptualizing Scottish music theory made this publication possible. Finally, I cannot thank Nir Cohen-Shalit enough for his meticulous work in proofreading the manuscript and creating the index.

I benefited greatly from the support of many dear friends while writing this book. I would like to extend my heartfelt thanks to the gimlet-eyed Caleb Mutch, who supplied many crucial intuitions about music theory as well as Presbyterian worship. Russell J. O'Rourke pushed me to write creatively and always to consider the stakes. Matthew Hall generously read the entire manuscript with his characteristic rigor and provided invaluable feedback. Lydia Goehr gave me excellent advice on eighteenth-century British philosophy and on writing in general, and Thomas Christensen provided warm encouragement and consistent inspiration. Lan Li lent her wonderful artistic skills to my initial study of Anne Young. Francesca Brittan, Elizabeth Lyon, and Marc Perlman kindly read chapters of the book and offered thoughtful and constructive critique. I am always grateful to Robert Trevino for his friendship and wisdom. And I would like to thank Yael Barolsky, Jamie Budnick, Leon Chisholm, Myke Cuthbert, Mi-Anh Duong, Lea Fink, Dan Hirschman, Sarah Godbehere, Yoel Greenberg, Elina Hamilton, Lester Zhuqing Hu, Racha Kirakosian, Pauline Larrouy-Maestri, Claudia Lehr, Céline Frigau Manning, Naila Makhani, John Muniz, Roya Pakzad, Lara Pearson, Steffi Probst, Shayna Silverstein, Nell Stanton, Michael Veal, and Alexander Wragge Morley for their support and friendship along the way.

There are, moreover, two scholars whose support was integral to the creation of this book: Jamie C. Kassler kindly shared many stimulating insights and ideas about British music theory over the past years with me and consistently provided valuable guidance on archival and intellectual matters. Our correspondence led to a friendship I treasure. This book perches on the shoulders of her two-volume magnum opus, *The Science of Music in Britain, 1714–1830*. My dear friend and colleague David E. Cohen read countless drafts of this manuscript, and in doing so generously shared his ideas and improved mine with his inimitable brilliance, originality, and insight.

xiv ACKNOWLEDGMENTS

This book would not have existed without him, and I owe him an enormous debt for helping me become a better scholar over the years.

Finally, I would like to thank my family: my mother Orna, my sister Mical and her family, my in-laws Rafi and Ruti, and my sister-in-law Nima and her family. Most of all, I would like to thank Nori and Uri for accepting John Holden into our family, and for their support, love, and inspiration.

Figures

Figure 1.1   The names listed on the wall facing the carillon within the Tolbooth Steeple. Photo courtesy of Historic Environment Scotland © Crown Copyright: HES. This image is not covered by the terms of the Creative Commons license of this publication. For permission to reuse, please contact the rights holder.   20

Figure 1.2   Isabella Holden's receipt. Glasgow, July 2, 1772: Received from Doctor Willson Ten Shillings owing by the College to John Holden for writing two Doctors Diplomas at five Shillings each. Isabella Holden. With permission of University of Glasgow Archives & Special Collections.   24

Figure 1.3   *Isabella Holden, nee Fawcett.* Miniature belonging to Jane Isabella Holden, John Holden's granddaughter. Image courtesy of Rotherham Museums, Arts and Heritage (Reference no. ROTMG: OP.1990.8).   37

Figure 1.4   Moorgate Hall, illustrated in Guest, *Yorkshire. Historic Notices of Rotherham,* 526. Image courtesy of Rotherham Museum, Arts and Heritage (Reference no. 942.741).   39

Figure 1.5   Richard Holden, *Review of Local Volunteers on Brinsworth Common, Rotherham, South Yorkshire 1804.* Image courtesy of Rotherham Museum, Arts and Heritage (Reference no. ROTMG:OP.1978.218).   39

Figure 2.1   An illustration of Holden's theory of visual grouping.   47

Figure 2.2   An illustration of Holden's analogy of the module as a measure and as a pitch. Left and center: the harmonics of a fundamental sound equivalent to the module represented as vibrations and as pitch; right: the same represented as a measure and equivalent rhythmic subdivisions.   54

Figure 2.3   Key and fifth divisions of the module.   58

Figure 2.4   Perceiving VII within the context of a scale means hearing it in relation to V.   61

Figure 2.5   We can compare hearing VII:V to the cognitive act of grouping that occurs when we perceive windows. If we see thirty windows, we view the five columns together as four groups (of six windows), with an extra six-window group included in the middle, analogous to Figure 2.1e.   62

xvi LIST OF FIGURES

Figure 2.6   Movement between scale degrees implies movement between fundamentals. The mind interprets the factor 5 in 3 x 2 x 5 (= 30) as a variant of the factor 4 in 3 x 2 x 4 (= 24), so that the bass g4 is projected at the major third below the b4.   62

Figure 2.7   Try moving from a division of a span into eight directly into a division of the same span into nine.   64

Figure 2.8a   The factorization of the descending scale, to be read as a directed graph starting from the left, following the arrows. The flawed option of taking II as the fundamental to IV is shown in brackets.   66

Figure 2.8b   The resulting descending scale. From top: ratios of the intervals between adjacent scale degrees; magnitudes of degrees relative to the module (key note) of 16; ratios of scale degrees to projected fundamentals; ratios between projected fundamentals.   66

Figure 2.9   Holden's ascending and descending scales. Holden, *Essay*, part 2, §44, 316.   68

Figure 2.10   Factorization and directed graph of the ascending scale (the *double emploi* at scale degree VI is indicated in dashed lines).   68

Figure 2.11   Holden's chart of the relationship of C to F and G major (Holden, *Essay*, part 2, §62, 338). The module of lines 1–4 is divided into 48; the module of lines 6–9 is divided into 64. The size of the interval between each scale degree is classified in one of five ways: the whole tone can be greater (*t. g.*), 9:8; lesser (*t. l.*), 10:9; or redundant (*t. r.*), or 8:7. The semitone can be proper (*s. p.*), 16:15; or deficient (*s. d.*), 21:20. These intervals arise from the derivation of scale degrees from the factors of 2, 3, 4, 5, and 7 and affect the size of larger intervals. In practice, Holden notes that the dominant seventh chords on the fifth scale degree of all three minor scales should be tempered to conform to the proportion of 4:5:6:7.   73

Figure 2.12   An illustration of the six different scales involved in the full chromatic gamut, with the temperings required for the minor mode. Roman numerals pertaining to minor scales are indicated by lower case letters. F major, D harmonic minor, and C major stand in proportion to a module of C taken as 48, while C major, A harmonic minor, G major, and E harmonic minor stand in proportion to a module of C taken as 64 (the raised leading tone G-sharp of A harmonic minor is shared with this module alone). Adjusted scale degrees in minor keys are indicated in rows 2, 7, and 12; all other pitches are shared with their relative major keys, and identical to the numbers in Figure 2.11. The fractions in these rows are an artifact of the low module values of this graphic representation (which range between 32 and 90 for the first module, and 48 and 120 for the second module), and reflect the aforementioned tempering of the minor triad and the dominant seventh chord built on V of minor.

	Members of the tonic triad of the minor keys are indicated in circles, and members of the dominant seventh chord of the minor keys are indicated in dashed squares.	75
Figure 2.13	Seventh chord with double fundamental. Holden, *Essay,* part 1, plate 4.	80
Figure 2.14	Implied difference tone generated by a minor third.	82
Figure 2.15	The implied sounds (shown in diamond note heads) of a perfect chord.	83
Figure 2.16	Inaudible implied sounds do not contribute to the consonance of dyads at the lower boundary of our hearing range.	84
Figure 2.17	A flowchart of Holden's theory, mapped onto the modern tripartite division of the cognitive faculties involved in audition, shown at the right. The gray shading illustrates the gradual boundaries between low-, mid-, and high-level processing.	88
Figure 3.1	Holden's setting of St. David's tune, with voice-leading infelicities indicated. Note that the melody is in the tenor. The diacritics indicate tuning adjustments and will be explained shortly. Holden, *Collection*, 4.	107
Figure 3.2	Ashworth's setting of the same tune (note the antiparallel octaves between the tenor and bass in measures 6 and 7 as well as measures 15 and 16). Caleb Ashworth, *A Collection of Tunes Suited to the Several Metres Commonly Used in Public Worship, with an Introduction to the Art of Singing and Plain Composition* (London: J. Buckland, 1762), 27.	108
Figure 3.3	St. David's tune as it appears in its original notation by Ravenscroft. Thomas Ravenscroft, *The Whole Booke of Psalmes: With The Humnes Evangelicall, and Songs Spiritual* (London, The Company of Stationers, 1621), 85.	111
Figure 3.4	St. David's tune as it appears in settings by Holden and by Ashworth. Holden, *Collection, 4*; Ashworth, *Collection, 27*.	111
Figure 3.5	My illustration of Holden's claim that we unconsciously regularize common-meter psalm tunes into phrases composed of four measures.	112
Figure 3.6	Compare (a), the "true" melody with a less intuitive fundamental bass with (b), which features the wrong melody with a more intuitive fundamental bass.	117
Figure 4.1	Some of Robertson's "Various Rhythms of Intervals." The Greek letter α marks the rhythms of the faster term; an ϑ marks the rhythms of the slower term. The Arabic numerals indicate the proportional size of the distances to their immediate left. Johnson Mus. d. 34, recto, Tenbury d. 36, p. 231, © Bodleian Libraries, CC BY 4.0 (http://creativecommons.org/licenses/by/4.0).	146
Figure 5.1	Stamitz, Symphony no. 6, mm. 1–5, violins. Throughout his "Essay," Walter merely cites works by name and bar number; I have included his musical examples in order to illustrate his ideas.	173

xviii   LIST OF FIGURES

Figure 5.2	Schematic of Walter's Account of the Development of "Rhythmical Measures."	182
Figure 5.3	Irregular bar structure in Stamitz's Symphony no. 11, fourth movement, mm. 1–47, first violin. Walter's interpretation is indicated with solid brackets; mine is shown in dashed brackets. The m. 32 upbeat is shown at the first asterisk; the m. 47 elision by two asterisks. (Note that a similar metrical structure occurs already at m. 24, which serves as an upbeat to m. 25, although this part of the work does not feature the changes of key, texture, and dynamic level that likely attracted Walter's attention to mm. 31–32.)	186
Figure 5.4	Mm. 1–17 of the last movement of Haydn's String Quartet op. 1, no. 1. The asterisk marks the "added" bar (m. 15).	187
Figure 5.5	Mm. 1–35 of Haydn's *Divertimento* in E-flat Major, Hob. II.6. The solid brackets indicate Walter's interpretation; dashed brackets show my interpretation. Note his "extra" bar at m. 23, indicated with an asterisk.	189
Figure 5.6	Violins 1 and 2, mm. 1–9 of the last movement of Haydn's *Divertimento* in E-flat Major, Hob. II:6. The solid brackets indicate Walter's interpretation; dashed brackets show my interpretation. Note his "extra" bar at the very beginning, indicated with an asterisk.	191
Figure 5.7	Haydn's String Quartet op. 2, no. 1 begins with two three-bar phrases.	192
Figure 5.8	Five-bar phrases in the last movement of Erskine's first overture op. 1, no. 1, mm. 66–75. The final sixteenth note (B) in m. 5 of the second violin is most likely a typographical error in the printed edition.	193
Figure 6.1	Anne Young's *Musical Games*.	202
Figure 6.2	*Captain Bosvill's March* by Lady Semphill. (Young, *Elements of Music*, 59).	207
Figure 6.3	*Miss Maxwell's Delight* by Lady Semphill. (Young, *Elements of Music*, 59).	208
Figure 6.4	*A Jigg Composed by a Young Lady of 7 Years of Age.* (Young, *Elements of Music*, 60).	209
Figure 6.5	Rowse, *A Grammatical Game in Rhyme, by a Lady* (1802). Image reproduced with permission of the Cotsen Children's Library (1292 Eng 18 / Board Games).	215
Figure 6.6	Rowse, *Mythological Amusement* (1804). Image reproduced with permission of the Cotsen Children's Library (34318 Eng 18 / Board Games).	216
Figure 6.7	Bryan, *Science in Sport, or the Pleasures of Astronomy.* Image reproduced with permission of the Cotsen Children's Library (40387 Eng 18 / Board Games).	217

LIST OF FIGURES xix

Figure 6.8	Mant, *The Study of the Heavens at Midnight*. Image reproduced with permission of the Cotsen Children's Library (17645-6 Board Games).	218
Figure 6.9	Anne's Chromatic Circle of Fifths. (Gunn (née Young), *Introduction*, facing page 64).	228
Figure 6.10	Burney's Circle of Fifths: the extreme sharp keys appear on the upper staff; an enharmonic shift is indicated in the seventh measure in the lower staff.	229
Figure 6.11	Crotch's examples 9 and 11, showing the accidentals involved in twenty-five major and minor keys.	229
Figure 6.12	Schultze, *Musical Balloon Containing 52 Different Keys*. Johnson Mus. d. 34, recto, © Bodleian Libraries, CC BY 4.0 (http://creativecommons.org/licenses/by/4.0).	230
Figure 6.13	Smith, *Musical Sacred Globe, Containing 50 Different Keys, & How to Transpose Them* (ca. 1814). From the British Library Archive (I.600.(15.)).	231
Figure 6.14	The face of the game board illustrated by the frontispiece of Anne's 1803 treatise. Note the extended circles of fifths along each margin.	233
Figure 6.15	Mm. 1–4 of Mozart's Sonata no. 12 in F Major, K. 332.	234
Figure 6.16	Gunn (née Young), *Introduction to Music* Plate XII, no. 3. The scale degree annotations are my own; the chord in question is likewise indicated by the addition of an arrow.	236
Figure 6.17	Frontispiece, Bryan, *A Compendious System of Astronomy* (1799). Photo courtesy of Dean Cooke.	244

Introduction

Our minds shape the nature of our musical experience. Indeed, there is an entire field, music psychology, that studies the ways in which our minds create music out of a stream of sounds. And yet the idea that our mental faculties are involved in processing music is a relatively recent one in Western intellectual history. Listeners from the Middle Ages through the seventeenth century considered music to be either "sounding number" or a physical stimulus of the affections, or in some way both at once—without, however, any real notice being taken of the active role played by the mind of the subject in the process of musical listening. To be sure, even prior to the advent of the modern view there were occasional inklings that faculties we would today regard as mental were implicated in music to various degrees; think, for instance, of St. Augustine's apparent assumption, in the *Confessions*, that our minds are able to measure and compare the relative durations of the syllables in metrical poetry.[1]

By and large, however, the West's Galenic heritage meant that music's effects were understood primarily as material, sounds being thought to impinge on us via our ears and correspondingly, and mechanically, alter our inner organs. According to this view, music was primarily a sensory, emotional, and ethical experience. Certain qualities, such as the sweetness of various intervals and their combinations, or the aptness of their rhythms, could be accounted for by appealing to the ideal simplicity of their integer ratios as in some way reflecting a profound cosmic order reflected in the human microcosm. When music was heard along with its accompanying text, such characteristics activated the listeners' emotions, "moving" the passions and causing material changes in their affections. The mind, at least as we understand it today, was bypassed, or merely brought in as an effect of that process.

The main subject of this book, a Glasgow-based merchant potter named John Holden (1729–1772), seems to have been the earliest musical thinker to theorize the perception of music as dependent on unconscious mental acts,

---

[1] See Augustine, *Confessions* 11.27.34–35.

Hearing with the Mind. Carmel Raz, Oxford University Press. © Oxford University Press 2025.
DOI: 10.1093/9780197786208.003.0001

2 HEARING WITH THE MIND

and to propose a detailed account of exactly how the human mind perceives music. The answer Holden offers to this very modern question in his *Essay towards a Rational Theory of Music* (1770) is striking in its non-modernity: he proposes a novel variation on the aforementioned theory, namely the notion that we are naturally moved and pleased by simple integer ratios. Yet Holden does not simply rehearse Pythagorean maxims about simplicity and perfection. Rather, he explicitly projects his speculative theory into our minds, hypothesizing that the act of perception itself entails an ongoing analysis of incoming sensory impressions—whether visual or auditory—into smaller groups of equally sized units limited in number to the small primes of 2 and 3 (and to a lesser extent, 5 and 7). "Where equal and equidistant objects affect our senses," he writes, "there is a certain propensity in our mind to be subdividing the larger numbers into smaller equal parcels; or as it may be justly called, compounding the larger numbers of several small factors, and conceiving the whole by means of its parts."[2] According to this view, our general preference for simple integer ratios is a property of cognition itself rather than an acoustic reality or a numerological prerequisite.

Holden's theory thus attempts to reveal the concrete mental process by which the mind accomplishes something very much like what Leibniz described when he wrote to the mathematician Christian Goldbach, "Music is an unobservable exercise of arithmetic by a soul unaware that it is counting."[3] Yet while Holden's theory depends on integer ratios, some of the oldest tools in the music theorist's toolkit, he applies them not merely to intervals but to tonal music in all its variety and complexity, proposing a dynamic theory capable of encompassing a broad range of the musical phenomena afforded by the music of his day, including tonality, harmonic functionality, and modulation.

To be sure, an early articulation of the idea that our experience of music is co-constructed by faculties which today we would ascribe to our minds can be found in Jean-Philippe Rameau's series of groundbreaking treatises published between 1722 and 1760. Rameau's notion of *le sous-entendu* enables him to refer to the listener's awareness of notes that are not physically present but are implicit in a given musical context. The doctrine of

---

[2] John Holden, *Essay towards a Rational System of Music* 3rd ed. (Edinburgh: Blackwood, 1807), part 2, §11, 288–289. All quotations in this book are taken from the third edition.

[3] Gottfried Wilhelm Leibniz, "Letter no. 6," Leibniz to Christian Goldbach, in "La correspondance de Leibniz avec Goldbach," ed. Adolf P. Juschkewitsch and Juri C. Kopelewitsch, *Studia Leibnitiana* 20 (1988): 182. I thank David E. Cohen for the translation.

le sous-entendu thus necessarily entails the existence of an acquired musical grammar that unconsciously contributes to the perception of music. However, as David E. Cohen demonstrates, Rameau locates the faculty by which this transpires not in the mind but rather in the *ear*. This is because for him, following Aristotle, and, to a certain extent, Descartes, mental acts are by definition conscious, if only in an attenuated way. Despite some striking similarities with subsequent psychological approaches, therefore, Rameau's account of music cognition remains vastly different from our present-day understanding. This is hardly surprising, since the philosophical and psychological ideas which would locate unconscious perception in the mind had not yet been fully realized.[4]

In his *Essay*, Holden provides a complete account of how we perceive music: not its emotional effects, but rather its intrinsic, intramusical relationships, the implicit, subconscious analysis that allows us to hear chord function or ascertain a given passage's key. In doing so, he invokes a number of familiar mental faculties, such as memory and attention, and also posits certain, more specialized, innate cognitive operations, such as the aforementioned activity of grouping, and describes how these work together in order to produce different experiences within us. This is a radically new account of the musical mind, one that would have no analogue in continental Europe until many decades later.

Holden's *Essay* was appreciated by many of the leading musical thinkers of his day, ranging from Charles Burney, John Wall Callcott, and Augustus Frederic Christopher Kollmann in London to Johann Nikolaus Forkel, Johann Georg Sulzer, and later, François-Joseph Fétis on the Continent.[5]

---

[4] See David E. Cohen, "The 'Gift of Nature': Musical 'Instinct' and Musical Cognition in Rameau," in *Music Theory and Natural Order from the Renaissance to the Early Twentieth Century*, ed. Suzannah Clark and Alexander Rehding (Cambridge: Cambridge University Press, 2001), 86–127.

[5] Charles Burney praised Holden as "no servile follower of any preceding writer; his precepts seem to arise from experience and reflection." Charles Burney, "Holden, John" in *The Cyclopaedia: Or, Universal Dictionary of Arts, Sciences, and Literature*, ed. Abraham Rees (London: Longman, Hurst, Rees, Orme, & Brown, 1811), 18:unpaginated; John Wall Callcott commended the "the Northern doctrines of Holden, whose 'Rational Essay' we have always esteemed," [John Wall Callcott], "Review of *Instructions for Playing the Musical Games*," *The British Critic* 21 (January, 1803), 45. Callcott's authorship of this review is confirmed by Kollmann in "A Chronological List of Periodical Musical Works," reproduced in Michael Kassler, *A. F. C. Kollmann's Quarterly Musical Register (1812): An Annotated Edition* (Farnham: Ashgate, 2008), 5. See also August Friedrich Christopher Kollmann, *An Essay on Practical Musical Composition, According to the Nature of that Science* (London: the Author, 1799), 75. Johann Nikolaus Forkel described Holden's *Essay* as "one of the best of its kind" (*unter die besten dieser Art*) in the *Allgemeine Litteratur der Musik, oder Anleitung zur Kenntniss musikalischer Bücher* (Leipzig: Schwickert, 1792), 418a, while Johann Georg Sulzer praised it as

4 HEARING WITH THE MIND

Despite this promising initial reception, however, Holden's thought, and the innovative tradition of music theory his *Essay* inspired, have not yet come in for sustained scholarly study.[6]

This book explores the emergence and the nature of Holden's proto-cognitive music theory as well as its afterlife in the writings of a pair of Scottish siblings, Walter (1745–1814) and Anne Young (1756–1813?). It utilizes methodologies drawn from the history of music theory, the history of science, social history, intellectual history, and feminist critique to embed their biographies and their productions within the social, musical, and philosophical contexts of the Scottish Enlightenment. By correlating innovative Scottish ideas about the perception of music with major transformations of European society, philosophy, religion, and culture, this book uncovers the ways in which the contributions of relatively marginalized figures in the endeavor of thinking abstractly about music reflect Britain's global entanglements in the rising age of empire.

"understandable and concise" (*verständlich und bündig*) in the second edition of his *Allgemeine Theorie der Schönen Künste* 3 (Leipzig: Weidemann, 1793), 457. Half a century later, François-Joseph Fétis praised the *Essay's* "philosophical spirit, which even today, renders it worth attention." François-Joseph Fétis, *Biographie universelle des musiciens* (Brussels: Meline, Cans et Compangie, 1839), 5:188. All translations are mine unless otherwise indicated.

[6] Louis Chenette provides a brief overview of Holden's conception of cadences, harmony, rhythm, and tuning in his dissertation; see Louis F. Chenette, "Music Theory in the British Isles during the Enlightenment" (PhD diss., Ohio State University, 1967), 346–366. In her dissertation, Jamie C. Kassler discusses Holden's theory in relationship to Scottish musical thought. Jamie C. Kassler, "British Writings On Music, 1760–1830: A Systematic Essay Toward a Philosophy of Selected Theoretical Writings" (PhD diss., Columbia University, 1971), 83–87. In her magisterial book, *The Science of Music in Britain, 1714–1830,* she elaborates on Holden's ideas in relationship to French music theory and Scottish common sense realism, while also providing new details about Holden's life, influences, and the first edition of the *Essay*. Jamie C. Kassler, *The Science of Music in Britain, 1714–1830: A Catalogue of Writings, Lectures, and Inventions* (New York: Garland Publishing, 1979), 1: 524–530. For a clear discussion of Holden's temperament see Owen Jorgensen, *Tuning: Containing the Perfection of Eighteenth-Century Temperament, the Lost Art of Nineteenth-Century Temperament, and the Science of Equal Temperament, Complete with Instructions for Aural and Electronic Tuning* (East Lansing, MI: Michigan State University Press, 1991), 118–128. A useful study of the relationship between the first part of Holden's treatise and Scottish common sense realism can be found in Leslie E. Brown, "The Common Sense School and the Science of Music in Eighteenth-Century Scotland," in *Essays in Honor of John F. Ohl: A Compendium of American Musicology,* ed. Enrique Alberto Arias (Evanston: Northwestern University Press, 2001), 122–132. More recently, David Damschroder has explicated aspects of Holden's harmonic system. David Damschroder, *Thinking about Harmony: Historical Perspectives on Analysis* (Cambridge: Cambridge University Press, 2008), 149–150, 166–186. Likewise, Justin London has described Holden's account of metrical accents as "remarkably prescient of [Bruno] Repp's work." Justin London, *Hearing in Time: Psychological Aspects of Musical Meter* (Oxford: Oxford University Press, 2004), 146.

INTRODUCTION 5

The Historical Context

The history of Scottish Enlightenment music theory is inextricable from the rise of the British empire in the eighteenth century. The Act of Union with England in 1707, a partnership to which Scotland—smaller, poorer, and religiously dissident—had only unwillingly acceded, brought about a range of new economic opportunities for enterprising Britons. Englishmen, many, like Holden, hailing from Northern England, began to settle in Scotland to take advantage of the country's growing economy and relatively free markets. Scottish trading companies embarked upon ambitious mercantile endeavors in British colonies in the Americas, West Indies, and the Far East, creating a significant influx of wealth. Scottish cities soon experienced a dramatic rise in size and prosperity. For example, Glasgow's population increased more than six-fold over the course of the eighteenth century, from ca. 14,000 in 1707 to 86,630 by 1806.[7] Its port on the banks of the river Clyde became the epicenter of the global tobacco trade, with imports increasing from 3.75 million pounds in 1726 to 25.6 million pounds in 1756.

The consequent increase in political and cultural exchange between Scotland and England meant that the ideology of Presbyterian asceticism that had dominated the former in the second half of the seventeenth century began to wane. A growing receptivity to English culture is evident in the attempts of Scottish landowners to improve their estates by importing English farming technologies and other forms of expertise.[8] Another example of this transformation is the decline of Scots in favor of English as the common tongue.[9] This cultural cross-pollination stands in a complex relation to the intellectual context of this book, the Scottish Enlightenment.

The astonishing accomplishments of the Scottish Enlightenment have been well studied in recent years, and need not be rehearsed here.[10] Suffice

[7] Statistical data taken from John Dougall, *The Modern Preceptor; Or, a General Course of Education* 2 (London: Vernon, Hood, and Sharp, 1806), 68.

[8] Ian D. Whyte, *Scotland before the Industrial Revolution: An Economic and Social History ca. 1050–ca. 1750* (Milton Park: Routledge, 2014), 299.

[9] See, e.g., Lilo Moessner, "The Syntax of Older Scots," in *The Edinburgh History of the Scots Language*, ed. Charles Jones (Edinburgh: Edinburgh University Press, 1997), 112. On the role of music and the notion "improvement" in Scottish society at the time see Andrew Greenwood, "Song and Improvement in the Scottish Enlightenment," *Journal of Musicological Research* 39 no. 1 (2020): 42–68.

[10] See, inter alia, Roy H. Campbell, Andrew S. Skinner, eds., *The Origins and Nature of the Scottish Enlightenment: Essays* (Edinburgh: John Donald Press, 1982); Michael A. Stewart, ed., *Studies in the Philosophy of the Scottish Enlightenment* (Oxford: Oxford University Press, 1990); Christopher Berry, *Social Theory of the Scottish Enlightenment* (Edinburgh: Edinburgh University Press,

6 HEARING WITH THE MIND

it to say that a particular brew of theological, economic, cultural, and social ingredients lent the ferment in Edinburgh, Glasgow, and Aberdeen a heady note.[11] Within the span of a few decades, these cities could boast foundational thinkers in the fields of philosophy (David Hume, Thomas Reid, Adam Smith, Adam Ferguson, Henry Home, Lord Kames), history (John Millar, William Robertson), aesthetics (Francis Hutcheson, Archibald Alison, James Beattie), economics (Smith again), geology (James Hutton), engineering (James Watt), literature (Robert Burns, Sir Walter Scott, Allan Ramsay, James McPherson), and many others. The work of most of these figures—and indeed the Scottish Enlightenment more broadly—is characterized by an emphasis on rational thought, simple explanations, and an interest in the "science of man," a nascent discipline that would later be known as the social sciences.[12]

Scottish Enlightenment music theory, as exemplified by the writings of Holden and the Youngs, is likewise characterized by a distinctly rational approach, one that links aspects of our mental and physical constitution to various musical behaviors and preferences. This school of thought draws directly on key philosophical discourses of the Scottish Enlightenment, and especially on Scottish common sense realism and conjectural history. It also engages with contemporaneous musical and pedagogical developments on the Continent: the writings of Jean-Philippe Rameau, Giuseppe Tartini, and Jean-Adam Serre, the musical dictionaries of Brossard and Rousseau, the music of Joseph Haydn, and the educational theories of Abbé Gaultier. The erudite and cosmopolitan orientation of the work of Holden and the Youngs can be taken as representative of the Scottish cultural blossoming which rendered England's northern neighbor for a time one of the major intellectual centers of Europe.

Other theories of music were, of course, written in Scotland in the second half of the eighteenth century. Notable examples include the *Essay on Tune*

1997); and Alexander Broadie and Craig Smith, eds., *The Cambridge Companion to the Scottish Enlightenment* (Cambridge: Cambridge University Press, 2019).

[11] With regard to the culture of intellectual socializing in the Scottish Enlightenment see, inter alia, Davis D. McElroy, "The Literary Clubs and Societies of Eighteenth-Century Scotland" (PhD diss., Edinburgh University, 1952); Mark C. Wallace and Jane Rendall, eds., *Association and Enlightenment* (Lewisburg, PA: Bucknell University Press, 2020).

[12] See e.g., Craig Smith, "The Scottish Enlightenment, Unintended Consequences and the Science of Man," *Journal of Scottish Philosophy* 7, no. 1 (2009): 9–28; Ryan Patrick Hanley, "Social Science and Human Flourishing: The Scottish Enlightenment and Today," *Journal of Scottish Philosophy* 7, no. 1 (2009): 29–46.

INTRODUCTION 7

(1781) by the Scottish landowner John Maxwell of Broomholm, the *Inquiry Into the Fine Arts* (1784) by the Presbyterian minister Thomas Robertson, and a number of works written by the Highland-born cellist and flutist John Gunn.[13] At Glasgow College and later at the University of Edinburgh, the physicist (and inventor of the siren) John Robison conducted research on acoustics, while more practical music manuals were published in Edinburgh by Italian immigrants such as Niccolò Pasquali and Domenico Corri, as well as by the music publishing impresario Robert Bremner. Some of these writers and their ideas will appear in minor roles throughout this book; I have chosen, however, to focus primarily on Holden and on the Youngs because their works, considered as a group, reveal a unique fusion of speculative musical thought with insights drawn from contemporaneous psychology.

We should keep in mind, however, that the daily exposure that Holden—an English merchant potter teaching church singing on the side in a provincial Scottish city—could have had to concert music would have paled in comparison with that of his counterparts in Paris, Padua, and Geneva. Indeed, as a member of the yeoman class and an amateur mathematician, Holden came to music theory as an outsider. He eventually titled himself "a teacher of church music" specifically, rather than a teacher of music more generally. A reticence to embrace the identity of professional musician is also reflected in the signature on the title page of his *Collection of Church Music* (1766), which reads "John Holden, φιλαρμονικος" (*philharmonikos*). The self-identification sometimes appears in the signature of various eighteenth-century authors, often as a pseudonym. It appears to be most closely analogous to the term "philomath" which, by middle of the eighteenth century, designated "a minor mathematical author, or a teacher of elementary mathematics," in the words of historian of science Olaf Pedersen.[14] Pedersen uses the term to characterize the contributors to a leading British amateur mathematical almanac, the *Gentleman's Diary or The Mathematical Repository*, a venue in which Holden also participated; and he further demonstrates that they also used the term to refer to themselves.

Holden's particular background suggests that his early musical experiences would have been almost exclusively formed by English sacred music: hailing from a tiny hamlet in Yorkshire, the only form of music he would have

---

[13] On the attribution of the *Essay on Tune* to Maxwell of Broomholm see Michael Kassler, "The Tuning of Maxwell's 'Essay,'" *Studies in Music* 11 (1977): 27–36.

[14] Olaf Pedersen, "The 'Philomath' of 18th Century England," *Centaurus* 8 (1963): 239, 257.

8 HEARING WITH THE MIND

regularly heard in childhood would have likely been the Anglican church service. Possibilities for musical development would have continued to be limited in scope after he embarked upon his apprenticeship, even had he done so in a major northern town such as Liverpool or Manchester. As an apprentice potter, he would have had relatively few opportunities to hear concert music on a regular basis, as such performances were typically held by closed associations and frequented almost exclusively by the wealthy at the time. After moving to Glasgow in 1757, his encounters with concert music would have continued to be scarce: while Scottish folk tune collections were beginning to come into vogue in the second half of the eighteenth century, and private musical clubs were flourishing (the Edinburgh Musical Society was founded in 1728, and the Aberdeen Musical Society twenty years later), no records of comparable musical associations exist in Glasgow.[15]

The curiously inept part writing evident in both of Holden's publications suggests that he did not have much, if any, practical training in composition, and that his experience with concert repertoire may have been relatively slight. Holden's decision to focus on psalmody in the *Essay* may thus have served not only to illustrate his claims with familiar tunes as he explicitly claims, but also (implicitly) to disguise his lack of musical experience. A similar conclusion was reached by Charles Burney, who praised Holden's theories but dismissed his compositional skills by observing that "a man that hears nothing but psalmody and national tunes, will never produce graceful elegant melody, or great effects in harmony."[16] This hypothesis may also explain why Holden's text eschews almost any philosophical discussion of aesthetics (arguably apart from two cursory paragraphs treating musical expression in the *Essay*).[17] Indeed, he seems entirely uninterested in the various debates on the nature of taste and beauty that are characteristic of the Scottish Enlightenment and found in the writings of figures such as Hutcheson, Hume, Beattie, Kames, or Reid.[18]

---

[15] On the Edinburgh Musical Society see Jennifer Macleod, "The Edinburgh Musical Society: its membership and repertoire, 1728–1797" (PhD diss., University of Edinburgh, 2001); on the Aberdeen Musical Society see David Johnson, "An Eighteenth-Century Scottish Music Library," *RMA Research Chronicle* 9 (1971): 90–95.

[16] Burney, "Holden, John" in *The Cyclopaedia,* unpaginated.

[17] On these two paragraphs see Leslie E. Brown, *Artful Virtue: The Interplay of the Beautiful and the Good in the Scottish Enlightenment* (New York: Routledge, 2016), 115–116. I thank Ben Piekut for raising the question of Holden's relationship to Scottish theories of aesthetics.

[18] On the major Scottish contributions to Enlightenment aesthetics in general see Jonathan Friday, ed., *Art and Enlightenment: Scottish Aesthetics in the Eighteenth Century* (Exeter: Imprint Academic, 2004) and Simon Grote, *The Emergence of Modern Aesthetic Theory: Religion and Morality in Enlightenment Germany and Scotland* (Cambridge: Cambridge University Press, 2017).

Holden's rustic origins thus seem to have been unpromising for a future music theorist. Yet his rather weak musical background may paradoxically have been an advantage—if not a prerequisite—in his creation of an innovative theory of music, insofar as his relatively modest musical training, shaped primarily by exposure to simple Anglican hymns, may have given him the audacity to envision a theory capable of accounting for all music based on a highly limited sample constituted by the music that he knew. At the same time, his advanced mathematical skills—which far exceeded the proficiency required for a merchant in his day—provided him the technical means for such an explanation. Finally, some explanation of the most essential and significant aspect of Holden's theory—its psychological foregrounding of mental acts, both conscious and unconscious—may be found in his exposure to the philosophy of Reid and the music-theoretical ideas of Rameau.

The six chapters that follow explore the nature and origin of Holden's ideas and the ways in which his near-contemporaries adapted his ideas about pitch perception to other modalities of musical knowledge. Chapter 1 reconstructs the biography of Holden himself, until now virtually unknown to modern scholarship. Interpreting primary documents ranging from tax rolls to shipping manifests, I examine the historical context in which he lived and worked, surveying his mercantile concerns, his mathematical activities, and his family history and social class. All of these aspects provide an intriguing window into the mind of a highly original thinker at a pivotal time of social mobility in Britain.

Chapter 2 explicates the speculative theories of Holden's principal theoretical treatise, his *Essay towards a Rational System of Music* (1770). I focus in particular on his notion of the *module,* a posited mental construct that explains how we hear pitches in relationship to each other and to a key, as well as on his approach to the faculties of memory and attention. As I demonstrate, Holden first defines explicit assumptions about the nature of perception, and he then derives his entire theory from a step-by-step elaboration of the conditions under which these first principles hold sway. Relying almost entirely on insights drawn from systematic introspection, the *Essay* depicts psychological phenomena that we today understand as grouping, subjective rhythmicization, and the interaction between top-down and bottom-up modes of perception. The chapter concludes with a comparison of Holden's implicit model of mental function to contemporary approaches to perception.

10   HEARING WITH THE MIND

The third chapter synthesizes the social history laid out in the first chapter with the music-theoretical approach described in the second in a close reading of the account of psalm singing that emerges from the pages of Holden's *Essay* and an earlier publication, *A Collection of Church Music* (1766). It does so by situating these writings within the context of sacred music in the Scottish Enlightenment, and in particular the movement known as the Monymusk Revival, which saw the restoration of part singing within Presbyterian congregations in Scotland throughout the 1750s. I conclude by examining the ways in which Holden's approach to psalm singing specifically, and to music more broadly, reflects his reception of the theories of Rameau.

In the fourth chapter I compare Holden's concept of attention with the ideas of the philosopher Thomas Reid (1710–1796), one of the founders of Scottish common sense realism and an important contributor to the emergence of the modern discipline of psychology. Reid and Holden were acquainted with each other in Glasgow, and the former's approach evidently inspired many of the latter's innovations. I also examine Holden's possible impact, in turn, on another Scottish philosopher and economist, Adam Smith (1723–1790). Reading Holden and Smith in conjunction with each other further suggests significant traces of influence, and I explore a potential intermediary between the two in the writings of a Scottish minister named Thomas Robertson.

In Chapter 5, I turn to an essay on rhythm published by the Scottish minister Walter Young. This relatively brief text, first delivered as a two-part lecture to the Royal Society of Edinburgh in 1784, clearly draws on Holden's ideas in its proposal of a cognitive approach to the subjects of musical rhythm and poetic meter. At the same time, it reflects its author's interest in the burgeoning "science of man," or incipient sociology. Contextualizing Walter's biography and analyzing his musical writings, I investigate how they respond to a larger problematic concerning universalism and relativism still hotly debated in the fields of psychology and music cognition today.

The last chapter explores a music-theoretical game set designed by Walter Young's sister, Anne, and its accompanying treatise that is part instruction manual, part theory textbook. Interpreting new archival evidence through the lens of feminist critique, I closely read her games and her writings to understand the music-theoretical, pedagogical, and social contexts from which they emerged. Anne's project offers, perhaps for the first time in the history of music theory, a novel synthesis of music theory rudiments with the dual

INTRODUCTION 11

constraints of antagonistic gameplay and juvenile pedagogy. Recovering the content, context, and reception of her work contributes to our understanding of the intricate relationships between didactic works and speculative music theory, while also widening our conception of the figures and traditions involved in the history of music theory.

Scottish Enlightenment Music Theory and the British Empire

With wealth and opportunity flowing into eighteenth-century Britain from its colonial holdings, enterprising working and middle-class Englishmen like Holden could, for perhaps the first time in their country's history, materially alter the social conditions into which they had been born through intellectual pursuits. Holden's relatively humble background (and his aspiration to transition into teaching) was not so unusual in a century that saw the meteoric rise of British men (and some women) from modest origins by dint of their intellect, imagination, and ambition.[19] These rising tides also affected native Scots like Walter and Anne Young.[20] The son of a schoolmaster, Walter was appointed to a comfortable living as the minister of Erskine, and he went on to attain most of his country's most distinguished academic honors. He participated in many of the discourses of the Scottish Enlightenment through his membership in social clubs limited to intellectuals of high status. Moreover, the Youngs were sufficiently well-off—and sufficiently well-connected—that Anne was able to develop a complex invention and file an expensive patent to protect her intellectual property.[21]

[19] Although absent from the first edition of his *Theory of Moral Sentiments* (1759), in his revised second edition of 1761 Adam Smith memorably observed that a man of "inferior rank" could hope to distinguish himself by "the labour of his body, and the activity of his mind ... he must acquire superior knowledge in his profession, and superior industry in the exercise of it. He must be patient in labour, resolute in danger, and firm in distress. These talents he must bring into public view, by the difficulty, importance, and, at the same time, good judgment of his undertakings, and by the severe and unrelenting application with which he pursues them." Adam Smith, *The Theory of Moral Sentiments* (London: A. Millar, 1761), 93.

[20] On the rising fortunes of Scotland in the context of the British empire, see John M. MacKenzie and T. M. Devine, eds., *Scotland and the British Empire* (Oxford: Oxford University Press, 2011).

[21] In his review of the *Musical Games*, the composer John Wall Callcott "reported that the patent, and other expenses attending the completion of the project, have amounted to nearly a thousand pounds." Callcott, "Review of *Instructions*," 45; a review of the *Musical Games* in *The Monthly Magazine* likewise noted that "the idea of [the whole apparatus] has, we learn, been realized at an expence little short of eleven hundred pounds." Anon., "Review of New Musical Publications: [Anne Young's] *Musical Game-Tables and Apparatus*," *The Monthly Magazine Or, British Register* 12, part 2, no. 80 (Dec 1, 1801): 428. This vast sum would have been far beyond the means of a music teacher at the time and must have been a loan or gift of some kind from a patron, friend, or family member.

12 HEARING WITH THE MIND

The design and execution of Anne's *Musical Games* further testify to the British empire's increasing affluence and global connectivity in the eighteenth century. The board and pieces of the *Musical Games* are made from commodities imported from the farthest reaches of the empire: mahogany from the West Indies, ebony and satinwood from East Asia or Africa, and ivory from West African elephants.[22] Moreover, the increasing financial means and growing size of the upper and middle classes in cities like Edinburgh and Glasgow resulted in new markets for instruction in general, and for refined accomplishments such as music in particular. These socioeconomic developments meant that women like Anne could now dream of alleviating the tedium of a teaching career through entrepreneurship.

So how can we account for the efflorescence of music-theoretical activity in late eighteenth-century Scotland by figures such as merchant potters, Presbyterian clergymen, and women, who were not traditionally involved as authors of works of music theory? Any answer to this question must reflect the fact that these three original thinkers—two of whom were siblings—were born in, or sufficiently near, Scotland at a time when the increased prosperity brought about by the rise of the British empire rendered it a uniquely fertile environment for the pursuit of their interests.

Archival sources shed no light on what Holden or the Youngs may have thought of the British empire, of its rapacious colonialism, or of its cruel exploitation of tens of thousands of Africans, native Americans, and other peoples during their lifetimes. (Other figures we will encounter in this book, such as Reid and Smith, publicly opposed slavery, and Holden's son Richard moved in abolitionist circles: he publicly supported the abolitionist William Wilberforce, and his father-in-law Samuel Tooker was a member of the Society for the Abolition of the Slave Trade).[23]

---

[22] I rely on the evaluators at the Bonhams auction house for their assessment of the materials from which the game was constructed; see https://www.christies.com/en/lot/lot-5067093. For more on the origins of these colonial materials see Chapter 6, note 3.

[23] Richard Holden also decried Liverpool's role in the slave trade in his private journal, observing in 1808 that the city had "made a disgraceful & ineffectual selfish resistance" to "the suppression of the slave Trade." Holden's account of a family trip to the lake district was serialized in six parts in *The Manchester Guardian* in 1953. See Richard Holden, "A Northern Tour in 1808: I—Manchester's 'Lower Orders,'" *The Manchester Guardian* (Thursday, September 10, 1953), 6. The reference to Richard Holden's support of Wilberforce can be found in Robert I. Wilberforce and Samuel Wilberforce, *The Life of Wm. Wilberforce, By His Sons* 2 (London: John Murray, 1838), 132. On Reid's attitude to slavery see Thomas Reid, *Practical Ethics: Being Lectures and Papers on Natural Religion, Self-government, Natural Jurisprudence, and the Law of Nations,* ed. Knud Haakonssen (Princeton: Princeton University Press, 1990), 46, 134–135; on Smith's attitude to the same see Adam Smith, *Lectures on Justice, Police, Revenue and Arms,* delivered at the University of Glasgow, ed. Edwin Cannan (Oxford: Clarendon Press, 1896), 94–104 and especially 99–100.

INTRODUCTION   13

The scanty records that we do possess, however, reveal some unexpected ways in which the lives of our three music theorists intersected with the long arm of the British empire. Holden, for instance, made his living initially by selling pottery, some of which he shipped to the American colonies on the distribution circuit run by the enterprising Glaswegian merchants who monopolized the highly lucrative tobacco trade at the time. At least one barrel of his goods was imported to the colonies on the "Charming Sally," a vessel which regularly transported enslaved people across the Atlantic.[24] Among the four named dedicatees of his *Essay* was the rector of Glasgow College, Adam Fergusson of Kilkerran; thanks to a family history written by a modern descendent, we know that Fergusson's family-owned plantations in Tobago and Jamaica that exploited the labor of hundreds of enslaved persons.[25]

Although far less directly involved in transatlantic commerce than Holden, Walter Young also maintained ties with the American colonies, corresponding for years with former parishioners who had joined the Scottish American Company of Farmers and founded the city of Ryegate in Vermont.[26] In 1784, moreover, he helped edit an anthology of "Highland Vocal Airs" collected by Joseph MacDonald, a childhood acquaintance who had died in Bengal while in service to the East India Company.[27] Anne Young's students in 1787–1788 included members of the Montagu family, whose great-grandfather, Duke John Montagu, had been granted a royal patent for the ownership of the Caribbean islands of St. Vincent and St. Lucia in 1722.[28]

---

[24] Between 1734 and 1758 "Charming Sally" made at least ten voyages carrying enslaved persons as cargo, departing from Britain to Africa and then to locations including Jamaica, Barbados, New York, and Charleston; see https://www.slavevoyages.org/voyage/database. Coincidentally, the Igbo abolitionist Olaudah Equiano describes being sold to the captain of the "Charming Sally" in 1762 in his autobiography, *The Interesting Narrative of the Life of Olaudah Equiano* (1789), a work which would become one of the most important sources for the British abolition movement. Olaudah Equiano, *The Interesting Narrative of the Life of Olaudah Equiano* (London: T. Wilkins, 1789), 176–177. Holden's contribution to the December 1757 voyage of the "Charming Sally" is discussed in Ann Smart Martin, "Scottish Merchants: Sorting out the World of Goods in Early America" in *Transatlantic Craftmanship: Scotland and the Americas in the 18th and 19th Centuries*, ed. Simon Gilmour and Vanessa Habib (Edinburgh: Society of Scottish Antiquaries, 2013), 34–36.

[25] No comparable project has been published with regard to the estate of the other university officials to my knowledge. For more on the history of the Fergusson estate see Alex Renton, *Blood Legacy: Reckoning with a Family's Story of Slaver* (London: Canongate, 2021).

[26] Edward Miller and Frederic Palmer Wells, *History of Ryegate, Vermont: From its Settlement by the Scotch-American Company of Farmers to Present Time: With Genealogical Records of Many Families* (St Johnsbury, CT: The Caledonian Company, 1913), 58 and 165.

[27] Walter's preface to this work is discussed in Chapter 5.

[28] John Montagu, 2nd Duke of Montagu (1690–1749), lost the islands almost immediately to the French. Later in life he assisted the celebrated musician Ignatius Sancho (ca. 1729–1780), who was in his employ, and helped secure the freedom of the enslaved African aristocrat Ayuba Suleiman Diallo (1701–1773). On Sancho's positive experiences with the Montagu family (whom Sancho described

14 HEARING WITH THE MIND

The ivory used to manufacture her *Musical Games* in 1801 was obtained by British traders via extractive trade practices in West Africa and transported to Scotland on slave ships. Even as the chapters that follow explore and often celebrate the accomplishments of these three musicians, we should keep in mind that the forces that enabled Scottish Enlightenment music theory to flourish came at an enormous cost around the world.

Medieval Aircraft: In Defense of Anachronistic Comparison

For sailing vessels can be built [that operate] without rowers, so that the largest ships, both river- and sea-going, may be borne along with but a single person in control, [moving] more swiftly than if they were full of people [i.e., rowers and sailors]. . . . Again, cars can be made that are moved, without [the force of] an animal, with an inestimable force, as we deem the force of the armed chariots used in battle in antiquity to have been. Again, flying craft can be fashioned so that a person would sit in the middle of one, turning a mechanism by means of which artificially made wings would beat the air, in the manner of a flying bird. . . . And an infinite [number of] such things can be made, for example, bridges over rivers without columns or any supports, and mechanical contrivances, and ingenious inventions not heard of before.[29]

---

as "one of the best families in the kingdom"), see Ignatius Sancho and Joseph Jekyll, *Letters of the Late Ignatius Sancho, an African: To Which are Prefixed, Memoirs of His Life* (London: J Nichols, 1782), vii–x; quotation at 96. Diallo's biography, written by the British judge Thomas Bluett, is dedicated at Diallo's request to Lord Montagu, in acknowledgment of Montagu's "great humanity and goodness to an unfortunate stranger." Thomas Bluett, *Some Memoirs of the Life of Job: The Son of Solomon the High Priest of Boonda in Africa; Who was a Slave about Two Years in Maryland; and Afterwards Being Brought to England, was Set Free, and Sent to His Native Land in the Year 1734* (London: Richard Ford, 1734), iv. On Anne Young's involvement with the Montagu family see Jeanice Brooks, "Staging the Home: Music in Aristocratic Family Life," in *A Passion for Opera: The Duchess and the Georgian Stage*, ed. Paul W. Boucher, Jeanice Brooks, Katrina Faulds, Catherine Garry, and Wiebke Thormählen (Boughton: Buccleuch Living Heritage Trust, 2019), 36.

[29] "Nam instrumenta navigandi possunt fieri sine hominibus remigantibus, ut naves maximae, fluviales et marinae, ferantur unico homine regente, majori velocitate quam si plenae essent hominibus . . . Item possunt fieri instrumenta volandi, ut homo sedeat in medio instrumenti revolvens aliquod ingenium, per quod alae artificialiter compositae aerem verberent, ad modum avis volantis . . . Haec autem facta sunt antiquitus, et nostris temporibus facta sunt, ut certum est; nisi sit instrumentum volandi, quod non vidi, nec hominem qui vidisset cognovi; sed sapientem qui hoc artificium excogitavit explere cognosco." Roger Bacon, *Epistola de secretis operibus artis et naturae, et de nullitate magiae*, in John S. Brewer, *Fr. Rogeri Bacon, Opera quaedam hactenus inedita* 1 (Oxford: Longman, Green, Longman, and Roberts, 1859), 533. I thank David E. Cohen for the translation.

INTRODUCTION   15

Ask any schoolchild about Roger Bacon (ca. 1220–1292), and chances are that if they have heard of the Franciscan friar, it is as a medieval prophet of science, one who predicted a great many of the technologies we use today. The futuristic aspects of Bacon's oeuvre, which had captivated readers for centuries, solidified his reputation in the Victorian era as a man ages ahead of his time.[30] Even today, Bacon's proleptic ideas continue to capture the contemporary popular imagination.[31] They interest us, I think, because *we* actually have many of the "ingenious inventions" of which he dreams: the suspension bridges, the motorized ships and cars, the flying machines and submarines. They were probably even more fascinating to the Victorians, who were themselves on the cusp of realizing the friar's visions.

Nearly eight hundred years after they were penned, Bacon's speculations prompt us to imagine medieval sea vessels, cars, and flying machines.[32] They also invite us to envision other technologies that could have been devised over the last eight centuries with the knowledge and materials available in various eras. Most importantly, they inform us about the horizons of possibility available to people in the thirteenth century: an ornithopter was imaginable; the internet, apparently, was not.[33]

In what follows, I sometimes consider the ideas of Holden and the Youngs rather as though they were akin to blueprints for a medieval airplane or

[30] This claim was asserted, for example, by the ballooning pioneer John Wise, who maintained that Bacon's ideas were "three or four centuries ahead of the age he lived in." John Wise, *A System of Aeronautics comprehending its Earliest Investigations, and Modern Practice and Art* (Philadelphia: Joseph A. Steel, 1850), 20. In the 9th edition of the *Encyclopedia Britannica* (1875), the Scottish philosopher Robert Adamson describes the prevailing view of Bacon thus: "Bacon, it is now said, was not appreciated by his age because he was so completely in advance of it; he is a 16th or 17th century philosopher, whose lot has been by some accident cast in the 13th century; he is no schoolman, but a modern thinker, whose conceptions of science are more just and clear than are even those of his more celebrated namesake [Francis Bacon]." See Robert Adamson, "Roger Bacon," in *Encyclopedia Britannica*, 9th ed. (Edinburgh: Adam and Charles Black, 1875), 3: 218–219. Adamson then immediately walks back this claim, remarking that "in this view there is certainly a considerable share of truth but it is much exaggerated." These sentiments do not appear in the 8th edition of the *Britannica* (1854).

[31] See, e.g., recent popular books on Bacon such as Brian Clegg, *The First Scientist: A Life of Roger Bacon* (London: Constable, 2003); Lawrence Goldstone and Nancy Goldstone, *The Friar and the Cipher: Roger Bacon and the Unsolved Mystery of the Most Unusual Manuscript in the World* (New York: Crown/Archetype, 2005). The Franciscan Friar is a character in the video game "Koudelka" as well as its sequel, "Shadow Hearts."

[32] Some scholars have argued that all or part of the *Epistola de secretis operibus artis et naturae* (1249–52) is a forgery; however, this conclusion is far from accepted. See, e.g., William R. Newman, "The Philosophers' Egg: Theory and Practice in the Alchemy of Roger Bacon," *Micrologus* 3 (1995): 93–94.

[33] A lovely meditation on medieval dreams of flight can be found in Peter Nilson, "Winged Man and Flying Ships: Of Medieval Flying Journeys and Eternal Dreams of Flight," trans. Steven Hartman, *The Georgia Review* 50, no. 2 (1996): 267–296.

16　HEARING WITH THE MIND

suspension bridge, and I point out some striking similarities between aspects of their ideas and later approaches in the cognitive sciences.[34] For example, Chapter 2 shows that Holden's investigation of music anticipates contemporary findings regarding the perception of just-noticeable difference, the limits of memory, and musical entrainment and anticipation, and it elucidates his theoretical model of music cognition in terms that we now call now call top-down and bottom-up processing. Chapter 5 takes a similar approach to Walter Young's examination of the role of nature and nurture in the development of individual differences, the influence of our embodied affordances on our rhythmic abilities, and the pleasure afforded by musical surprise. Chapter 6 argues that Anne Young's *Musical Games* represent a very early ludic program of music-theory pedagogy.

Such a use of anachronistic comparisons runs the risk of distortion: the contemporary lens that invites us to disregard Bacon's discussion, a few pages later, of the philosopher's stone and to instead celebrate his technological predictions obscures the fact that the friar's quest for the former was far more emblematic of his day and age—and of his interests—than the latter.[35] And the issues raised in the *Epistola* are not at all characteristic of the friar's work, most of which focused on linguistics, optics, and Aristotelian commentary.

Insofar as my study of Holden and the Youngs makes occasional use of anachronism, it does so, I believe, in a way that does not lead to distortion or misinterpretation. First of all, I take all of their extant music-related writings into account. Then too, I complement my application of anachronistic concepts by exploring the ways in which their ideas respond to contemporaneous discourses in philosophy and music theory, as well as to the specific social and cultural contexts of their time. Most importantly, however, I employ these anachronistic comparisons as clearly-identified hermeneutic maneuvers designed to bring out ideas that are demonstrably present in the analyzed texts themselves.

The readings I offer here lead, I hope, to a new and more comprehensive understanding of the horizons of possibility in late-eighteenth-century British thought regarding the role of the mind in the perception of music.

---

[34] As it happens, suspension bridges were independently invented in South America and East Asia long before the thirteenth century. See Leonardo Fernández Troyano, *Bridge Engineering: A Global Perspective* (London: Thomas Telford, 1997), 50–51.

[35] Bacon treats the philosopher's stone (or egg) in the last three chapters of his *Epistola de secretis operibus artis et naturae* (1249–1252).

They also shed light on the nature and the pre-history of one of our widely held and fundamental views regarding the perception of music: the idea that our mental faculties are essential in constructing our musical experience—that we hear music not only with our ears, but also with our minds.

1

Potter, Musician, Merchant, Scribe

The Many Lives of John Holden (1729–1772)

Writing in *Rees's Cyclopædia* (1811), Charles Burney characterized John Holden's *Essay towards a Rational System of Music* as "excellent" and regretted that it was "much less noticed by the public than it deserves."[1] However, he admitted, he was completely "unable to give a biographical account of this ingenious author."[2] François-Joseph Fétis's entry on Holden in the *Biographie universelle des musiciens* (1839) likewise praised the treatise while lamenting its author's inexplicable absence from the historical narrative:

> The silence of the English biographers on this musician, or rather the indifference of all music historians and compilers of musical anecdotes from Britain, is even more astonishing given that Holden was author of the best musical treatise published in England for over a century.[3]

He then goes on to (erroneously) surmise that Holden was born in Scotland and employed as a professor of music at Glasgow University.[4]

Half a century later, the music historian James Love attempted to piece together Holden's biography for his *Scottish Church Music: Its Composers and Sources* (1891). Examining Glasgow city records, Love found the following:

> Holden, John, probably a native of England, seems to have settled in Glasgow about 1757, and carried on business as a potter. He was made a Burgess and Guild-brother of the city of Glasgow by purchase, July 8, 1757, and is described on the Burgess Roll as a merchant.[5]

---

[1] Burney, "John Holden."
[2] Ibid.
[3] François-Joseph Fétis, *Biographie universelle des musiciens* (Brussels: Meline, Cans et Compangie, 1839), 5:188.
[4] Ibid.
[5] James Love, "John Holden," in *Scottish Church Music: Its Composers and Sources* (Edinburgh: W. Blackwood, 1891), 171.

Hearing with the Mind. Carmel Raz, Oxford University Press. © Oxford University Press 2025.
DOI: 10.1093/9780197786208.003.0002

This account appears to be the first to include a theory as to Holden's profession. However, three years later, David Baptie's entry on Holden in *Musical Scotland, Past and Present: Being a Dictionary of Scottish Musicians* mentions that Fétis had mistakenly believed Holden to be a professor at Glasgow University, and it exclaims that "it is a matter of regret that the biography of this writer is so obscure!"[6] Baptie makes no mention of Holden's profession, despite acknowledging a reliance on Love's *Scottish Church Music* in his preface, possibly owing to the incongruity of a potter having authored a work such as the *Essay*.[7]

More recently, in her magisterial book, *The Science of Music in Britain, 1714–1813* (1979), Jamie C. Kassler lists four John Holdens active in Glasgow around 1770: a guild member, a potter, a bell ringer, and the father of a student who matriculated in 1772. She proposes that "any one or more of these could be the John Holden in question."[8] Relying on newly discovered archival material, this chapter demonstrates that Kassler's four Holdens were indeed the same person, and that he can be conclusively identified with the music theorist who authored the *Essay towards a Rational System of Music* (1770).

A Bell-Ringing Merchant

Let us begin with the simplest part of this tale. The inner wall of the Tolbooth Steeple in Glasgow features an engraved list of people who have held the position of carillonneur since the bells were first installed in 1738, a list which extends to the present day. An image of this wall is shown in **Figure 1.1**.

We can be confident that the "J. Holden" inscribed on the wall refers to "John" because a Glasgow council minute from March 17, 1772 records the appointment of one Joshua Campbell as the successor to John Holden, deceased, suggesting that Holden held the position for about seven years.[9]

---

[6] David Baptie, "John Holden," in *Musical Scotland, Past and Present: Being a Dictionary of Scottish Musicians from about 1400 Till the Present Time, to Which is Added a Bibliography of Musical Publications Connected with Scotland from 1611* (Paisley: J. and R. Parlane, 1894), 82.

[7] Ibid, iv. Holden's profession does not appear in a miniscule biographical entry by William Hume in the 1895 edition of *Grove's Dictionary of Music and Musicians*, which simply states that he was not a professor at Glasgow; the first notice of his profession appears in a slightly expanded entry in the 1906 edition. Compare William Hume, "John Holden," *Grove's Dictionary of Music and Musicians*, ed. Sir George Grove (Philadelphia: Presser, 1895), 4:678; to William Hume, "John Holden," *Grove's Dictionary of Music and Musicians*, ed. Sir George Grove (New York: Macmillan, 1906), 2:419.

[8] Kassler, *The Science of Music in Britain*, 1:524–525.

[9] Ibid.

20 HEARING WITH THE MIND

Figure 1.1 The names listed on the wall facing the carillon within the Tolbooth Steeple.

Indeed, the city of Glasgow lists the death of John Holden, merchant, from consumption on March 3, 1772, a date that aligns with the appointment of his successor two weeks later.[10] Thus, we can be fairly sure that the Holden who was ringing the bells was also the merchant of the same name.

In the first half of the eighteenth century, smaller Scottish cities generally imported their bell-ringers from England or from Edinburgh, as ringing carillons and church bells involved training that was relatively rare in Scotland due to the Scottish Reformation's prohibition on the use of musical instruments in religious worship, a topic we will revisit in Chapter 3.[11] We learn from the Glasgow Council minutes, for example, how the first ringer of the Tolbooth Chimes, a Scot named Rodger Redburn, travelled to Edinburgh

---

[10] Family records indicate that the actual date of death was March 2, 1772; this is stated in John Holden Jr.'s will, dated March 2, 1799 and confirmed by Isabella Holden's tombstone, which states that she was widowed on March 2, 1772. John Guest, *Yorkshire. Historic Notices of Rotherham: Ecclesiastical, Collegiate, and Civil* (Rotherham: White, 1879), 264.

[11] The association of bell-ringing skill with England, rather than Scotland, preceded the English Civil War. In 1636, the English traveler William Brereton reported that "there are but few bells in any steeple [in Edinburgh], save in the Abbey Church steeple, which is the king's palace. Herein is a ring of bells erected by King Charles immediately before his coming into Scotland, anno Dom. 1635, but none here knew how to ring or make any use of them, until some came out of England for that purpose, who hath now instructed some Scotts in this art." William Brereton, *Travels in Holland, the United Provinces, England, Scotland, and Ireland, M. DC. XXXIV–M. DC. XXXV* (Manchester: Chetham Society, 1844), 1:111.

POTTER, MUSICIAN, MERCHANT, SCRIBE   21

at the city's expense in order to learn the craft.[12] A minute of April 22, 1765 suggests that this position entailed considerable musicianship:

> The Council are of opinion that the foresaid office [of Tolbooth bell ringer] should be bestowed on a person learned in the parts of music, and recommend the Magistrates to cause [to] intimate in the public newspapers, that any person skilled in playing on bells, as well as on the violin, spinnet, or harpsichord, and well versed in church music, will meet with good encouragement.[13]

The position is recorded as having gone to "Collett of London"—possibly the English composer John Collett (1735–1775)—who does not seem to have lasted very long, as his appointment commenced on October 9, 1765 and concluded that same year.[14] Figure 1.1 further reveals that the dating of Collett's end and Holden's start differs from the other entries, suggesting that the two men shared the position for part of 1765.[15] Holden's abilities as a carillonneur suggests that he was not a native of Glasgow, and as such they lend support to Love's suggestion that he was of English extraction.

Yet, a letter penned in 1773 by the Scottish philosopher Thomas Reid (1710–1796) describes John Holden in a different light. Addressing himself to the trustees of a scholarship at Glasgow College on behalf of Holden's orphaned son, also named John,[16] Reid writes:

> The Young Man I would recommend is John Holden son to the Deceased John Holden who was a Teacher of Musick, Writing, & Mathematicks in

---

[12] In 1736, the city of Glasgow also purchased a set of expensive practice bells to enable Redburn to hone his craft. John Oswald Mitchell, "Four Old Glasgow Bells," in *Publications of the Regality Club, Glasgow* (Glasgow: James Maclehose and Sons, 1893), 2:38–42.

[13] John McUre, *Glasghu Facies: A View of the City of Glasgow* (Glasgow: Tweed, 1872), 391.

[14] Although hailing from London, Collett ended his days in Edinburgh. For more on the English composer see Jürgen Schaarwächter, *Two Centuries of British Symphonism: From the Beginnings to 1945. A Preliminary Survey* (Hildesheim: George Olms Verlag, 2015), 2:43–46. Schaarwächter does not locate Collett in Glasgow; that is my own hypothesis.

[15] It is perhaps conceivable that both Collett and Holden had tenures of less than one year (Collett in 1765, Holden in 1772), but I have found no report of a gap in bell ringing during the period from 1765–1772. Moreover, as Holden died on March 2, 1772 from consumption, we can assume that the illness would have debilitated him some time before his demise (and would have likely prevented him from climbing up the many stairs required to reach the musician's loft in the bell tower). It thus seems unlikely that his name would have been inscribed for a few weeks' worth of bell ringing, if indeed his health had permitted him to do so. I thank Matthew Hall for discussing this point with me.

[16] This letter was first published in 2002. See Thomas Reid, *The Correspondence of Thomas Reid*, ed. Mark Wood (University Park: Penn State Press, 2002), 74. This letter is also reproduced in Brown, "The Common Sense School," 131–132.

22   HEARING WITH THE MIND

Glasgow. The Father was really an ingenious literary Man as well as of an excellent Character. He wrote a Treatise on the theory of Musick which does him much honour in the Opinion of good judges. He commonly had some Students of better Rank at the University who boarded in his house. He wrote our Records and Diploma's, and Directed the Church Musick in the College Chappel, so that he was much connected with the College & much respected by the Masters[.] He had been some time a teacher as I have heard in an English Academy & both he & his wife were English People although the Young Man I recommend was born in Scotland. . . . The Father died two years ago and left his family in straitned [sic] Circumstances. The Son entered to the College in the beginning of this Session and is a boy of very good hopes.[17]

Having student lodgers, as Reid reports, was not indicative of poverty, but rather a sign of prestige and involvement in university life: well into the nineteenth century, Glasgow college students regularly roomed with their professors and other college employees.[18] This would have meant that Holden lived in a spacious house, one that was suitable for hosting the off-spring of well-to-do and well-connected families.[19] Reid notes that Holden was deeply involved with Glasgow college in various capacities, and that he was held by the academic community in high esteem. Crucially, he also mentions that Holden and his wife were of English extraction.

The identification of Holden the teacher with the bell-ringing merchant requires us to wonder why a successful businessman would want to write records and diplomas for relatively small sums (records indicate that he was paid five shillings per diploma, and just under two pounds for a year's worth of minute keeping).[20] The information supplied in Reid's letter suggests that

---

[17] Ibid.

[18] As the Scottish antiquarian and Glasgow college historian David Murray (1842–1928) would later write, "It was the practice for Principals and Professors to receive students into their houses as boarders, and it continued down to my day." David Murray, *Memories of the Old College of Glasgow: Some Chapters in the History of the University* (Glasgow: Jackson, Wylie & Co., 1927), 2:379.

[19] Indeed, the Glasgow window tax rolls reveal that a "John Holden, potter" was living in a house with seven windows or less in Glasgow in 1764, but that by 1769 he had moved to spacious quarters—a house with 16 taxable windows—close to Glasgow College, and listed his profession as that of teacher. For Holden's entry in 1764 see National Records of Scotland, window tax vol. 175/108; for 1769 see ibid., vol. 176/113.

[20] A minute of 1767 states: "An account of John Holden for writing the record of the minutes of the rector, dean of faculty, and principals meetings for the year preceding the 14th June 1767 was given unto the meeting. It amounts to 1 pound 10 shillings 7 pence sterling. And the account being judged reasonable, the principal is appointed to grant a precept for pay of it upon the factor." University of Glasgow Senate Minutes for December, 1767, Glasgow University Archive vol. 26643:129.

this may have been a way for Holden to develop his academic connections. According to the faculty minutes of Glasgow University, Holden began "instructing the band" in the chapel in 1765 and began writing the records of the rector, dean of faculties, and principal a year later. This latter task would have presumably involved his presence at meetings where sensitive university matters would have been discussed.[21] Holden would have thus been a familiar and trusted figure at the college.

We can learn more about Holden's activity at the college from the Preface to his *Collection of Church Music . . . Designed for the Use of the University of Glasgow,* first published in 1766. The text's unusual inclusion of complex music-theoretical content (discussed in Chapter 3) suggests that it may have also been designed to impress other members of the college, perhaps with the ultimate aim of improving its author's standing in the academic community. Indeed, the university senate minutes of January 23, 1766 record a decision by the chapel committee to "purchase thirty copies of Mr. Holden's music book," with the purchase being made by the chapel fund. Holden thus ensured that his colleagues would have access to his new work.[22]

Holden seems to have enthusiastically promoted music at Glasgow college. The faculty minutes reveal that Holden directed the music in the chapel as a labor of love, and that he regularly donated his annual salary of five pounds back into a fund he set up for the encouragement of music at the college.[23] These minutes further record that in 1769, he donated the sum of fifteen pounds as "a present . . . for buying music for the college,"[24] and that a year later, he gave twenty-one pounds, ten shillings, and ninepence "for the encouragement of music in this university."[25] The establishment of such gifts by professors was not unusual: the same page that lists Holden's second gift also catalogues a donation of twenty pounds made by the wealthy Professor John Anderson "for the encouragement of natural philosophy and good

---

[21] Holden would later dedicate his *Essay* to these three men by name: "Sir Adam Fergusson of Kilkerran, Baronet, Lord Rector; The rev. Mr. John Corse, dean of faculties, and Dr. William Leechman, principal." The first dedicatee is William Duke of Montrose, chancellor of the University of Glasgow, and the last line of dedication is to "all the professors of Glasgow College, patrons of useful and polite literature."

[22] Minutes of the University of Glasgow Senate for January 23, 1766, Glasgow University Archive, vol. 26643:81.

[23] Minutes of the University of Glasgow Senate for June 10, 1782, Glasgow University Archive, vol. 26692:191–192.

[24] Minutes of the University of Glasgow Senate for June 1, 1769, Glasgow University Archive, vol. 26644:93.

[25] Minutes of the University of Glasgow Senate for May 18, 1770, Glasgow University Archive, vol. 26644:179.

elocution among the masters of arts at this university."[26] It appears, therefore, that (certainly unlike most professional musicians) Holden was sufficiently well-off that he could not only work for free but even make modest philanthropic gifts to support his passion. Such a pattern of behavior suggests that he was very keen to establish himself amongst the members of the college, initially by performing relatively mechanical tasks such as writing records and copying diplomas, and later by making gifts of a scope that matched those of his professorial colleagues.

Curiously, a receipt signed by Isabella Holden that I found in the University of Glasgow archives dated July 2, 1772, a few months after Holden's death, describes a payment of ten shillings for two doctors' diplomas copied by her husband (see **Figure 1.2**). This receipt, as well as a record in the university minute books from March 17, 1772 of the sum of seven pounds, thirteen shillings, tenpence, and one halfpenny to the "late Mr. Holden for writing college records and other papers" suggests that Holden continued to copy records and diplomas until the end of his life.[27] The size of this sum

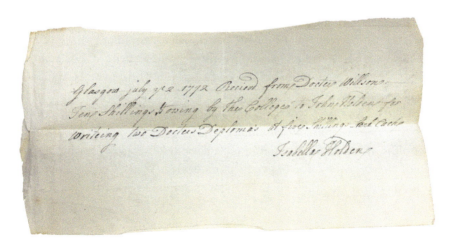

Figure 1.2 Isabella Holden's receipt. Glasgow, July 2, 1772: Received from Doctor Willson Ten Shillings owing by the College to John Holden for writing two Doctors Diplomas at five Shillings each. Isabella Holden.

---

[26] Ibid.
[27] Minutes of the University of Glasgow Senate for March 17, 1772, Glasgow University Archive, vol. 26690:59–60.

POTTER, MUSICIAN, MERCHANT, SCRIBE   25

and its singular appearance in the university minutes—Holden's only other documented bill for writing records was in 1767—suggests that he may have declined to invoice the university for his work, and that these payments were made in retrospect, as various odds and ends relating to Holden's university business would have been tidied up following his death.

More importantly, Isabella Holden's name enables us to definitively link the four John Holdens that Kassler described. An entry from 1760 in the registry of St. Andrew's Episcopal Church in Glasgow includes John Holden "Englishman," and his wife, Isabel [*sic*] Faucet, and describes Holden as a potter.[28] The names of John Holden and Isabella Fawcett also appear in the Episcopal Register of Baptisms at Glasgow: from 1760 to 1765, three children were born to "John Holden, potter and Isabella Faucet his Spouse," including Hannah (b. 1760), John Jr. (b. 1762), whom we encountered earlier as Reid's "boy of very good hopes," and Agnes Maria (b. 1765). However, Richard, baptized in 1768, appears as "Son to John Holden Teacher of Church Music, formerly Potter and Bella Faucet his Spouse," while Eleanor, baptized in 1771, is the daughter of "Mr John Holden Teacher of Music" and Bella Faucet. Yet the municipal record of the deaths of three of his children—Elizabeth, †1767; Bridget †1770; and Agnes Maria, †1770—as well as the record of Holden's own death in 1772 consistently list his profession as "merchant," suggesting that he retained some involvement in mercantile activities until the end of his life.[29] Thus, we can be certain that the same Holden was a guild member, a potter, a bell ringer, and the father of a student at Glasgow College.

Although the younger John Holden (henceforth John Holden Jr.) is not included in the university's (very) partial matriculation lists, the faculty minutes reveal that he was awarded a scholarship in 1775.[30] He would go on to attend Sidney Sussex College in Cambridge on a scholarship, attaining the

---

[28] Love, "John Holden," 171. Cornelius Hallen, *The Scottish Antiquary: Or, Northern Notes & Queries* (Edinburgh: Constable, 1891), 91. The first Episcopal church in Glasgow, the congregation of St. Andrew's-by-the-Green would have consisted primarily of English and Irish expatriates.

[29] See the *Register of Baptisms at Musselburgh Dalkeith & Glasgow by the Rev. John Falconer 1754–1793*, Glasgow City Archives, The Mitchell Library, Glasgow. This teaching career seems to have extended beyond music as well: as mentioned earlier, Reid described Holden as having been "a Teacher of Musick, Writing, & Mathematicks in Glasgow," while the admissions register of Sidney Sussex College, Cambridge describes the younger John Holden as the son of "John, teacher of mathematics at Glasgow." John A. Venn, *Alumni Cantabrigienses; A Biographical List of All Known Students, Graduates and Holders of Office at the University of Cambridge, from the Earliest Times to 1900*, part 2, vol. 3 (Cambridge: The University Press, 1947), 409.

[30] Minutes of the University of Glasgow Faculty for November 9, 1775, Glasgow University Archive, vol. 75:335.

26 HEARING WITH THE MIND

notable distinction of "second wrangler," or the second-best student at the university in mathematics, in 1784. He was subsequently engaged as tutor of mathematics at his alma mater and honored with a symbolic appointment as preacher at the Chapel Royal at Whitehall.[31] Further details about his background emerge from the records of the diocese of Winchester, where he assumed the curacy of the town of Boldre in 1785. These documents indicate that one John Holden, born in 1762 in Glasgow to John Holden Sr. and Isabella Faucet, was ordained at Cambridge in 1785. They also list Holden Sr.'s roots as Westhouse, Thornton, in Yorkshire.[32]

This information in turn can help us definitively identify the Glasgow-based merchant-potter-musician John Holden with John Holden of Westhouse, a frequent contributor to *The Gentleman's Diary or The Mathematical Repository* in 1753 and 1754. Published annually, the *Diary* included both an almanac and a repository of mathematical problems and riddles. It reflected a burgeoning culture of popular practical mathematics, one that linked a community of like-minded individuals throughout England. Contributors to the *Diary* included a number of women as well as teachers, tradesmen, and even schoolboys.[33]

In 1753 and 1754, both John Holden (then at Westhouse) and a Francis Holden, of "Settle, near Westhouse"—probably his younger brother, whom I shall henceforth term Francis (II), with John's father as Francis (I)—sent in a number of solutions, and in some cases, posed mathematical questions to the readers of the almanac.[34] Holden's questions include the following:

> A Curate borrowed of his Rector 20£. giving him his Bill for the same, with Interest at the Rate of 5 per Cent. A Year after which the Curate begun to pay to the Rector, 12d. per Week, in Part of the aforesaid Bill, I desire to

---

[31] Denys A. Winstanley's history of early Victorian Cambridge describes the younger John Holden as "an extremely successful tutor of [Sydney Sussex College at Cambridge] [who] had seemed destined to become its master in due course." Denys A. Winstanley, *Early Victorian Cambridge* (Cambridge: Cambridge University Press, 1940), 8.

[32] See Arthur J. Willis, *Winchester Ordinations, 1660–1829, from Records in the Diocesan Registry, Winchester* (Hambleden, Lyminge, Folkestone, Kent: A. J. Willis, 1964), 1:68.

[33] See Shelley Costa, "The 'Ladies' Diary': Gender, Mathematics, and Civil Society in Early-Eighteenth-Century England," *Osiris* 17 (2002): 49–73. See also Frank J. Swetz, *The Impact and Legacy of The Ladies' Diary (1704–1840): A Women's Declaration* (Providence: MAA Press, an imprint of the American Mathematical Society, 2021).

[34] This Francis Holden, as well as his nephew Francis (III), about whom more forthwith, should not be confused with Francis Holden, a precentor in Barony, near Glasgow, who was the father of Smollett Holden, a music seller, and the grandfather of Francis Holden, a compiler of Irish tunes. I have not been able to ascertain any connection between the Yorkshire Holdens and this other family, which hailed from Dublin.

know how many such weekly Payments will be required to clear off the whole Debt, the Interest due at each Time of Payment being always first accounted for?[35]

At what Time of the Year 1753 was the Night (exclusive of Twilight) longer at York, than it was either at London or Edinburgh?[36]

These two questions, as well as various problems that Holden is listed as having answered, reveal that he had considerable facility with algebra. The first, also designated in the *Diary* as a "prize question" (that is, an exceptionally difficult question for which the compiler awarded prizes to subscribers sending in correct answers), is a variant of the well-known "yield approximation problem," which attracted considerable attention from the late seventeenth century and well into the nineteenth century. The complier of the *Diary* printed five solutions to the problem, including Holden's own, with answers ranging from 533.937 weeks to 540.4 (these deviations probably have to do with the day of the week on which the payment is calculated to be due).[37] Answering Holden's second question requires familiarity not only with trigonometry, but also with latitudes and longitudes, in a calculation that would have been intimately familiar to the compliers of the *Diary,* which was of course itself an almanac. While we do not know more about Holden's training in mathematics, his engagement with the *Diary* reveals that his ability in that domain extended far beyond the skill necessary for his profession of merchant, and it would have been far more characteristic of a teacher or some other form of professional mathematician.

Holden's music-theoretical discussions in the *Essay* similarly reveal a facility with both practical and advanced mathematics. For example, he briefly explains how a mainstay of mercantile calculation at the time, the "Rule of Three," can be applied to musical ratios, including a brief exposition presumably aimed at refreshing the memories of his readers, all of whom would

---

[35] Thomas Simpson, ed. *The Gentleman's Diary or the Mathematical Repository* (London: Company of Stationers, 1754), 14:24.

[36] Ibid., 43. The answer to this question appears in the 1755 edition of the almanac, which provides two methods of arriving at the answer (February 9, 1753 and October 31, 1753) sent in by different contributors; see 29–31.

[37] Holden does not provide any further detail about his (presumably fictional) curate and rector. However, such small-scale loans backed by personal connections would have been extremely common at the time. There may also be an implied connection here to the "Court of the First Fruits and Tenths," a tax which clergy were required to pay upon assuming new ecclesiastical positions, and which consisted of the sum of their annual salary. Twenty pounds would have been a reasonable annual salary for a curate to have earned in the 1750s. I thank Will Deringer for communicating these insights to me in a private conversation, November 16, 2021.

28 HEARING WITH THE MIND

have encountered the rule during their schooling.[38] The Rule of Three, as historian of mathematics Caitlin Rosenthal has emphasized, was "essentially an algorithm designed to be run by a user who lacked training in mathematics."[39] By spelling out how it applied to musical intervals, Holden was making sure that all of his readers could precisely follow his mathematical arguments.

Holden's reference to a Gunter's Line a few paragraphs later, however, suggests that he also had a more sophisticated mathematical audience in mind. A Gunter's Line was an early seventeenth-century forerunner of the slide rule, which would, in its nineteenth-century form, become a ubiquitous calculating tool until it was displaced by pocket calculators in the 1970s. In Holden's day, it was typically a two-foot-long ruler inscribed with intervals marking the logarithms of trigonometric functions, which "enabled the addition or subtraction of distances on the scale, by which, according to the properties of logarithms, products or quotients of numbers could be found."[40] After explaining the logarithmic relation between the sizes of intervals and their ratios (or as Holden prefers to say, their quotients, that is, the number of times a ratio's smaller term divides its larger term), he observes:

> The progression of intervals, and their correspondent quotients, exactly answers to that of Logarithms, and their correspondent natural numbers; as those who understand the nature of logarithms will immediately perceive from the above. The line of numbers, called Gunter's Line, is still more easily applied to the use of attaining proper ideas of magnitudes and proportions of intervals: for here the real-magnitude of every interval is represented by the distance between the two numbers which express its terms; so that nothing could be devised more suitable for this purpose.[41]

---

[38] Holden remarks that he will "lay down a few rules for the management of numbers, in computation relative to musical intervals," and proceeds to observe that just as "when we say in the Rule of Three: as the first term is to the second, so is the third term to the fourth, we mean that in what manner soever the whole or any assignable part of the first term is contained in the second, in the same manner must the whole or the like part of the third term be contained in the fourth: or in other words, we mean that the quotient of the first term divided by, or dividing, the second, must be the same with the quotient of the third term divided by, or dividing, the fourth: and when this is the case, we say the ratio of the first to the second is the same as the ratio of the third to the fourth term." Holden, *Essay*, part 2, §49, 327. On the ubiquity of the rule of three in eighteenth-century education see Caitlin Rosenthal, "Numbers for the Innumerate: Everyday Arithmetic and Atlantic Capitalism," *Technology and Culture* 58, no. 2 (2017): 529–544.

[39] Ibid., 533.

[40] Florian Cajori, *A History of the Logarithmic Slide Rule and Allied Instruments* (New York: Engineering News Publishing Company, 1909), 2.

[41] Holden, *Essay*, part 2, §50, 329.

A few pages later, Holden turns to Roger Cotes's *Harmonia mensurarum* (1722), and specifically to proposition 1, scholium 3, which offers an algorithm "to find the measure of any ratio whatsoever," that is, its logarithm.[42] Holden translates this passage, originally written in concise Latin for an audience of mathematical experts, into English and demonstrates how it would apply to the case of musical ratios.[43] Elsewhere in the *Essay*, he refers to William Emerson's *Doctrine of Fluxions* (1743), and specifically section III, proposition 2 ("to find the motion of a musical string vibrating and very small distances"), and to Robert Smith's *Harmonics, Or the Philosophy of Musical Sounds* (1749) in order to discuss the precise number of vibrations of a string. These three aforementioned sources indicate that he was fluent in calculus.[44]

It appears that Holden's interest in mathematics was shared by his family. Confirmation of his Westhouse origins further allows us to fully ascertain his relationship to Richard (1718–1775) and George Holden (1723–1793), who produced the *Holden Almanack and Tide Tables*, the "first high-quality, publicly-accessible tide tables in the UK" from 1770 on.[45] These tide tables, historian Philip Woodworth writes, were ubiquitous, "reproduced in newspapers, printed as calendars for offices, and, most famously, published as the main component of a diarylike almanac for each year."[46] The mathematical secret of the tables, which was based on a new application of the Bernoulli method to a dataset of Liverpool tide measurements contributed by the English mariner William Hutchinson, was kept within the family for nearly a century.[47]

---

[42] The Latin is "Invenire Mensuram Rationis cujuscunque propositae." Roger Cotes, *Harmonia mensurarum, sive analysis et synthesis per rationum et angulorum mensuras promota* (Cambridge, 1722), 4. The first part of the book, an essay entitled *Logometria,* had been published in the *Philosophical Transactions of the Royal* Society in 1714; however, Holden only refers to the later (posthumous) collection of Cotes's writings.

[43] On the obscurity of the original and its intended audience of specialists, see Ronald Gowing, *Roger Cotes: Natural Philosopher* (Cambridge: Cambridge University Press, 2002), 23. On the terseness of Cotes's Latin, see Christa Jungnickel and Russell McCormmach, *Cavendish: The Experimental Life* (Lewisburg: Bucknell, 1999), 149.

[44] Holden, *Essay,* part 2, §47, 323–324.

[45] Philip L. Woodworth, "Some Further Biographical Details of the Holden Tide Table Makers," *Proudman Oceanographic Laboratory Report* 58 (July 2003): 3–20. The first Holden tide table came out in Liverpool in 1770. As mentioned earlier, Holden's Westhouse roots are also confirmed in Willis, *Winchester Ordinations,* 68; they also appear in an obituary for his son, John Holden Jr. See "John Holden," in *The Annual Register, Or, A View of the History, Politics, and Literature for the Year 1806,* ed. Thomas Morgan (London: Otridge and Son, 1806), 546.

[46] Philip L. Woodworth, "Three Georges and One Richard Holden: The Liverpool Tide Table Makers," *Transactions of the Historic Society of Lancashire and Cheshire* 151 (2002): 20.

[47] See ibid. for Woodworth's reconstruction of this secret.

30   HEARING WITH THE MIND

In his investigation of Richard and George Holden, Woodworth provides the baptismal records of their siblings, where we learn that John was born in Clapham, Yorkshire, in 1729 (Francis [II], probably the contributor to the *Diary*, was born in 1737).[48] Holden's father, Francis (I), appears in the parish records as a "yeoman," or freeholding farmer—that is, the class below the gentry and above the tenant farmers. Francis (I) owned a farm at Westhouse and other properties, and he left these in his will to his sons George and Thomas, with his other sons John, Proctor, Richard, Francis (II), Christopher, and Robert inheriting just five shillings each.[49] However, many of the brothers achieved subsequent intellectual and mercantile success: Richard, Thomas, and Christopher went on to own businesses in Liverpool, while George took up various clerical positions, culminating in the perpetual curacy of Tatham Fells, and Proctor became a schoolmaster (Richard later joined him in the profession as a teacher of mathematics). Francis (II)'s aforementioned contributions to the *Diary* indicate that he was a promising mathematician in his own right, but I have found no trace of him after 1755. John Holden's family tree is provided in the Appendix on page 257.

It thus appears that many members of the Holden family were exceptionally well educated. This point also comes up in a different document that corroborates John Holden's relationship to the Westhouse Holdens and to Glasgow: the English historian and abolitionist William Roscoe (1753–1831) left a lengthy biographical sketch of his closest childhood friend, Francis Holden (1752–1788), the son of Proctor Holden, John Holden's brother, not to be confused with his uncle or his grandfather. Roscoe's sketch describes this younger Francis, whom I term Francis (III), as "descended from a family, many members of which had distinguished themselves by their progress in scientific pursuits."[50] Francis (III) tutored Roscoe in ancient and modern languages, having also taught mathematics and languages at his uncle Richard's Liverpool academy.[51] In early 1772, he moved to Glasgow to study at the university, writing to Roscoe that his uncle [John] had "died the morning after I arrived. He declared all along that he only lived to see

---

[48] See Woodworth, "Some Further Biographical Details," 3.

[49] Francis Holden, "Will and Testament," March 13, 1741, Clapham, Yorkshire. Registered probate, vol. 87, f.557 MF 1007.

[50] Henry Roscoe, *The Life of William Roscoe* (London: T. Cadell, 1833), 1:18.

[51] Francis is mentioned in Proctor Holden's obituary from 1810: "Aged 83, Mr. Proctor Holden, formerly master of the free grammar-school at Westhouse, near Ingleton, Yorkshire. He was brother to the late Rev. G. Holden, calculator of the tide-table for Liverpool, and father of that universal scholar, the late Francis Holden, of Trinity College, Cambridge." "Proctor Holden," *The Monthly Magazine* 29 (June 1810): 502.

me."[52] Francis (III) would end up lodging with "the widow of his uncle, John Holden, well known by his learned treatise on music."[53]

Finally, John Holden's lineage is again verified by a footnote in a death notice of a music-loving Liverpudlian named John Runcorn, whose favorite book, we learn in an obituary, was Holden's treatise, his copy having "most of the blank spaces filled up, and many places interleaved with curious and learned MS. Illustrations."[54] The author of this memorial adds in a footnote:

> The Holdens have been a family of genius; the author above mentioned, whose name was John, arrived to a title of honour in the University of Glasgow. To complete his work, such was the enthusiasm of Mr. H. that he studied more than one language, to enable him to peruse the works of some particular authors, who had written on his favourite subject. Richard H. (brother to the above, and an intimate friend of the deceased,) was well known to many in the [Liverpool] neighbourhood of Walton, and for several years kept a mathematical school in Liverpool, and in the latter part of his life established an academy at Rainford, which, at this time, is conducted with much credit by two nephews. Another brother, with the assistance of his son, are the calculators of the Tide-table, published annually under their name. If an account of this ingenious and large family was collected, and given to the world, is would make an entertaining and interesting memoir.[55]

Glaswegian Burgess

I have found no documentary trace of Holden between his last mathematical contribution to the *Diary* in 1754 and his reappearance in Glasgow three years later as a merchant potter. Holden's decision to open a pottery in Glasgow would have been a sensible one at the time. It was only in the late 1740s that the city saw the founding of its first industrial pottery—the

---

[52] Roscoe, *The Life of William Roscoe,* 20.

[53] Ibid., 20.

[54] Edward Cave, ed., "John Runcorn," in *The Gentleman's Magazine and Historical Chronicle* 57, no. 2 (1787): 1195. Runcorn's annotated copy of the 1770 edition of the *Essay* would later appear in the sale catalogue of the English antiquarian John Sidney Hawkins's library in 1843. Anon., *Catalogue of the Extensive and Valuable Library of the Late John Sidney Hawkins, Esq. F.S.A. . . . sold by Auction by Mr. Fletcher* (London: Fletcher, 1843), 109.

[55] Cave, ed., "John Runcorn," 1195.

32 HEARING WITH THE MIND

famous Delftfield pottery company, whose workers were all Englishmen—by local merchants who regarded pottery as a promising investment given the city's access to American colonies via Scottish ports. Moreover, the nascent Scottish pottery scene attracted some noteworthy characters: in 1768, for example, the polymath inventor James Watt joined Delftfield as a managing partner, where he introduced many innovations and improvements in the manufacturing process, enjoying a stream of regular income that enabled him to fully transition into his career as an engineer.[56]

By 1757, Holden appears to have set up shop in this profession, having joined the city of Glasgow's merchant guild as a burgess by purchase on July 8 of that year. Holden wasted no time: by December, he had already sent an entire hogshead—a barrel with a volume of approximately 300 liters—containing hundreds of items of pottery to Port Tobacco, Maryland, under the auspices of John Semple, a Scottish merchant based in Maryland.[57] These included relatively upscale articles: enameled and carved sauce boats, for example, as well as teacups and saucers, all of which were aimed at a well-to-do clientele.[58] The presence of these relatively expensive and labor-intensive items suggests that Holden was running a substantial operation which involved at least a dozen workers.[59]

Holden's payment of guild-brother and burgess fees—memberships amounting to seven pounds, eighteen shillings, and threepence in the municipal guild, which allowed him to set up business in the city of Glasgow—and the speed at which he was able to embark upon transatlantic trade make it highly probable that he brought some capital with him and purchased an

---

[56] Richard L. Hills, "James Watt and the Delftfield Pottery, Glasgow," *Proceedings of the Society of Antiquaries of Scotland* 131 (2001): 375–420.

[57] See Martin, "Scottish Merchants," 34–36. Semple would later correspond at length with George Washington about his initiative for developing the Potomac; see David C. Skaggs, "John Semple and the Development of the Potomac Valley, 1750–1773," *The Virginia Magazine of History and Biography* 92, no. 3 (1984): 282–308.

[58] Martin, "Scottish Merchants," 35. Semple's cargo manifest also includes a delivery from the Delftfield pottery company of cruder ceramics at lower prices. Ibid.

[59] In his 1795 account of the Staffordshire earthenware industry, John Aikin noted that the manufacture of any "single piece of ware, such as a common enamelled tea-pot, a mug, jug, &c. passes through at least fourteen different hands before it is finished, viz. The slipmaker, who makes the clay; The temperer, or beater of the clay; The thrower, who forms the ware; The ballmaker and carrier; The attender upon the drying of it; The turner who does away its roughness; The spoutmaker; The handler, who puts to the handle and spout; The first, or biscuit fireman; The person who immerses or dips it into the lead fluid; The second, or gloss fireman; The dresser, or sorter in the warehouse; The enameller, or painter; The muffle, or enamel fireman. Several more are required to the completion of such piece of ware, but are in inferior capacities, such as turners of the wheel, turners of the lathe, &c. &c." John Aikin, *A Description of the Country from Thirty to Forty Miles Round Manchester* (London: John Stockdale, 1795), 534–535.

extant business.[60] This inference is likewise embraced by historian of Scottish crafts Michael Donnelly—the only researcher I am aware of who has studied Holden's pottery in any depth—in a two-page contribution to the Scottish Pottery Society's 1978 journal.[61] Examining two letter fragments from a cache of eighteenth-century business records from the very early 1760s, in which John Bucknall, a partner in a major Staffordshire pottery firm, asks after Holden and his wares, Donnelly surmises that Holden may have spent time as "a partner or at least a manager of a Staffordshire pot-house prior to his arrival in Glasgow."[62] I have not been able to confirm whether this was indeed the case.

In view of the fact that the skills involved in running a pottery on an industrial scale were not to be found in Scotland at the time, Donnelly's hypothesis of Holden's earlier professional experience in Staffordshire, the capital of British pottery, is certainly plausible.[63] However, given the success of Holden's brothers in establishing their own businesses in Liverpool— Richard owned and operated a foundry before opening a mathematical school, while Christopher owned a foundry across the street from Thomas's brewery—it is also possible that Holden could have acquired his training in that city, which boasted a lively community of potters in the Shaw's Brow neighbourhood.

We might well wonder where the fourth son of a yeoman—who, like most of his brothers, had inherited very little from his father—could have obtained sufficient capital to purchase his guild membership and set up a successful pottery in a new city. Perhaps initial support for Holden's mercantile endeavours was supplied by the family of his wife, Isabella Fawcett of Dent, Yorkshire.[64] The Fawcetts and the Holdens had a family history of close

---

[60] This sum appears in George Crawfurd, *A Sketch of the Rise and Progress of the Trades' House of Glasgow* (Glasgow: Bell & Bain, 1858), 142.

[61] In *Scottish Pottery*, Fleming spends a few lines on Holden's establishment, mistakenly dating it to the early eighteenth century. John A. Fleming, *Scottish Pottery* (Glasgow: EP Publishing, 1973), 132. As Donnelly writes, "Of all the Glasgow potteries listed by Arnold Fleming, Holden's in Gorbals remains the most obscure." Michael Donnelly, "John Holden's Gorbals Pottery, 1762–1786," *Scottish Pottery: The Journal of the Glasgow Branch of the Scottish Pottery Society* 1 (1978): 2–3. More recently, Ann Smart Martin describes Holden as a "little known potter . . . [he] and his business appear to be a cipher." She does, however, cite the biographical information found in Love's aforementioned *Scottish Church Music*. Martin, "Scottish Merchants," 34–35.

[62] Donnelly, "John Holden's Gorbals Pottery," 2.

[63] The dearth of Scottish potters is evident in the protracted legal battle between the Delftfield Company and its English manager, John Bird. A detailed summary of this case is provided in Jonathan Kinghorn and Gerard Quail, *Delftfield: A Glasgow Pottery 1748–1823* (Glasgow: Glasgow Museums and Art Galleries, 1986).

[64] The village of Dent, about ten miles north of the hamlet of West House, where Holden had grown up, was, in the eighteenth century, a lively bastion of the wool trade, its inhabitants famously

34 HEARING WITH THE MIND

ties: another Isabella ("Bella") Fawcett, evidently a relation of John's wife Isabella, was married to John's brother Christopher.[65] Christopher and Bella joined John and Isabella in Gorbals, Scotland around 1761, but by 1766 this branch of the Holden family appears to have moved to Liverpool (we find public records of Christopher Holden, a brass founder in Liverpool through 1783).[66] Available records indicate that Christopher Holden benefitted from his wife's family connections in the form of a significant loan, which his brother-in-law, the wealthy attorney James Fawcett, intended to forgive in favor of his nieces at some point in the future.[67]

Isabella Holden appears to have maintained close ties with her family in the Yorkshire Dales, and in 1782, her son Richard would be articled to this same James Fawcett as a clerk as part of his qualification as an attorney.[68] We might speculate that Holden, too, could have had some kind of business dealings with his wife's extended family. This relationship would have likely begun following the couple's engagement, as John and Isabella were married in Dent on October 7, 1759, two years after Holden had moved to Glasgow and set up his pottery.[69]

Holden's mercantile success can be further ascertained from an announcement in a 1761 issue of the *Glasgow Journal* that confirms that he was renting commercial space from "Dreghorn" (probably the wealthy architect Allan Dreghorn, a prominent Glaswegian property owner) in one of the city's main

---

specializing in knit caps and stockings that were exported throughout Britain. As we learn from the Holdens' marriage bonds, John's bondsman—typically a close friend of the groom—was "James Fawcett, hosier," evidently Isabella's relative. On Dent, see Adam Sedgewick, *A Memorial by the Trustees of Cowgill Chapel with a Preface and Appendix, on the Climate, History and Dialects of Dent* (Cambridge: University Press, 1868).

[65] See the will of James Fawcett of Kirkby Stephen, gent made January 2, 1777, in WD U/Box 59/3/T 22–23 Cumbria Archive Centre, Kendal, as confirmed by comparison with the family tree of the Holden family of Westhouse & Clapham, ref 17/18, Cambridgeshire Archives, Ely.

[66] With regard to Fawcett's business dealings see, e.g., the shopkeeper Abraham Dent's mortgage of his lands, described in Thomas Stuart Willan, *An Eighteenth-Century Shopkeeper: Abraham Dent of Kirkby Stephen* (Manchester: Manchester University Press, 1970), 136.

[67] The will of James Fawcett of Kirkby Stephen (not to be confused with the James Fawcett of Dent, hosier, who was John Holden's bondsman), a wealthy attorney and landowner in the region and the brother of Christopher's wife Bella, mentions that the "mortgages and securities from brother-in-law Christopher Holden with principal sum of £550 due thereon and also further sum of £200 to be raised as after-mentioned from real estates" are to be managed by James's brother John "in maintenance and education of testator's nieces ... until age of 21 years in such manner as he shall think most proper and beneficial for them." See the will of James Fawcett of Kirkby Stephen.

[68] Articles of Clerkship for Richard Holden, articled to James Fawcett (1782), CP 5/121/1, The National Archives, Kew.

[69] The marriage bonds lists John Holden as a merchant from Glasgow, Scotland. See "England, Lancashire, Marriage Bonds and Allegations, 1746–1799."

avenues, Stockwell Street, alongside the Scottish merchants James Scott, Robert Donald, and David Leitch.[70] While the lack of archival evidence does not allow us to ascertain the nature of his relationship with these men, they seem to have been well-established and prosperous: David Leitch was a merchant whose family had significant dealings in the American colonies, while Robert Donald (1724–1803) was one of the Glasgow Tobacco Lords and would go on to become Lord Provost of Glasgow. By 1763, as an advertisement in the *Glasgow Journal* indicates, Holden operated a pottery and glass store on Stockwell, where he sold "earthen ware, in several branches never before attained to in North Britain" and also offered "Newcastle glass of the usual kinds . . . sold on commission from Messrs Airey, Cookson and co."[71]

Additional evidence suggesting that Holden's pottery was a profitable endeavour can be gleaned from an advertisement in the *Glasgow Mercury* of April 19, 1781, which informs the public of the impending sale of the "pottery adjoining the Gorbals Church, which belonged to the deceased Mr. John Holden, and presently possessed by Mr Andrew Boag. If any person incline to make a private bargain, they will please apply to Mrs Holden, at Burrel Hall, any time before the roup [public auction]."[72] Boag, Donnelly informs us, was "one of Glasgow's major china and glass dealers," which suggests that the Holden pottery was sufficiently profitable so as to be worth operating a decade after its founder's demise.[73] More indicative of Holden's business acumen is the fact that nearly a decade after his decease Isabella Holden was still living at an upscale address: Burrel Hall, a mansion in central Glasgow.[74]

Nonetheless, we can clearly establish that Holden's widow and children experienced some financial insecurity following his death, as alluded to by Thomas Reid's aforementioned letter nominating the promising young John for a fellowship on account of his recent bereavement.[75] Moreover, from

---

[70] An advertisement in the *Glasgow Journal* of 1761 indicates that "By publick roup [auction], within the Exchange Coffee-house, Glasgow, upon Wednesday the 14th of October next, at 12 of the clock mid-day, to be sett for the space of seven years, from Whitsunday next, the several lodgings in Stockwell of Glasgow, belonging to Mr. Dreghorn, presently possess'd by Mr. James Scot, Mr. Robert Donald, Mr. David Leitch, and Mr. Holden." Advertisement, *Glasgow Journal,* no. 1049, September 3–10, 1761.

[71] Advertisement, *Glasgow Journal,* no. 1127, March 3–10, 1763.

[72] Robert Reid, *Glasgow, Past and Present: Illustrated in Dean of Guild Court Reports* (Glasgow: David Robertson, 1884), 3:85.

[73] Donnelly, "John Holden's Gorbals Pottery," 2.

[74] George Eyre-Todd, *The Book of Glasgow Cathedral: A History and Description* (Edinburgh: M'Farlane & Erskine, 1898), 392. We cannot rule out the possibility that Isabella Holden merely suggested Burrell Hall as a convenient meeting point.

[75] Reid, *Correspondence,* 74.

36 HEARING WITH THE MIND

Holden Jr.'s registration records at Sidney Sussex College in Cambridge in 1780 we learn that he was admitted as a "sizar," a student receiving financial aid in conjunction with certain menial duties. The reasons for the Holdens' fiscal difficulties may have had not only to do with the passing of the family's main breadwinner (and much of his wealth presumably being locked up in his business), but also with Holden's having also embarked upon the career of teacher just a few years before his death. Various expenses relating to this change of career—ranging from the acquisition of a spacious home suitable for student boarders to the substantial donations made to the college chapel fund in hopes of solidifying Holden's status as a music teacher—did not have time to bear their intended fruit.[76] The university minutes from 1782, a full decade after Holden's death, reveal that the university agreed to return the annual salary of five pounds, along with the accumulated interest, which Holden had habitually donated back to the university during his tenure as director of the chapel band, to his widow and children. [77] (Isabella Holden did, in fact receive the sum of fifty-six pounds, eleven shillings, and ninepence from the faculty in April 1783).[78]

In embarking upon a teaching career, then, Holden would have left at least some of the day-to-day running of his pottery business in the hands of others. Some of the work would have been delegated to apprentices—we know of at least one such employee, Alexander Grunton, who became a merchant burgess and guild brother of the city of Glasgow in 1780, in consequence of "serving apprentice with John Holden, merchant and potter."[79] Moreover, as the archival evidence we have seen indicates, Isabella Holden (**Figure 1.3**) could read, write, and would later negotiate the sale of her husband's pottery,

---

[76] On the spaciousness of Holden's home (with its sixteen taxable windows), see note 19 of this chapter.

[77] The minutes state: "As the fund called the music fund, in the chapel accepts began by Mr Holden giving the 5 pounds which he was allowed annually, out of the chapel fund as precentor in the college chapel. The faculty, on account of his great merit as a musician, agrees to give unto his widow and children the total of the five pounds which he was entitled to receive. Together with the interest thereof. And the chapel committee, together with Mr. Anderson who is guardian to the children, is appointed to ascertain that sum and to report." Minutes of the University of Glasgow Faculty for June 10, 1782, Glasgow University Archive, vol. 77:191–192. The Mr. Anderson described in the report can be identified with John Anderson (1726–1796), professor of natural philosophy at Glasgow College and the founder, through a bequest, of the Andersonian Institution in 1796, an educational establishment aimed at the working classes, now the University of Strathclyde.

[78] Minutes of the University of Glasgow Faculty for April 11, 1783, Glasgow University Archive, vol. 77:233–234.

[79] James R. Anderson, ed., *The Burgess & Guild Brethren of Glasgow, 1751–1846* (Edinburgh: Skinner, 1935), 163.

Figure 1.3 *Isabella Holden, nee Fawcett*. Miniature belonging to Jane Isabella Holden, John Holden's granddaughter.

suggesting that she had some business competence of her own.[80] The historian of pottery Gerard Qual affirms that the social history of eighteenth- and nineteenth-century Glasgow includes "several stalwart potters' widows, who, at a period when women were not accustomed or encouraged to hold positions of authority, competently carried out the daily task of management of their family pottery on the demise of their loved one."[81]

Evidence pertaining to the lives of Holden's sons suggests that they prospered in spite of the hardship they experienced from the loss of their father at an early age. In the obituary of the eldest, we learn that "Mr. H. in his youth visited the better part of Europe, and returned to his own land improved; having joined an acquaintance with the living tongues to his Greek and Latin attainments."[82] The fact that John Jr. could travel abroad—even had he done so as a tutor or secretary in the employ of a nobleman—suggests

---

[80] More possible details about what her education may have been like can be ascertained from the geologist Adam Sedgwick's 1868 recollections of life in Dent in the eighteenth century: "All the women with very rare exceptions learned to read; and the upper Statesmen's [prominent families] daughters could write and keep family accounts. They had their Bibles, and certain good old-fashioned Books of Devotion; and they had their Cookery Books; and they were often well read in ballad poetry, and in one or two of De Foe's novels. And some of the younger and more refined of the Statesmen's daughters would form a little clique, where they met—during certain years of the last century—and wept over Richardson's novels." Sedgwick, *A Memorial*, 69–70.

[81] Gerard Quail, "The Millroad Street Pottery Calton: The Background—Thomas Wyse and his Stoneware Shop, 1799–1814," *Scottish Industrial History* 18 (1996): 37.

[82] Edward Cave, ed., "Obituary of the Rev. John Holden," *The Gentleman's Magazine and Historical Chronicle* 76, no. 2 (1787): 878–879.

38 HEARING WITH THE MIND

that he must have come from a position of relative social and financial privilege.[83] We also learn from the same document that "his family connections are above mediocrity," and that "his conversation proved him the accomplished scholar, his manners the polished gentleman."[84] This glowing account of John Jr.'s pedigree and comportment suggest that he was regarded as a gentleman, no mean achievement given his circumstances in early life.

If anything, the slope of Holden's younger son Richard's (1768–1809) ascent within society was even steeper. An attorney by profession, he married an heiress in Rotherham, Yorkshire and lived at Moorgate Hall (**Figure 1.4**), on an estate inherited from his wife's uncle.[85] During his lifetime, Richard Holden mingled in high society, as evident in his correspondence with William Wilberforce (the noted abolitionist), Francis Ferrand Foljambe (the M.P. for Yorkshire), and George Osborne (the Duke of Leeds).[86] He is remembered in Rotherham today as a talented amateur painter (**Figure 1.5**), and many of his paintings are still extant in the local museum.[87]

The family's upward mobility was in many ways characteristic of their day.[88] Other members of Holden's intimate circle experienced a similar dramatic rise in their fortunes, notably his brother Richard, who sold his foundry in 1760 and opened a successful school for mathematics in Liverpool; his brother George, who, as we saw earlier in this chapter, devised

---

[83] Ibid. Travel abroad in the eighteenth century was no trivial matter: Adam Smith, for example, resigned his professorship at the University of Glasgow in order to tutor the young Duke of Buccleuch during his travels abroad in 1764–1766.

[84] Ibid.

[85] "Mr. [Richard] Holden combined with his profession of the Law the somewhat incongruous accomplishment of that of artist; and, in his day, whilst taking the lead in his profession, Moorgate Hall was rich in family portraits painted by him." Guest, *Yorkshire. Historic Notices of Rotherham*, 528.

[86] Wilberforce and Wilberforce, *The Life of Wm. Wilberforce*, 132; Richard Holden, "Letter to Francis Ferrand Foljambe of Hertford Street, London," of June 1, 1794, The National Archives, Kew, HO 42/31/15 Fol. 59–60; Richard Holden, "Letter from Mr. Holden to his Grace the Duke of Leeds, Read November 26, 1795," in *Archaeologia: or, Miscellaneous Tracts Relating to Antiquity* 12 (1796): 207–208.

[87] In 1953, furthermore, the *Guardian* printed excerpts from his account of a six-week family trip to the Lake District, in which he described various Northern cities, among them Liverpool, which he critiqued for having made a "a disgraceful & ineffectual selfish resistance," to the "suppression of the slave Trade." Richard Holden, *Diary of a Journey to the Lake District* (1808), reprinted as "A Northern Tour in 1808," in *The Guardian*, September 10, 1953, 6. Other excerpts from the diary appear in the editions of September 11, 14, and 16, 1953.

[88] This social transformation would later be digested in fiction—we might think here of the many startling class distinctions arising from advantageous or ill-favored marriages in Jane Austen's novels, as well as her depiction of ambitious lower middle-class characters such as the plucky sailor William Price in *Mansfield Park* (1814), "distinguishing himself and working his way to fortune and consequence" by serving in the West Indies and the Mediterranean. Jane Austen, *Mansfield Park: Authoritative Text, Contexts, Criticism*, ed. Claudia L. Johnson (New York and London: Norton, 1998), 162.

Figure 1.4 Moorgate Hall, illustrated in Guest, *Yorkshire. Historic Notices of Rotherham,* 526.

Figure 1.5 Richard Holden, *Review of Local Volunteers on Brinsworth Common, Rotherham, South Yorkshire 1804.*

40 HEARING WITH THE MIND

and sold the lucrative *Holden Tide Tables* based on Richard's computations; and his aforementioned nephew Francis (III), who opened a school in Cambridge.[89] To give a few comparable examples from beyond Holden's immediate family: Thomas Simpson (1710–1761), the editor of the *Diary* which we encountered earlier in this chapter, was a weaver by day and taught himself mathematics at night; eventually, he became a teacher of mathematics and joined the Royal Society in 1745; and Charles Hutton (1737–1823), whom we will meet briefly in Chapter 6 as a supporter of the science writer Margaret Bryan, worked as a coal hewer in his youth, and he later became a teacher of mathematics, joining the Royal Society in 1774.

Holden's carefully executed professional transition from running a pottery to teaching church music suggests that he regarded music theory as a conduit to the more prestigious world of learning and letters. In that, too, he was not unusual: no less a figure than Rousseau (whose *Dictionnaire* Holden cites) hoped that approval of his new system of musical notation by the Académie des Sciences would win him a great reputation and thereby make his fortune.[90] Holden's fellow Glaswegian, the inventor of the steam engine James Watt, began his career as a musical-instrument manufacturer and dabbled in music theory (along with the Glasgow College physicist John Robison) in the 1760s.[91] Another famous example of a middle-class artisan who publicized his activities in music theory is the English clockmaker John Harrison, who published his invention of the marine chronometer along with a theory of tuning (with references to his experience leading psalmody) in *A Description concerning such Mechanism as will afford a nice, or true Mensuration of Time; together with*

---

[89] See Richard Holden, "Advertisement," *The General Evening Post (London)*, June 12–14, 1760, 4.

[90] In his *Confessions* VII.2.1 (1782), for example, Rousseau writes that he departed "regretfully for Paris ... counting on my system of musical notation as though on a fortune that was already assured" (Je me suis laissé, dans ma première partie, partant à regret pour Paris ... et comptant sur mon système de musique comme sur une fortune assurée), and notes a few pages later that, "engrossed in my musical system, I concentrated on using it to try and effect a revolution in that art, and thus to achieve the sort of celebrity that in the fine arts in Paris always goes hand in hand with success." (Concentré dans mon Système de musique, je m'obstinois à vouloir par lui faire une révolution dans cet art, et parvenir de la sorte à une célébrité qui dans les beaux arts se conjoint toujours à Paris avec la fortune.) Jean-Jacques Rousseau, *Confessions*, trans. Angela Scholar (Oxford: Oxford University Press, 2000), 271, 278; Jean Jacques Rousseau, *Les Confessions* in *Œuvres completes I: Les confessions; Autres textes autobiographiques*, ed. Bernard Gagnebin, Marcel Raymond, and Robert Osmont (Paris: Gallimard, 1959), 280, 286.

[91] See Michael Wright, "James Watt: Musical Instrument Maker," *The Galpin Society Journal* 55 (2002): 104–129; for an account of Watt's music-theoretical interests, see John Robison, "A Narrative of Mr Watt's Invention of the Improved Engine by Professor John Robison," in *The Life of James Watt: With Selections from his Correspondence*, ed. James Patrick Muirhead, 46–47 (London: John Murray, 1858).

Some Account of the Attempts for the Discovery of the Longitude by the Moon; and also An Account of the Discovery of the Scale of Musick (1776).[92]

Upon arriving in Scotland in the 1750s, therefore, Holden benefited from a confluence of related developments: the new economic opportunities afforded by the British empire, the growing city of Glasgow, the rise of the Scottish Enlightenment, and, as we will see in Chapter 3, the consequent weakening of Presbyterian strictures against part singing in church. As a music-loving English businessman, he had the dexterity to take advantage of the moment, bringing desirable mercantile and musical skills to a relatively open market. The same general trend of increased opportunity continued throughout the lives of his sons, who enjoyed far more formal education than their father, and who were regarded as gentlemen. Although most of the family's activities took place in Glasgow, Rotherham, and Cambridge, the Holdens' rising fortunes nonetheless exemplified the new economic opportunities afforded by the expanding British empire at the time.

---

[92] Holden's brother Richard refers to "Mr. Harrison's time-keeper" in describing his method in an advertisement in *Williamson's Liverpool Advertiser & Mercantile Chronicle* (February 3, 1764): 2–3, cited in Woodworth, "Some Further Observations," 16. On Harrison's music theory see Loren Ludwig, "J. S. Bach, the Viola da Gamba, and Temperament in the Early Eighteenth Century," *BACH: Journal of the Riemenschneider Bach Institute* 53, no. 2 (2022): 260–300.

2

An Eighteenth-Century Theory of Musical Cognition?

John Holden's *Essay towards a Rational System of Music* (1770)

In 1863 the Englishman Charles Isaac Stevens (1835–1917) submitted a doctoral dissertation in absentia to the Georg-August-Universität Göttingen. Titled "An Essay on the Theory of Music," the work derived a speculative theory of music perception from the hypothesis that the mind has an innate tendency to group all stimuli into small units of equal size—an essentially cognitive approach foreshadowing experimental work on rhythm perception from the late nineteenth century, as well as key findings in Gestalt psychology. The author of this forward-looking work, however, was not in fact Stevens. He lifted the text verbatim, wholesale, and without attribution from a work published in Glasgow nearly a century earlier: John Holden's *Essay towards a Rational System of Music* (1770). As far as I can tell, Stevens's plagiarism was never detected. He lived out the rest of his life in London, eventually assuming leadership positions in the Free Protestant Episcopal Church of England. Living well before the era of searchable text, Stevens had little reason to fear that his deception would ever come to light.[1]

For Stevens's examiners, the dissertation's exploration of music perception may have seemed attuned to contemporaneous interest in the physiology and psychology of audition, evidenced in works such as Hermann von Helmholtz's *Die Lehre von den Tonempfindungen als physiologische Grundlage für die Theorie der Musik* (1863) and Hugo Riemann's dissertation *Über das musikalische Hören* (1874), which as it happens was submitted to the same institution.[2] Even if we grant that these readers might have simply

---

[1] Stevens plagiarized only the first three chapters of the second part of Holden's *Essay* and cannily removed all references and citations that could have dated the manuscript.

[2] Riemann's dissertation was published as *Musikalische Logik: Hauptzüge der physiologischen und psychologischen Begründung unseres Musiksystems* (1874). The philosopher Hermann Lotze evaluated the dissertations of both Stevens and Riemann, as well as that of his own

---

Hearing with the Mind. Carmel Raz, Oxford University Press. © Oxford University Press 2025.
DOI: 10.1093/9780197786208.003.0003

rubber-stamped the English-language dissertation, the very fact that Stevens considered a century-old treatise worth pirating attests to the novelty of its ideas to European musical discourse. Holden's comprehensive exploration of cognitive preferences, manifested in a system of harmony as well as various aspects of rhythm and pitch perception, is unlike any other eighteenth-century theory of music. Stevens's act of plagiarism demonstrates the extent to which the *Essay* anticipated, by nearly a century, developments in philosophical, psychological, and musical thought.

Still, my subject here is neither Stevens nor the state of nineteenth-century music theory. Rather, this chapter focuses on the speculative theory of Holden's treatise itself. Perhaps the most exciting aspect of the *Essay* for modern readers lies in its explicit ambition to explain musical practice by means of a limited set of psychological first principles. Relying chiefly on introspection, the work describes phenomena that we today conceptualize as grouping, chunking, subjective rhythmicization, and just-noticeable difference. The *Essay* also recognizes the distinct contributions of the mental faculties of memory and attention to our musical experience. The originality and scope of Holden's theory are thus exceptional both within the Enlightenment and beyond.

Holden seeks, by investigating the operations of the mind, to elucidate the nature of our inborn mental tendencies and their role in our experience of hearing music. His methodology therefore prioritizes human behavior over the laws of counterpoint or physics. Criticizing not only mathematicians who "have bestowed their labour on wrong materials, for want of sufficient practice in music" but also musicians who have "adopted erroneous principles, without submitting them to accurate trial,"[3] he declares that his *Essay* will examine "other properties which have hitherto passed almost unobserved, and which seem to constitute a very essential part of the theory of music."[4] This aim may explain his reversal of the customary order of the first and second parts of a musical treatise: rather than initially presenting abstract theories that will subsequently be applied to the acquisition of practical skills, the *Essay* first expounds the rudiments of musical practice, often

---

advisee Carl Stumpf, although Stumpf's dissertation does not deal with music. See William R. Woodward, *Hermann Lotze: An Intellectual Biography* (Cambridge: Cambridge University Press, 2015), 468–472.

[3] Holden, *Essay*, part 2, §1, 284.
[4] Ibid., part 2, §2, 284.

44 HEARING WITH THE MIND

with experiential description, and then distills them into psychological principles. It therefore seems clear that Holden understood his focus on the perception of music as diverging from both the speculative and the practical traditions of his day.

Holden's explanation of how we perceive music relies chiefly on the actions of posited mental faculties including attention, memory, imagination, and expectation. These concepts allow him to develop detailed speculations about a range of conscious and unconscious mechanisms of perception. The key difference between Holden and every other eighteenth-century music theorist is his comprehensive grounding of musical experience in a hierarchy of explicitly mental actions that are largely independent of any affective or mimetic response. His explicit focus on the psychological first principles of our minds can thus be understood as an early attempt at what we today would recognize as a cognitive theory of our mental processing of music. Seen in this light, Holden's approach is best characterized as synthesizing the rationalist, cognitively oriented approach to music theory ascribed to Jean-Philippe Rameau with Scottish philosophy of perception, in an effort to explain precisely how the human mind perceives, grasps, and enjoys music.

Isochronous parcels and the analogy to rhythm

In line with the ideals of eighteenth-century rationalism, Holden begins the theoretical portion of his *Essay* with a first principle, one that he regards as primary and foundational, and which he introduces as a means of answering a fundamental question about music: What is the criterion according to which "the ear chuses and refuses" particular pitches as constituents of a musical context, such as a scale or a piece? "We want, if possible," he writes, "to know *why* the ear chuses and refuses as we find it does, and for this purpose, some new principle must be adopted."[5] He goes on:

> The new principle we have here to propose, as being that whereby the various choices of a musical ear are best accounted for, is that of our distributing the vibrations of musical sounds by *isochronous*, or *equal*

---

[5] Ibid., part 2, §9, 288; my italics. Holden raises the question by way of a critique of Rameau, who (he believes) had tried to answer it in his *Génération harmonique* (1737) but had arrived at an answer that was, in Holden's view, merely circular ("this is only proving that the ear chuses such sounds *because it chuses them*," ibid.; original italics.

timed parcels, something very similar to the distributions we find naturally to be made among quavers, or other short notes, in the timing of music.[6]

Holden here asserts that our perception of both pitch and rhythm in music is structured by an innate strategy of grouping sounds into small equal units, which he terms *isochronous parcels*. Our preference for certain rhythms and pitch relationships is thus a consequence of this single far-reaching imperative. According to Holden, it is the role of the ear to determine the isochronism of successive parcels of vibrations and thereby to conclude whether the sound has a recognizable pitch. It is guided in this by the same principles that govern musical rhythm.

Holden's emphasis on the workings of the ear allows him to avoid having to claim that we are able actually to count, and thereby determine the frequency of, sound vibrations.[7] "It is one thing," he writes, "for the sense to be pleased with a number, and another thing to count it."[8] While individual vibrations may be too fast for us to grasp consciously, there is no need for them to be counted, for the ear is capable of simply assessing incoming sounds regarding the equality of their consecutive vibrations. Thus Holden compares the experience of sensing a given pitch to that of hearing a drum roll, where "the quick succession renders it impossible to count the single pulses; all we can here do is to judge of, and I may say—feel—the isochronism of each."[9] The essential element in our hearing of pitch and rhythm, then, is the ear's recognition of isochronous parcels of sound impulses—at slower time scales for rhythm and faster ones for pitch.

Whereas *the ear* gauges the isochrony of sounds, Holden entrusts *the mind* with a higher task: "among the isochronous single vibrations of musical sounds the mind naturally seeks to constitute isochronous compound parcels."[10] These isochronous compound parcels are cumulative or nested groupings (groups of groups), and Holden argues that our musical preferences result from this cognitive strategy of equal grouping. Again, this approach is shared across a range of perceptual domains:

---

[6] Ibid., part 2, §10, 288.

[7] As David E. Cohen has shown, Rameau's theory assumes the "musical ear" extracts the consequences of the *corps sonore*—including the diatonic succession, the fundamental bass, and the perfect (authentic) cadence—from the music itself. Cohen, "The 'Gift of Nature,'" 68–74. However, Rameau does not attempt to provide an account of the mechanism of unconscious processing involved in music perception.

[8] Holden, *Essay*, part 2, §10, 288.

[9] Ibid., part 2, §15, 293.

[10] Ibid., part 2, §15, 292.

46 HEARING WITH THE MIND

> Where equal and equidistant objects affect our senses ... there is a certain propensity in our mind to be subdividing the larger numbers into smaller equal parcels; or as it may be justly called, compounding the larger numbers of several small factors, and conceiving the whole by means of its parts.[11]

Holden thus regards our tendency to understand large numbers as compounded of smaller factors as an innate cognitive strategy: on sensing that a number of similar objects are equal in size, the mind immediately groups them into equal sets. The sensing of the ear is thereby implicitly distinguished from the operations of the mind, which is charged with conceiving the whole by means of synthesizing its parts. The mind thus imposes structure onto a sensed object in order to be able to grasp it: the information we extract from it is ultimately only that which we ourselves are able to project onto it.

Holden clarifies this point by appealing to a familiar visual experience:

> When we cast our eyes on nine equidistant windows in a row, they are no sooner seen than subdivided into three times three: eight appears at first to be two fours, and each of these fours, two twos; seven we conceive as two threes disjointed, and one in the middle; six most naturally divides itself into two threes; but if seen along with nine, or immediately after it, we then trisect it, in conformity with nine, and it appears three twos: five becomes two twos disjoined and one in the middle; four becomes two twos, and single three or two need no subdivision.[12]

An illustration of this passage is shown in **Figure 2.1**. Holden asserts that the mind instantly arranges larger sets of similar objects into smaller equal portions factorized by the small primes 2 and 3. This grouping is internally hierarchical, with our minds subdividing each of two groups of four into two groups of two (**Figure 2.1b**). He further asserts that the primes 5 and 7 divide into equal groups plus a remainder of one. Finally, in ambiguous cases, such as the grouping of six (**Figure 2.1d**), Holden maintains that subdivisions are determined not only by the current state but also by the context of the grouping of the previous state. These claims prove essential to his theory of harmony.

---

[11] Ibid., part 2, §11, 288–289.
[12] Ibid., part 2, §11, 289.

JOHN HOLDEN'S *ESSAY* 47

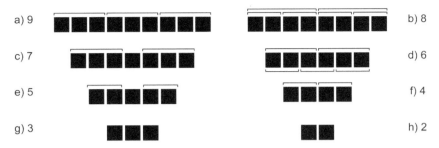

Figure 2.1 An illustration of Holden's theory of visual grouping.

Describing visual experiences, Holden observes that we obtain gratification from encountering "a certain symmetry which strikes us immediately with delight, in the prospect of a regular and well-designed piece of architecture."[13] Regular proportions between different sounds in music, like proportions between different parts of such a building, "are the very things whence our pleasure is derived."[14] However, whereas in vision we can access an entire quantity at once, a different approach is required for sound, which occurs in time. Returning to the topic of isochronous compound parcels in music, Holden first addresses rhythm, claiming:

> The mind insists that all our notes be made up, as it were, into isochronous parcels, which we call bars, or measures, and that the number of equal short notes which constitutes each measure, be a number some way compounded of the small factors two and three multiplied together, and rarely admits any larger factor than these.[15]

In perceiving music, therefore, our mind constantly groups incoming sounds into smaller equal units of twos and threes and similarly organizes these into groups of groups, that is, hierarchical structures. This mental activity, by which we comprehend larger wholes as comprising groups of twos and threes, is for Holden the principle of rhythm. Duple and triple divisions

---

[13] Ibid., part 2, §15, 293. The statement is reminiscent of contemporary eighteenth-century analogies between music and architecture, e.g., Charles-Étienne Briseux, *Traité du beau essentiel dans les arts* (Paris: 1753).
[14] Holden, *Essay*, part 2, §15, 293.
[15] Ibid., part 2, §12, 289. Holden uses the term *rhythm* in a broad sense to denote what we today define as both meter and rhythm. Notably, the word *meter* does not appear a single time in his treatise, although he uses the term *mood* to indicate the various types of duple or triple time signatures.

48 HEARING WITH THE MIND

govern the mood, or metric character of a measure, as well as the surface rhythm, which subdivides further into twos or threes at the beat level. Holden notes that subdivisions of five and seven are also occasionally encountered at the beat level but emphasizes a difference between them: whereas the effort it takes to perceive groups of five can be attributed to lack of familiarity, groups of seven stand at the very limits of the ear's capacity and challenge our mental faculties.[16]

Given an undifferentiated series of beats, Holden asserts, the listener will "naturally parcel them by 4 and 4 together, by giving a greater regard to every fourth beat; by which regard these beats would to him acquire a kind of emphasis or accent."[17] This accent, which arises from involuntary acts of attention ("a greater regard"), is explicitly a mental phenomenon: it exists in our minds, not (necessarily) in acoustic reality, and "would be the very same with a hearer if an inanimate machine were made to beat the drum."[18] This subject is also treated in the first part of the *Essay*, where Holden suggests that the aspiring student acquire a sense of rhythm by counting off duple and triple groupings of watch ticks. In this case, "we imagine the pulses which we count, to be really stronger than the intermediate ones, which we pass over. The superior regard which we bestow on the counted pulses is, here, the sole cause of these imaginary accents."[19]

Holden's remarkable description of our involuntary mental impulse to metricize duple or triple groups of beats as arising from a regular, recurrent increase in attention appears to be the earliest depiction of a crucial element in the phenomenon of meter as we understand it today, namely, what we now call metrical accent. His recognition of this behavior, known today as subjective rhythmicization, predates Johann Philipp Kirnberger's account of accentual grouping in *Die Kunst des reinen Satzes* by six years.[20]

Holden observes that changes within the domain of rhythm can at times require our minds to shift between comparable isochronous compound

---

[16] Ibid., part 2, §14, 290.

[17] Ibid., part 2, §14, 294.

[18] Ibid.

[19] Ibid., part 1, §95, 83.

[20] Johann Philipp Kirnberger, *Die Kunst des reinen Satzes in der Musik* (Berlin and Königsberg: Decker und Hartung, 1776), 2:114–115. For more on Kirnberger's theory of rhythm see Roger M. Grant, *Beating Time and Measuring Music in the Early Modern Era* (Oxford: Oxford University Press, 2014), 93–125. The earliest report of psychological experiments pertaining to subjective rhythmicization can be found in Thaddeus L. Bolton, "Rhythm," *American Journal of Psychology* 6, no. 2 (1894): 185; see also Ernst Friedrich Wilhelm Meumann, "Untersuchungen zur Psychologie und Aesthetik des Rhythmus" (Habilitationsschrift, Universität Leipzig, 1894), 26.

parcels. He argues that, given that any music that includes triplets will feature repeated changes between triple and duple subdivisions, we generally prefer to retain the smallest (fastest) invariable beat division common to the changeable subdivisions, in order to shift smoothly between different groupings.[21]

Next, scaling up from beats and measures, Holden proposes that division and grouping by twos and threes also determine the lengths of musical phrases, though these are subject to the constraints of our memory:

> The mind extends its view, and as far as the memory can be supposed distinctly to retain, goes on to constitute some number of measures into isochronous phrases, or strains of a tune; and these strains may contain a greater number of measures in quick time than in slow, because of the inability of the memory; but here, as before, the number of measures in a strain must always be either two or three, or some product of these numbers: for here five bars in one strain is not used, and seven proves much more intolerable.[22]

Thus the action of creating isochronous groups or chunks (i.e., groups of groups) does not arbitrarily stop at the measure but extends to phrase lengths in terms of measure groupings, that is, numbers of measures, which, like rhythm, generally comprise multiples of twos and threes. This comparison is not naive, as Holden is intent on establishing grouping as an innate principle of our perception. Therefore he argues that, in addition to influencing the domain of sight, the strategy of grouping into the powers of twos and threes affects our hearing, specifically the perception of rhythm and form along a temporal continuum, subject to the constraints of our memory.

Holden now makes a striking analogy between our perception of rhythm and form and that of pitch and tonal relationships:

> The faculty of remembering the key note [i.e., the tonic] and the constant expectation of returning to it at the conclusion, which is so remarkably perceived by the musician, resolves immediately into that of retaining the idea of a small portion of time, divided and subdivided in some eligible manner, by the vibrations of the same key, or its octaves; and agrees exactly

---

[21] Holden, *Essay*, part 2, §17, 295–296.
[22] Ibid., part 2, §14, 290.

50 HEARING WITH THE MIND

with the remembrance of the length of one bar or strain, and of the mood of time, in the timing of music.[23]

Holden here asserts that the faculty at work in our perception of rhythm (and meter) is the very same as that which operates in our perception of pitch and scale degrees. The opening terms of his analogy—the retention of the tonic and the expectations associated with a sense of a key—revisit a theme that he treated at length earlier in the *Essay*, namely, the roles of memory and syntactical expectation. Our ability to fix a tonic in our minds implies the presence of features that distinguish it from other pitches, and in the very first chapter of the *Essay* Holden takes the unusual step of observing that not only the tonic, but indeed each of the scale degrees, is characterized by specific qualities:

> The key note is remarkably bold and commanding; the third and seventh have something supplicative in them. . . . The sixth is a kind of plaintive sound; the fourth, as observed before, is grave and solemn; the fifth partakes of the nature of the key, and the second is not unlike to the sixth.[24]

Holden emphasizes that these properties depend on the tonal context in which the notes occur, thereby identifying them as relational rather than inherent in the pitches themselves or in their physical means of production. That is, "if we were only to consider musical sounds singly, without any regard to their relations to, or dependencies upon each other; no such properties as these could be attributed to any one sound more than another."[25]

Given that the particular effect of each scale degree arises solely out of its relationship to the other degrees, and thus to the key, Holden argues, we must "keep the key note constantly in view during the whole course of a tune; and to consider all the other notes of the scale, chiefly with regard to their several relations to the key [note]."[26] (By "key" Holden typically means

---

[23] Ibid., part 2, §20, 298–299.

[24] Ibid., part 1, §21, 14–15. Something very like this appears in Walter Young's preface to Patrick MacDonald, *A Collection of Highland Vocal Airs Never Hitherto Published. To Which Are Added a Few of the Most Lively Country Dances or Reels of the North Highland and Western Isles: And Some Specimens of Bagpipe Music* (Edinburgh: the Author, 1784), 5–6; see note 85 in Chapter 5.

[25] Holden, *Essay*, part 1, §21, 15. Holden writes, "We might indeed ascribe different properties to the sounds of different instruments; thus, the trumpet is bold, the violin is chearful, the bassoon is solemn, etc." Ibid.

[26] Ibid., part 1, §28, 23.

the "key note" or tonic, as opposed to the key as we now understand it.) The fact that all pieces of music end on the key note is further proof that we must be preserving it in our minds throughout, as "if ever [the listener] were supposed to have totally forgot it, what then could hinder him from being as fully satisfied with a final close, upon some other note?"[27] Consequently, the act of sustaining the tonic in our awareness gives rise to the familiar phenomenon of musical expectation. Perceiving pitches in relation to a key note, he concludes, is thus "both necessary for the purposes of music and natural to the human mind."[28]

Holden summarizes this claim as follows:

> In the practice of music, the key note is constantly kept in mind; and all other notes which are admitted, are some way compared with the key. These comparisons, and the consequent perceptions, are, indeed, the very essence of music. It is impossible for us to hear two different sounds, either together, or in succession, without attempting to make some comparison, either between one and the other, or between each of them, and some third sound, with which our mind may previously be possessed, and which we regard as a key note.[29]

According to this theory, our unconscious retention of the tonic is the basis for the comparisons of sounds that are essential to the perception of music, giving rise to our experience of music as being in a key, and consequently, to tonal expectation. It is, in short, the principle and cause of the phenomenon that we call tonality.

Beyond identifying the crucial role played in tonality by our mental retention of the tonic, Holden also offers an ambitious attempt to explain that retention itself, to say exactly *how* we are able to accomplish it, and on that basis to explain as well our ability to relate other pitches to the tonic as a central point of reference. The foundation of his account of these matters is a concept which directly exploits his analogy between the perception of time and of pitch: that of the *module*.

---

[27] Ibid., part 1, §29, 25.
[28] Ibid., part 1, §32, 28.
[29] Ibid., part 1, §29, 25.

52 HEARING WITH THE MIND

The module

The module, according to Holden, is "the small portion of time, by which we suppose the vibrations not only of the key [note], but also of every other sound which we admit into our music, while we retain that key, to be measured and distributed into isochronous parcels."[30] That is to say, it is a temporal quantity abstracted from the acoustical signal of the key note and retained by the mind, where it functions as a standard unit in reference to which other pitches are measured and thus perceived.[31] As we shall shortly see, this quantity stands in a simple proportional relation to the frequencies of all the pitches of the diatonic scale belonging to the key represented by the module (where "simple" means, in effect, defined by one of the first four prime numbers, 2, 3, 5, or 7). The module, then, effectively transforms frequency relations from physical events into unconscious mental acts.

We can therefore regard Holden's module as an innovative adaptation of that well-known music-theoretical heuristic, the monochord. The unit of the monochord—the full vibrating length of the string—is divided and subdivided to produce a set of intervals which together represent a given pitch collection, such as the diatonic scale. The module, too, constitutes a unit that is divided and subdivided to produce a set of intervals that represents a diatonic scale. But whereas the monochord is (at least in principle) a concrete physical object that exists out in the world, the module represents an introjection of the monochord's quantitative model within the mind.[32]

Holden's concept of the module depends on the familiar experience of octave equivalence, whereby octave-related pitches are perceived as variants of a single pitch-class, and on the acoustical fact that all octave-related pitches are proportionally related to each other by powers of two. (Thus 2:1 = one octave; 4:1 = two octaves; 8:1 = three octaves; and so on.) Octave equivalence allows the mind to take any specific pitch as a representative of all other pitches octave-related to it, while the simple proportional relation makes the frequencies of all octave-related pitches easily derivable from any of them.[33]

---

[30] Ibid., part 2, §21, 299.

[31] The term "module" might have been familiar to Holden from the domain of architecture: the Roman architect Vitruvius coined the term (*modulus*) in his *De architectura* (15 C.E.) to denote the proportionally determined unit of measure endemic to the plan of a specific building, used to determine the dimensions or positions of various architectural elements. See Pollio Vitruvius, *De architectura*, 3.3.7.

[32] I thank David E. Cohen for sharing this observation with me.

[33] Holden regards octave equivalence as fundamental to his analogy between rhythm and pitch, claiming that just as "the division of a crotchet or other short note, into 2, or 4, or 8, makes no

Holden posits that, for any given pitch, the mind can, unconsciously and instantaneously, imagine its frequency reduced, by powers of two, down to a (sub-audible) frequency many times slower: so slow, in fact, that at this new frequency a mere two vibrations occupy a duration which contains the hundreds of vibrations that define the frequency of the actually given pitch.[34] Now any piece of music, Holden assumes, has a key note or tonic: not a single specific, registrally determined pitch, but a set of octave-related pitches, or rather, frequencies, to which the mental operation just described is unconsciously and instantaneously applied. The duration defined (that is, measured off) in this way, by the mind's application of that mental operation of proportional reduction to the piece's key note, is what Holden calls a "module." Holden further assumes that the frequencies of all the other notes belonging properly to the key (that is, the diatonic scale) of the piece are, by definition, proportionally commensurable with that of the key note, and therefore also commensurable with the module, which thus divides them into spans of equal duration—what Holden calls "isochronous parcels"—the duration being, of course, that of the module. In setting out a system of related pitches such as the diatonic scale, Holden, following the time-honored procedure of the monochord theorist, simply selects that numerical representative of the key note (16, say, or 32) that allows him to locate the numbers of all the required pitches using only ratios of whole numbers.

Holden compares the module to the musical measure insofar as it is a durational window in relation to which shorter durations stand as fractional divisions and are thereby measured, as a centimeter is measured as a defined fraction of a meter. (Unlike the measure, however, the module cannot participate in groupings that would be of greater duration than itself.) I have illustrated the (partial) analogy between a module and a measure in **Figure 2.2**.[35]

---

alteration in the nature of the time in music, so the doubling and redoubling the number of vibrations in a parcel, makes no alteration in the effect of such parcel on our sense." Holden, *Essay*, part 2, §19, 298.

[34] Holden uses the terms *vibrations* and *pulses* nearly interchangeably while observing the distinction that two vibrations entail three pulses. A single pulse cannot determine a module, nor can the two pulses that constitute a single vibration; as Holden notes, "three pulses only determine two of the small intervening particles of time; and till we have perceived two or more of these particles, we can form no judgment concerning their equality; that is, we cannot estimate the tone [i.e., pitch] of the sound." Ibid., part 2, §25, 302.

[35] Ibid., part 2, §22, 299.

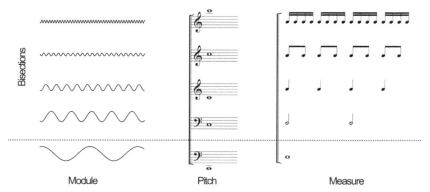

Figure 2.2 An illustration of Holden's analogy of the module as a measure and as a pitch. Left and center: the harmonics of a fundamental sound equivalent to the module represented as vibrations and as pitch; right: the same represented as a measure and equivalent rhythmic subdivisions.

Holden observes that for any module there is an infinite number of pitches whose vibrations divide that module in an infinite number of ways. Most of these, he asserts, are not "of use" because they are not proportionally related in an acceptable way to the pitch of the key note that determines the module. In the context of a musical work, therefore, we include only the pitches whose various frequencies "divide the module in certain simple and intelligible manners, and such we say have *eligible parcels*."[36] These simple and intelligible kinds of division of the module, which constitute the eligible parcels that produce usable pitches, are those meeting Holden's requirement that the number of parts into which any module is divided be one of the powers of 2 and/or 3 or their products. Divisions involving fives or sevens are also possible at times, he notes, but only after an initial division into two or three. Such units are "admitted, but with more difficulty, and generally only as dependents, or *harmonics* ... [while] those whose parcels involve any of the higher prime, or uncompounded numbers 11, 13, 17, etc. are totally rejected."[37]

Holden suggests that we normally establish the module for a piece of music on the basis of the piece's first pitch or chord, since we assume that to be the key note:

---

[36] Ibid.
[37] Ibid., part 2, §23, 300. Holden uses the term *harmonics* here (in the sense of harmonic partials) to indicate a further quintuple or septuple division of a primary duple or triple division of the original frequency.

Supposing our ear to be entirely unbiassed, and not retaining the least impression of any former heard sound, when a musical sound is first proposed: in this case, we shall most naturally regard this first heard sound as a principal key note, and parcel its vibrations by continual reduplication, or by the powers of 2, rather than in any other practicable manner; and thus we shall constitute a module divided and subdivided by continual bisection, like the measure in common time.[38]

In the absence of other contextual cues, therefore, our minds constitute a module from the vibrations of the first sound we hear, which, Holden asserts, is generally the tonic, or notes that are "nearly related" to the key, typically by third or fifth.[39] (Holden is thinking here chiefly of a-capella hymnody, where the choir is given its starting note by the leader.)[40]

Importantly, even though Holden was familiar with the concept of frequency, he chose to use relative rather than absolute values to construct the module.[41] His decision to rely on a perceptual measure, rather than a unit anchored to absolute time, arises from the nature of the module, which grounds a cognitive framework against which other pitches are hierarchically organized and perceived.

Thus, in a passage in which Holden attempts to determine the likely length of a module in relation to the frequency of the key note, his observations regarding the limits of our ability to perceive pitch lead him to an interesting hypothesis as to the cause of those limits. The entire range in which we can perceive pitch, he observes, comprises approximately seven octaves from F1 to F8.[42] He therefore proposes the following hypothesis: since F1, being the

---

[38] Ibid.

[39] Holden explains that when this is not the case, and when "we sometimes begin with another note of the scale, it must be one of those which are more nearly related to the key, viz. the third or the fifth and the initial passages must be such, as will sufficiently point out, to an experienced hearer, the real key of the following piece." Ibid., part 1, §30, 25–26.

[40] I explore this important point in the next chapter.

[41] Holden mentions experiments reported by William Emerson in *The Doctrine of Fluxions* (London: J. Bettenham, 1743) and by Robert Smith in *Harmonics; or, The Philosophy of Musical Sounds* (Cambridge: J. Bentham, 1749) and compares their measurement of frequency to refute the possible argument that the fastest rhythmical subdivisions used in music might influence our conception of the module. Holden, *Essay*, part 2, §47, 323–324.

[42] Ibid., part 2, §25, 302. Holden's estimation of the physiological limitations of our hearing range resembles a claim made by Leonhard Euler in *Tentamen novae theoriae musicae* (1739), namely, that the ear can perceive sounds within the range of eight octaves corresponding to thirty and 7,552 vibrations per second. Leonhard Euler, *Tentamen novae theoriae musicae ex certissimus harmoniae principiis dilucide expositae* (St. Petersburg: Academiae scientiarum, 1739), 8. However, Euler and Holden use different methods and provide distinct results. Euler bases his conclusion on the sounding frequency, i.e., the number of vibrations per second, whereas Holden chooses to examine idealized numbers. Additionally, Holden's hearing range is more limited, comprising "more than six,

56 HEARING WITH THE MIND

lowest audibly identifiable pitch, has the longest module, and F8, being the highest, divides that module into the maximum possible number of parts, the latter represents "the smallest mental subdivisions which we can make in a [i.e., any] given module, [and this] may be the cause why no sound still acuter can be admitted into our music."[43] The size of the parcels therefore reflects constraints on our cognitive capacities. He thus effectively portrays the physiological limits of human hearing as a consequence of the challenges involved in the mental act of subdividing, which is to say grouping.

As an idealized span of time, the module exists in the mind rather than in the physical world. It thereby joins a host of other phenomena that theorists have posited as existing only in our imagination, such as the aforementioned case of subjectively rhythmicized metrical accents, as well as Rameau's fundamental bass and implied dissonances, which may be absent from a given sonority but are entailed by the musical context.[44] Unlike these constructs, however, Holden proposes that the module is constantly retained over long spans of time—or at least until the mind can no longer parse sounds according to a given module, at which point it selects another. The module thus appears to be that instrument of the mind, so to speak, which affords the mental acts of comparison required to hear pitches as participating in distinct relations to a tonic.[45] To clarify this matter, I turn to his account of the major scale.

The Ascending and Descending Major Scale

The major scale, Holden asserts, "has always been, and will always be the same in all ages and countries: it seems to have been one of those laws which the great Author of Nature prescribed to *himself*, in the formation of the

---

but less than seven successive octaves." Holden, *Essay,* part 2, §25, 302. Holden never cites Euler nor mentions his ideas.

[43] Ibid.

[44] See e.g., Cohen, "The 'Gift of Nature,'" 69–71.

[45] In the first part of the *Essay* Holden depicts the experience of hearing a melody that remains in a single key (his term is *connected*) as follows: "A whole tune is often in reality no more than a kind of division or breaking upon the key note as fundamental. The key might be held on in the bass from first to last, as in the musette and bagpipe music; and although it should not actually be so held on, yet it is undoubtedly always kept in mind in all connected pieces" (part 1, §321, 278). The equivocation in this passage, whereby we could retain the tonic note as a drone, although we do not actually do so, bolsters the interpretation of the module as the representation of the tonic that is sustained in our minds throughout a single diatonic piece.

human mind, that such certain degrees of sound should constitute music."[46] We "chuse" certain pitches as the notes of the scale, then, while "refusing" others because the scale embodies his first principle of music—the mental strategy of grouping equally proportioned stimuli into nested groups of small-prime cardinality. The notes of the scale are the sonic manifestations of proportionally related frequencies unconsciously selected by our minds in accord with that strategy. The pitches that we perceive as members of a given major scale are thus simply those that we hear as belonging to the tonal context constituted by the key note and the module that represents it.

As we shall shortly see, Holden regards movement between scale degrees as involving perceptual shifts among groupings of pulses of differing eligible cardinalities. For example, a move from, say, the key note (tonic) up to the fifth degree of the scale, involves a re-division of the module in 2:3 ratio, as the duple division of the module that produces the tonic is replaced by a triple division to produce the pitch a fifth higher. This means, again, that each degree must stand in proportion to an initial duple or triple division of the module by some combination of the prime factors 2, 3, 5, 7, and 4 (the last included for the sake of the ratio 5:4, the major third). He justifies the inclusion of 4, 5, and 7 by appealing again to the case of musical rhythm: in the case of 4, because 4/4 time differs metrically from 2/4; in the case of 5 and 7, because, as he observes, when we perform a quintuplet or septuplet we usually feel them as slightly faster variants of the more familiar divisions into four and six.

More broadly, Holden's explanation rests on a fundamental analogy between visual grouping and the experience of hearing scale degrees in relation to a key. He proposes that in hearing, just as in vision, we grasp directly the simple factors of 2, 3, and 4, while the higher primes of 5 and 7 can be perceived only in reference to these simple factors. Just as, when we see a row of five or seven equally spaced windows, "we readily conceive 5 by its affinity to 4, and 7 by its affinity to 6,"[47] so we perceive 5 as a variation of 4 by the addition of 1, and 7 likewise with respect to 6 (3 × 2). As a consequence, the higher primes of 5 and 7 are "admitted, but [only] with more difficulty,"[48] because they impose a greater strain on our cognitive abilities. Scale degrees whose grouping cardinalities include 5 or 7 as factors must therefore be combined with smaller and simpler factors (2, 3, or 4), as we shall shortly see in more detail.

---

[46] Ibid., part 1, §38, 37. Original italics. I critique this claim in the conclusion to this book.
[47] Ibid., part 2, §29, 305.
[48] Ibid., part 2, §23, 300.

Holden's scale derivation also relies on a further property of our perception. Once heard, a pitch influences the way in which we perceive subsequent sounds in the present. As a result of this protensive feature, Holden, as we will see, assigns scale degree IV different divisions of the module according to whether the scale ascends or descends.[49]

To illustrate Holden's approach to the scale, let us again consider moving from scale degree I to scale degree V.[50] This is illustrated in **Figure 2.3**, which depicts the module as the duration of a measure of 4/4.

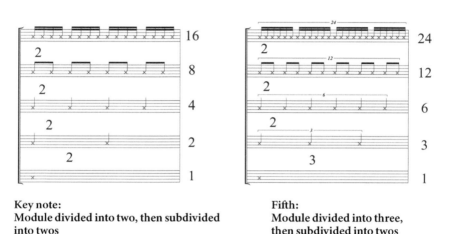

Key note:
Module divided into two, then subdivided into twos
2 × 2 × 2 × 2 = 16

Fifth:
Module divided into three, then subdivided into twos
3 × 2 × 2 × 2 = 24

Figure 2.3 Key and fifth divisions of the module.

In Figure 2.3, the note values in the left column stand to those in the right column in the proportion of 2:3. As any module occupies only the miniscule amount of time required for the minimal number of vibrations it includes—a duration far briefer than could be realistically noted in rhythmic notation—we can regard the note values in this illustration as idealized frequencies, and their proportions as intervals.

The proportions representing the scale degrees of Holden's descending scale are listed in **Table 2.1**, along with each degree's corresponding division of the module into the relevant combination of the factors 2, 3, 4, 5, and 7.

---

[49] Following Holden's practice, I use roman numerals to denote scale degrees, regardless of whether they signify chords, chordal roots, or merely notes.
[50] Ibid., part 2, §27, 304.

JOHN HOLDEN'S ESSAY 59

Table 2.1 The descending scale represented as divisions of the module.

Note	Scale degree	Division of module
C	Key: 32	$2 \times 4 \times 4$
B	Seventh: 30	$3 \times 2 \times 5$
A	Acute sixth: 27	$3 \times 3 \times 3$
G	Fifth: 24	$3 \times 2 \times 4$
F	Grave fourth: 21	3×7
E	Fundamental great third: 20	$2 \times 2 \times 5$
D	Second: 18	$3 \times 2 \times 3$
C	Key: 16	$2 \times 2 \times 4$

For example, since the factor 3 produces an upper fifth and thus scale degree V, and since scale degree II is a fifth above V, therefore II is represented by 3×3 (additionally multiplied by 2 to bring it into the same octave register as the key note, represented here by 16). Note that scale degree IV ($21 = 3 \times 7$) is a "grave" fourth (21:16), which is to say, a note class equivalent to the harmonic seventh (7:4) of the dominant scale degree below the key note (3:4), and is therefore lower by a comma of 64:63 ($3:4 \times 21:16$) than the usual "perfect" fourth (4:3), which, as we will soon see, occurs as scale degree IV of the ascending scale. Holden describes the resulting discrepancy between the two kinds of fourths as his own invention, remarking that as this "small interval, not having formerly been supposed to exist among musical sounds, has no established name; we shall therefore call it a bearing; not chusing to borrow from any other language the name of an interval which a Briton first introduces."[51]

---

[51] Ibid., part 2, §44, 317. The ratio 64:63 (later termed the Archytas and the Leipziger comma by George Secor and Siegfried Karg-Elert respectively) is precisely the amount by which the ratio of the major whole tone, 9:8, is exceeded by the larger whole tone with ratio 8:7 that lies between V and the grave fourth. Thus 72:64 = 9:8, while 72:63 = 8:7. These relationships are further detailed in Holden's chart, reproduced in Figure 2.11. In the "Conjecture sur la raison de quelques dissonances généralement reçues dans la musique," Euler likewise makes use of 64:63, claiming that the proportions of the dominant seventh chord are 36:45:54:64 but that they are heard as 36:45:54:63 and thus reducible to 4:5:6:7. Leonhard Euler, "Conjecture sur la raison de quelques dissonances généralment reçues dans la musique," *Histoire de l'académie royale des sciences et des belles-lettres de Berlin année 1764* (Berlin: Haude et Spener, 1766), 172. I thank Roger Grant for this observation. It seems unlikely, however, that Holden was influenced by the "Conjecture," as he issued a compilation of twenty-four new hymn settings advertising his forthcoming *Essay* the same year. In the preface to that work Holden claims that his only predecessor in using 7 as a musical ratio was William Jackson (in *A Scheme Demonstrating the Perfection and Harmony of Sounds*, 1726). See John Holden, *A Collection of Church-Music Consisting of New Setts of the Common Psalm-Tunes ... for the Use of the University of Glasgow* (Glasgow: Watt and McEwan, 1766), 1–12.

60 HEARING WITH THE MIND

In order to fully understand how Holden conceives of his descending scale, let us now take for example the transition from scale degree VIII (32), a purely duple division of the module, to scale degree VII (30), which decomposes into the factors of 2, 3, and 5. In Holden's words, "we take the seventh instead of the fifth; that is, we take 30 vibrations instead of 24; and this presents a module divided, like the fifth, into 3; but each of these subdivided into 5, instead of 4; or 10, instead of 8."[52] Holden here proposes that we regard scale degree VII ($3 \times 2 \times 5$) in relationship to scale degree V ($3 \times 2 \times 4$), as these degrees share two of three factors. Therefore, "as we conceive of 5, by its affinity to 4, so, while we sound the seventh of the scale, the fifth is essentially implied, and is our fundamental."[53]

Holden thus understands the fundamental bass as a consequence of the mind's attempt to grasp sounds in terms of various duple and triple primary divisions of the module. When we move from VIII to VII, he argues, our imagination instantly projects their corresponding fundamental basses, namely, I and V, respectively, an act that essentially reflects how our mind perceives tonal relationships.[54] We can compare this to the experience of sight: recall Holden's claim that when we view five windows, our minds involuntarily sort them into two groups of twos and a remainder of one by perceiving them as two twos with a surplus, as illustrated in Figure 2.1e. The relationship between scale degree VII and its fundamental, V, is illustrated in **Figure 2.4** by an analogy with rhythm and in **Figure 2.5** by the analogy with vision.

To reiterate, moving from scale degree VIII ($2 \times 4 \times 4$) to VII ($3 \times 2 \times 5$) involves simultaneously moving from fundamental bass VIII or I ($2 \times 4 \times 4$ or $2 \times 2 \times 4$) to V ($3 \times 2 \times 4$), as shown in **Figure 2.6**. (Again, VIII [$2 \times 4 \times 4$] and I [$2 \times 2 \times 4$] are treated interchangeably, as they are members of the same note class.)

Holden thus posits that the listener's mind automatically and unconsciously supplies an appropriate accompanying fundamental bass note below each melodic scale degree heard. His theory is reminiscent of Rameau's, but with significant differences. For Rameau, as David E. Cohen has shown, the

---

[52] Holden, *Essay*, part 2, §33, 308.

[53] Ibid.

[54] Holden's presentation of the scale as composed of scale degrees accompanied by fundamentals subject to constraints on their possible motion responds, of course, to Rameau, who repeatedly addressed this subject, starting with the *Traité* (1722), which Holden cites in the *Essay*, though not on this point.

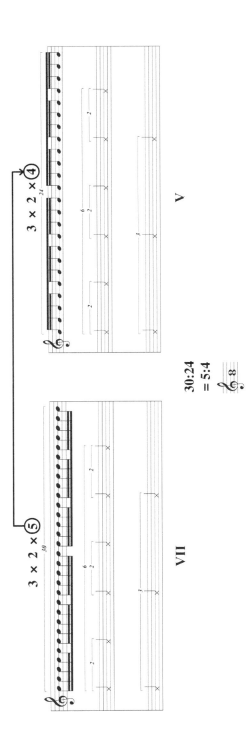

Figure 2.4 Perceiving VII within the context of a scale means hearing it in relation to V.

62 HEARING WITH THE MIND

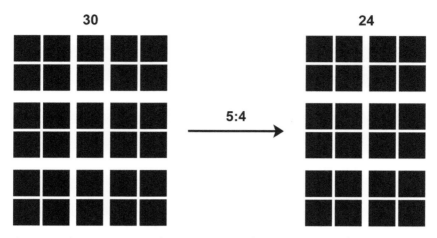

Figure 2.5 We can compare hearing VII:V to the cognitive act of grouping that occurs when we perceive windows. If we see thirty windows, we view the five columns together as four groups (of six windows), with an extra six-window group included in the middle, analogous to Figure 2.1e.

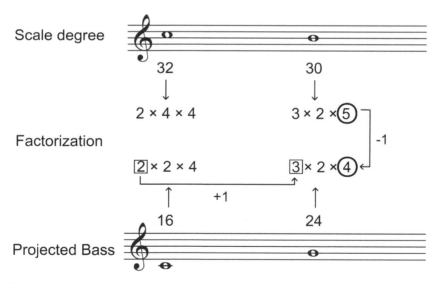

Figure 2.6 Movement between scale degrees implies movement between fundamentals. The mind interprets the factor 5 in 3 x 2 x 5 (= 30) as a variant of the factor 4 in 3 x 2 x 4 (= 24), so that the bass g4 is projected at the major third below the b4.

agency within the listener responsible for this activity is not the mind but a faculty that he calls "the ear" and later "instinct."[55] However, the detailed operations of this faculty and the specific means by which it achieves its results remain occult. In contrast, Holden assigns the generation of fundamental basses to the mind rather than the ear and provides a coherent explanation of how we accomplish this task by proposing that the fundamental bass is an artifact of our overarching cognitive tendency toward grouping certain kinds of (auditory and visual) stimuli into nested hierarchies of small primes. Thus, while both Rameau and Holden assume that our projection of fundamental basses is unconscious, Holden's theory provides a detailed speculative account of how exactly the precise pitches are generated within the mind.

A number of consequences follow from Holden's conception of the relationship between the scale degrees and the module as mediated by projected fundamentals. Given that we are dealing with factors of 2, 3, 4, 5, and 7, the factorization of the number corresponding to any scale degree will differ from that of its projected fundamental by either 0 or 1.[56] In other words, the relationship between fundamentals and scale degrees will be one of either note class identity or a superparticular (or sub-superparticular) ratio. We can describe the constraints that are implicit in his harmonization of the ascending and descending diatonic scales as follows:

1. The numbers representing fundamental bass notes have as their factors only 2 and 3.[57]
2. The tonic and the fifth scale degree take a member of their own respective note classes as their fundamental, as does the fourth of the ascending scale.[58]
3. A scale degree with a 5 or a 7 among its factors can only be conceived as bearing a superparticular ratio to its fundamental.[59]

---

[55] Cohen, "The 'Gift of Nature,'" 71–72.

[56] Thus, for example, scale degree V (24, or $3 \times 2 \times 4$) is its own fundamental, whereas scale degree VI (27, or $3 \times 3 \times 3$) accepts a fundamental bass of II (18, or $3 \times 3 \times 2$), or 3:2. The complete inventory of scale degrees and fundamentals is illustrated in Figure 2.7.

[57] "The real fundamentals themselves include only the numbers 2 and 3 in the composition of their parcels" (Holden, *Essay*, part 2, §39, 311). Again, I treat 2 and 4 interchangeably for the reasons discussed above.

[58] Holden writes, "The next note is the fifth itself, which we naturally take as a fundamental." Ibid., part 2, §36, 309.

[59] According to Holden, "In the descending scale ... the sounds which include the number 5 in the composition of their parcels naturally become great thirds [to their fundamentals], and the sound which includes the number 7, will be either a less third, or an added seventh [to its fundamental]." Ibid., part 2, §39, 310–311.

64   HEARING WITH THE MIND

4. Because we directly perceive only groups of 2 or 3, we are only able to shift directly between such groups: higher-order groups are perceived as multiples of 2 or 3, and a shift between higher-order groups must be mediated by a shift between the underlying groups of 2 and 3.[60]

Remarkably, rule 4 effectively prohibits stepwise motion between fundamentals. The rule states that direct cognitive comparison is possible only between groups of two (or four) and groups of three. Now, triple division of a module produces the pitch a twelfth (which reduces to a fifth) higher than the module's fundamental, and a second triple division produces the pitch a twelfth (fifth) above that. That second pitch, of course, is a ninth—which reduces to a major-second step—above the fundamental. Rule 4, by stipulating that only one such division can be carried out at a time, effectively mandates that one can only attain that major-second step above the fundamental by way of the fifth above the latter. (In effect, the rule asserts that the major second must be divided into two fifths.) In this way, the step progression between the fundamental and the second fifth is prohibited. Thus fundamental-bass progression by step, which Rameau regarded with disfavor, is prohibited as a consequence of Holden's cognitive-perceptual principles.

In a brief aside, Holden notes that the same holds true for rhythm; indeed, the reader is invited to attempt to move directly from a measure divided into eight eighth notes to one of the same duration divided into nine eighth-note tuplets, as shown in **Figure 2.7**[61]:

The four rules produce the descending major-scale harmonization shown in **Figure 2.8a**. The figure collates the transitions between scale degrees (and their corresponding fundamentals) that are compatible with rules 1–4 and

Figure 2.7 Try moving from a division of a span into eight directly into a division of the same span into nine.

---

[60] Holden writes that, "after the module has been heard divided simply into 3 parts, we may, at the next step, proceed to trisect each of its third parts; and, by this means, introduce a division into 9, which corresponds with the second of the scale: but we cannot easily reconcile the making of two such trisections at one step, that is, we can substitute 3 vibrations instead of 4, and contrariwise, 4 vibrations instead of 3, but we cannot immediately substitute 9 vibrations instead of 8 nor 8 instead of 9, which is exactly the case in the timing of music also." Ibid., part 2, §35, 308.

[61] See previous footnote.

JOHN HOLDEN'S *ESSAY*   65

displays them as a directed graph, tracking the course of the descending scale shown in the top row. The columns show by means of check marks or x's the compatibility or incompatibility of each scale degree with each of the available fundamentals, listed in the leftmost column. VIII, V, and I are compatible only with themselves (i.e., with pitches one or more octaves below them), and VI and II are compatible only with II and V, respectively. III is compatible only with I, and VII only with V; in both of these the factor 5 must be understood in terms of 4. A more challenging case arises in regard to scale degree IV, which, having been assigned the pitch number 21 (3×7), is theoretically compatible with both the fundamentals II ($18 = 3 \times 6 = 3 \times (2 \times 3)$) and V ($24 = 3 \times 2 \times 4$), either of which would be eligible according to rule 3. Recall that, according to his aforementioned grouping principles, we typically perceive seven in reference to six, so that IV would normally be comprehended in reference to II as its fundamental. Indeed, he writes, this *would* have been the case

if the succession [i.e., the fundamental-bass progression] would have admitted it; but perceiving that the third of the scale is next to follow, which having 5 vibrations for 4 of the key, will inevitably require the key for its fundamental. . . . [Therefore] [i]f we here take the second as fundamental, we cannot take the K*f* [i.e., the tonic] immediately after it [because to do so would result in a fundamental bass motion by descending step]: we therefore refer the grave fourth, and its implied second to the fifth fundamental which existed in the preceding note, and to which the fourth becomes an added 7th, as the second is its 5th.[62]

As Figure 2.8a shows, therefore, instead of taking II as our fundamental, we conceive of IV as an added seventh to the preceding scale degree (V) and retain its fundamental from the preceding step of the scale. The remaining descent from scale degrees III to II to I is straightforward. Note the elegance of this solution: the descending scale is fully compatible with the four rules above, and no other possible solution fulfills all these conditions. (Note also how this necessarily produces the mandatory resolution of the seventh of the V7 chord.) The ratios between adjacent scale degrees, their successive fundamentals, and each scale degree and its fundamental are shown in **Figure 2.8b.**

Holden next introduces the ascending scale by remarking that one can also derive the scale degrees by continual trisection of the module; this generates

---

[62] Ibid., part 2, §37, 310. K*f* here refers to the fundamental of the key, or I.

66 HEARING WITH THE MIND

Possible Fundamentals	Scale Degree and Factorization							
	C (VIII) 2x4x4	B (VII) 2x3x5	A (VI) 3x3x3	G (V) 3x2x4	F (IV) 3x7	E (III) 2x2x5	D (II) 2x3x3	C (I) 2x4x2
C (I / VIII) 2x4x2/2x4x4	✓	✗	✗	✗	✗	✓	✗	✓
G (V) 3x2x4	✗	✓	✗	✓	✓	✗	✓	✗
D (II) 2x3x3	✗	✗	✓	✗	(✓)	✗	✗	✗
A (VI) 3x3x3	✗	✗	✗	✗	✗	✗	✗	✗

Figure 2.8a The factorization of the descending scale, to be read as a directed graph starting from the left, following the arrows. The flawed option of taking II as the fundamental to IV is shown in brackets.

Figure 2.8b The resulting descending scale. From top: ratios of the intervals between adjacent scale degrees; magnitudes of degrees relative to the module (key note) of 16; ratios of scale degrees to projected fundamentals; ratios between projected fundamentals.

a scale degree III of 81 and a VII of 243.[63] He suggests that a trained musician can opt to perform III:I and VII:V as major thirds in the ratio 5:4 "in order to confine his fundamental progression more closely to the original key," or he can perform III:VI and VII:III as perfect fifths in 3:2 ratio, which is "no more than what nature suggests to the most uncultivated singer."[64] He therefore includes alternate variants of certain scale degrees as possible options within his system, depending on context.

---

[63] Assuming octave equivalence, the difference between the third scale degree generated according to Holden's given ratio of 5:4, which renders a major third represented by the number 80 (= 5 × 2^4), and a major third generated by four stacked fifths resulting from four trisections of the module, represented by 81 (3^4), is the syntonic comma, 81:80. The same comma occurs between his seventh scale degree (15:8) and a seventh generated by continual trisection (3^5), or 243, which, compared to 240 (the 15 of 15:8 multiplied by 2^4), reduces to 81:80.

[64] Ibid., part 2, §42, 314. These proportions, Holden observes, are found in Scottish folk tunes, which (he adds) are difficult to harmonize.

Referring to musical practice, Holden emphasizes that we frequently replace a given sound with another that is a comma away without noticing the discrepancy. Moreover, he observes that we can substitute not only sounds (pitches) but also idealized mental groupings: "These different ways of conceiving the third and seventh of the scale may effectually take place, although no alteration be made in the real pitch of the sounds."[65] This is possible because our senses generally cannot detect the difference when we vary the division of our module to constitute, for instance, 80 versus 81. (In other words, intervals such as the syntonic comma are too small for us to perceive.) Holden makes an analogy to musical timing, where "one bar may be, and always will be some very small matter longer than another, perhaps much more than one eighteenth part of the whole bar, and yet we are sensible of no impropriety."[66]

Holden appeals again to his earlier distinction between the sense of hearing and the mind. Recall that the sense conveys the scale degree to the mind by assessing the equality of successive pulses, while the mind, once it has received a sound, groups the pulses into small groups and determines their proportional relationship to the module by means of a projected fundamental. Holden argues that "the mind allows not the least deviation from the proper method of dividing and subdividing each parcel; but the equality of the whole successive parcels, being determined by the sense, is not so perfectly estimated."[67] As the senses are responsible for simply determining the isochrony of the pulses that constitute a given sound, pitches that are close enough to the proportions of the scale "produce the same effects as if they were perfectly proportioned, because we adapt our module to them without any *sensible* inequality."[68] (This also explains why an expert musician is more sensitive to intonation than a beginner: because "he is more critical in regard to the equality of the time of successive modules.")[69]

It is the fact that such small intervals normally go unperceived that allows Holden to maintain that a module derived from the tonic is actively retained in the mind and used to evaluate successive pitches, such that they reflect the specific ratios characteristic of the diatonic scale, even when, in acoustical reality, there is a discrepancy between the actual pitches and idealized divisions of the module (e.g., as is the case in fixed keyboard instruments). The mental action of extracting

---

[65] Ibid., part 2, §42, 314–315.

[66] Ibid., part 2, §42, 315. While descriptions of our mind's tendency to approximate pitch discrepancies so as to conform to integer ratios can be found in a letter by Leibniz to the mathematician Christian Goldbach in April 1712 and in Euler's "Conjecture," Holden's generalization of this principle to encompass both pitch and timing appears to be unique.

[67] Ibid., part 2, §42, 315.

[68] Ibid.

[69] Ibid.

68 HEARING WITH THE MIND

a module from the sounding tonic and sustaining it in our minds thus allows us to perceive the various scale degrees as the different divisions of the module, which is precisely the feature that lends them their distinct tonal qualities.[70]

Figure 2.9 displays Holden's ascending and descending scales side by side. These are identical with a single exception: the former features a perfect fourth (64) and the latter a grave fourth (63).[71] Although Holden does not make this point explicitly in the second part of the treatise, he discusses *double emploi*, which he terms "double meaning," in the first part of his treatise.[72] As shown in **Figure 2.10**, coming from V, scale degree VI can be

Ascending Scale		Descending Scale	
key,	96	key,	96
seventh,	90	seventh,	90
acute sixth,	81	acute sixth,	81
fifth,	72	fifth,	72
perfect fourth,	64	grave fourth,	63
fund. great third,	60	fund. great third,	60
second,	54	second,	54
key,	48	key,	48

Figure 2.9 Holden's ascending and descending scales. Holden, *Essay*, part 2, §44, 316.

Possible Fundamentals	Scale Degree and Factorization								
	C (I) 2x4x3x2 48	D (II) 3x3x2x3 54	E (III) 2x5x2x3 60	F (IV) 2x4x4x2 64	G (V) 3x4x3x2 72	A (VI) 2x4x2x5 80	A (VI) 3x3x3x3 81	B (VII) 3x5x3x2 90	C (VIII) 4x4x3x2 96
F (IV) 2x4x4x2	✗	✗	✗	✓	✗	✓	✗	✗	✓
C (I / VIII) 4x4x3x2 / 2x4x3x2	✓	✗	✓	✗	✓	✗	✗	✗	✓
G (V) 3x4x3x2	✗	✓	✗	✗	✗	✗	✗	✓	✗
D (II) 3x3x2x3	✗	✗	✗	✗	✗	✗	✓	✗	✗
A (VI) 3x3x3x3	✗	✗	✗	✗	✗	✗	✗	✗	✗

Figure 2.10 Factorization and directed graph of the ascending scale (the *double emploi* at scale degree VI is indicated in dashed lines).

---

[70] This notion anticipates Riemann's assertion that our "tonal imagination" (*Tonvorstellungen*) leads us to perceive various pitches and chords in ways that sometimes significantly deviate from hypothetical tuning differences. See Hugo Riemann, "Ideen zu einer 'Lehre von den Tonvorstellungen,' " *Jahrbuch der Musikbibliothek Peters* 21, no. 22 (1914): 21.

[71] The descending scale in Table 1 was fixed between 16 and 32; however, here, to obtain a full comparison with the ascending scale, Holden multiplies these ratios by 3, as discussed earlier in this chapter.

[72] See Holden, *Essay*, part 1, §212, 190–191, for Holden's discussion of "double usage"; for more on Rameau's concept of *double emploi*, see Thomas Christensen, *Rameau and Musical Thought in the Enlightenment* (Cambridge: Cambridge University Press, 1993), 193–195.

interpreted as 80:64, or 5:4 of IV, while in relation to VII it is heard as 81:64, or 3:2, of II. Holden's harmonization of the scale refers to both harmony and melody and is identical to Rameau's *règle de l'octave* harmonization from *Génération harmonique*, which Holden cites elsewhere in the *Essay*.[73]

To summarize, retaining a module in our minds and using it to interpret successive pitches is what enables us to retain and attend to the tonic in our minds throughout a piece; indeed, this is precisely what it *means* to have "a sense of key." Because, moreover, hearing different pitches as scale degrees requires us to mentally carry out distinct divisions of the module, the sense of key is maintained even when the actual pitch is not phenomenally present. Significantly, the various scale degrees can be conceived—and consequently heard—differently according to context. These variations, which are reflected in the differing module divisions and fundamental bass progression involved in each case, help inform us about their tonal function.

Memory, Attention, and Modulation

Our very perception of pitches as diatonic—that is to say, as standing in some readily cognizable relation to a key and key note—requires us to retain the module and to use it as a measure of every pitch we hear. Holden calls the mental faculties operative in this activity "memory" and "attention." Both, it should be noted, typically (though not necessarily) function without our conscious awareness. Memory enables us to retain the key note, and consequently its module, even when the latter is not phenomenally present to our awareness.[74] According to Holden, we retain the key throughout a tune unless we encounter sounds that imply a different key, thereby requiring a change in the size of our module. The faculty or activity whereby we perform this mode of evaluative listening is what Holden calls attention. He writes:

> When a person has fixed his attention on one sound as a key note, if other
> sounds accidentally intrude upon his ears, which belong to a different key,

[73] Compare Holden's Plate 6 Example XLII to Rameau's Plate 15 in *Génération harmonique*. Holden does not provide the generation of the ascending scale in the second part of the *Essay*. John Holden, *Essay towards a Rational System of Music* (Glasgow: Urie, 1770), part 1, facing page 76. Jean-Philippe Rameau, *Génération harmonique, ou Traité de musique théorique et pratique* (Paris: Prault fils, 1737), 129.

[74] The faculty of memory corresponds to our innate capacity for "retaining the impressions of musical sounds, for some considerable time after they cease to be heard, [it] is purely natural, and requires no improved abilities at all." Holden, *Essay*, part 1, §31, 28.

70 HEARING WITH THE MIND

they have often a most disagreeable effect.... Before he can hear such incompatible sounds, with satisfaction, he must quit his own former key, and
turn his attention altogether on that to which they properly belong.[75]

Thus attention is the faculty at work not only in our experiencing music as
being in a key, but also in our ability aurally to comprehend changes of key.

Attention plays a significant role in the *Essay*. Throughout Part 1, for example, Holden repeatedly invokes the degree of mental effort required to sustain the attention as reflecting the tonal function of the pitches we encounter.
He proposes that a beginner can identify the mode of a piece as major or
minor by attempting to hear which third scale degree is compatible with its
tonic. This exercise succeeds, he writes, "because it is very difficult to turn
our attention from one sort of scale to the other, while we retain the same
key note."[76] The transition between major and minor thus requires a "change
of attention," in which previous sounds "must be entirely disregarded, and
quite new ideas substituted in their stead."[77]

On the other hand, when moving between the most closely related keys,
we can largely rely on our recollection of the original key. This is because
"there is such an affinity between the key and its fifth or fourth that we can
for a little while turn our attention upon either of these as an occasional new
key, without entirely relinquishing our principal key."[78] The experience of
hearing non-diatonic pitches thus gives rise to a change in attention, and this
serves as an indication that the music may be in the process of modulating
to a key closely related to the home key. (Holden uses the term "modulation"
in the eighteenth-century sense whereby it signifies movement both within
a key and toward a secondary key—one that is typically closely related to the
home key.)

Holden also speaks of "divided" or "distracted" attention. This occurs
under three circumstances: (1) when the current module does not sufficiently account for a given pitch; (2) when there are multiple potential
explanations of a given pitch or combination of pitches; or (3) when the
music has changed keys. The implications of regarding the faculty of attention as participating in a broader system of music perception are fully revealed in Holden's conceptualization of the full chromatic system, which he

---

[75] Ibid., part 1, §31, 27.
[76] Ibid., part 1, §35, 32.
[77] Ibid., part 1, §35, 33.
[78] Ibid., part 1, §31, 27.

obtains by juxtaposing the diatonic scale of a major key with scales built in the identical fashion on IV and V, plus the leading tone of each of these three keys' relative minors. That is, a tonic key (with its relative minor) shares one module with the key of its adjunct fifth (and its relative minor) and another with the key of its adjunct fourth (and its relative minor), and these two modules stand in a 3:2 (or 4:3) relationship to each other. In other words, each module is shared by four keys, of which two—the tonic key and the relative minor key—have the same scale, except for the leading tone of the latter, which lies a major third (5:4) above scale degree III. Jointly, these two modules supply double versions of two scale degrees: first, two versions of scale degree IV—one a perfect fourth (4:3), and the other a "grave fourth (21:16) above the key note—and second, justly tuned (5:3) and Pythagorean (27:16) versions of scale degree VI. These options allow the mind to actively select which one to perform or mentally project.

Holden asserts, "In every passage of music our attention is partly divided, either between the key and its adjunct fifth, or between the key and its adjunct fourth, as fundamentals."[79] When we listen to a passage of music, therefore, our minds constantly retain and project—that is, attend to—two possible modules, each divided so as to afford the scale degrees of two fifth-related major keys and their relative minors, as explained a moment ago. Jointly, the two modules provide all twelve chromatic pitches. (To do so via a single module would require its division into 3,072 [48 × 64] parts, that being the least common denominator of all the ratios present in Holden's complete chromatic scale.) In short: by taking the same C to equal 48 units of one module and 64 of another, we can obtain whole-number module values for all degrees of the full chromatic gamut.[80] Shifting between these modules and "conceiving the principal key occasionally in one or the other of these ways" thus generates the entire range of chromatic pitches.[81] In a sense, we can think of moving between these two pairs of scales as representing a cognitive grouping strategy that enables us to comprehend highly complex proportions.

---

[79] Ibid., part 2, §45, 317.

[80] The numbers 48 and 64 represent the smallest division of two modules standing in 3:4 proportion that afford the construction of the chromatic scale. Holden maintains that we cannot grasp this gamut directly, as he places an upper perceptual limit on possible divisions of the module at just under seven octaves, which means that the most detailed divisions of the module that we could still comprehend equal 3×2^7, or 384 (part 2, §25, 302). Thus "the highest power of 3, which we can admit in the parcelling of musical vibrations, is its fifth power 243, for the sixth power, which is 729, is unquestionably beyond our limits." Ibid., part 2, §41, 312.

[81] Ibid., part 2, §45, 317.

72   HEARING WITH THE MIND

Holden encapsulates this theory in a chart titled "Scheme of the System of Modulation of C," shown in **Figure 2.11**. The column on the left labels each row with an Arabic numeral purely for ease of reference, while the roman numerals across the top and at lines 1, 5, and 9 represent scale degrees (diatonic or altered) with reference to C as tonic. The table is most easily interpreted starting from line 5, labeled "Scale of the Principal Key," which contains the Roman-numeral representation of the scale degrees of C major. Line 6, labeled "Final," contains the numerical proportions for each scale degree taking C as 64,[82] while line 4, termed "Medial," likewise contains the numerical proportions for each scale degree but starting by taking C as 48.[83] The proportions taken here are those of the descending scale, which is to say that the fourth in both cases is grave (21:16). That is, we have here two identical representations of the descending scale of C that stand to each other in the proportion of 64:48, that is, 3:2.

Moving to lines 1–3, line 1 contains the scale degrees in relation to C major as generated by the descending scale of its lower fifth, F (hence including B-flat), labeled as the "Scale of the Adjunct Fourth," that is, the scale of F major but with its degrees numbered with reference to C as tonic; line 2 contains the proportions of this scale according to the same module we encountered in line 4; and line 3 contains their letter names. However, here the descending scale has been built on F, rather than C, and hence the grave fourth is located between F and B-flat, at VII-flat of C, and the fourth itself is a "bearing" higher than the fourth we encountered in the scale built on C, shown in lines 4 and 6; that is, it is a perfect fourth above the tonic.[84] This can be seen if we compare the numerical value of F in row 4 (63) and row 2 (64). Likewise, the A that is scale degree VI in C major (81) is a syntonic comma higher than the A that is scale degree III in F major (80). Lines 7–9 represent an analogous scenario in mirrored order: line 9 contains the scale degrees of C major and the label "Scale of the Adjunct Fifth"; line 7 contains the letter

---

[82] That is, taking as reference pitch a C with a module divided into 64 equal parts.

[83] Holden terms the relationship between the key and its adjunct fourth *medial* and that of the key and its adjunct fifth *final*. He uses these terms to refer to a key's degree of "perfection," that is, conclusiveness (see part 1, §146, 130). He argues that we cannot conclude on a medial chord or key, and at times he even uses the terms *medial* and *inconclusive* interchangeably (part 1, §197, 173). In contrast, he regards the key of the adjunct fifth as more perfect than that of the adjunct fourth, not only because every piece must conclude with a movement forming a perfect authentic cadence from V to I (part 1, §191, 167) but also because he maintains that it is more closely related to the key (part 1, §242, 221).

[84] I discuss the interval of the bearing (64:63) in the previous section (on "The Ascending and Descending Major Scale"); see also part 2, §44, 317.

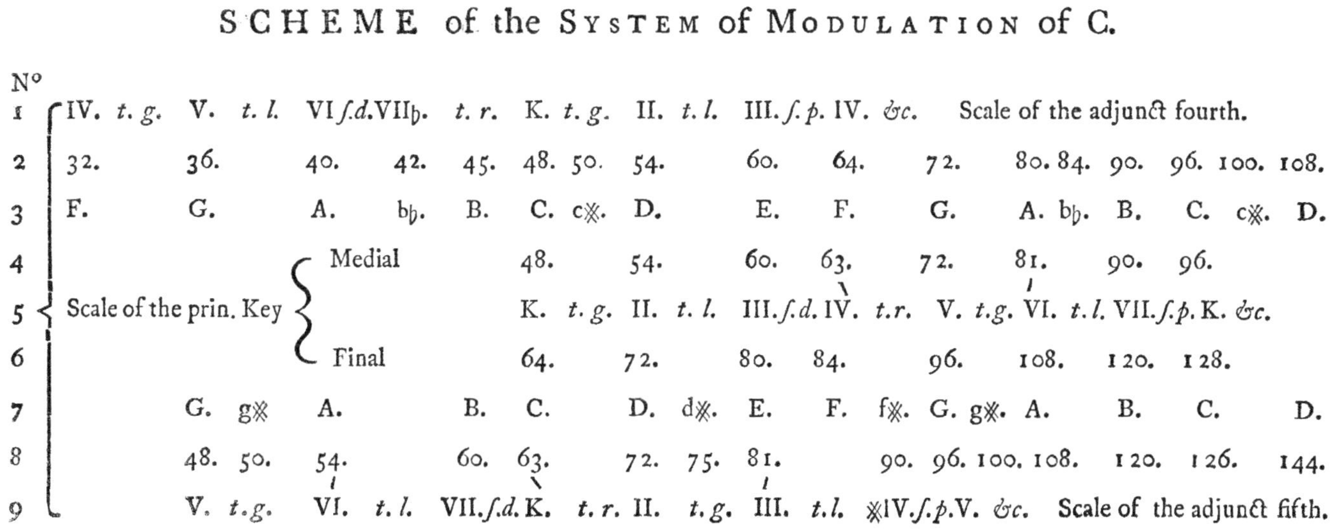

SCHEME of the SYSTEM of MODULATION of C.

Nº																
1	IV. *t. g.*	V. *t. l.*	VI *ſ.d.* VII♭.	*t. r.*	K. *t. g.*	II. *t. l.*	III. *ſ. p.* IV. &c.			Scale of the adjunct fourth.						
2	32.	36.	40.	42.	45. 48. 50.	54.	60.	64.	72.	80. 84.	90.	96.	100.	108.
3	F.	G.	A. ♭♭. B.	C. c𝄪.	D.	E.	F.	G.	A. ♭♭. B.	C. c𝄪.	D.
4			Medial	48.	54.	60.	63.	72.	81.	90.	96.
5	Scale of the prin. Key			K. *t. g.*	II. *t. l.*	III. *ſ.d.* IV. *t.r.*	V. *t.g.*	VI. *t. l.*	VII. *ſ.p.* K. &c.
6			Final	64.	72.	80.	84.	96.	108.	120.	128.
7		G. g𝄪	A.	B.	C.	D. d𝄪.	E.	F. f𝄪. G. g𝄪. A.	B.	C.	D.
8		48. 50.	54.	60.	63.	72.	75.	81.	90. 96. 100. 108.	120.	126.	144.
9		V. *t. g.*	VI. *t. l.*	VII. *ſ.d.* K. *t. r.* II. *t. g.*	III. *t.l.* 𝄪IV. *ſ.p.* V. &c.		Scale of the adjunct fifth.

Figure 2.11 Holden's chart of the relationship of C to F and G major (Holden, *Essay*, part 2, §62, 338). The module of lines 1–4 is divided into 48; the module of lines 6–9 is divided into 64. The size of the interval between each scale degree is classified in one of five ways: the whole tone can be greater (*t. g.*), 9:8; lesser (*t. l.*), 10:9; or redundant (*t. r.*), or 8:7. The semitone can be proper (*s. p.*), 16:15; or deficient (*s. d.*), 21:20. These intervals arise from the derivation of scale degrees from the factors of 2, 3, 4, 5, and 7 and affect the size of larger intervals. In practice, Holden notes that the dominant seventh chords on the fifth scale degree of all three minor scales should be tempered to conform to the proportion of 4:5:6:7.

74 HEARING WITH THE MIND

names; and line 8 contains the proportions of the descending scale on G, built according to the module we encountered in line 6. In this case, the grave fourth is located between G and C; hence the C in line 8 is a bearing lower than the C in line 6, and the syntonic comma is located between scale degree III in C major and the E that is VI in G major. Taking all these elements together, we have C major, flanked by G major below and F major above. Furthermore, each of these three diatonic pitch collections is augmented with sharp V, the raised leading tone of its relative minor, and thus we have in effect six keys represented, and hence the full chromatic gamut.

Figure 2.12 offers my alternative representation of Holden's "Scheme of the System of Modulation of C," and its derivation of the 12-pitch gamut by means of the scales of F, C, and G major and their harmonic minor relatives. Unlike Holden's chart, which depicts the relationship in terms of the scale degrees of C major, Figure 2.12 displays the degrees of each scale with reference to its own tonic. As shown in the figure (and as necessarily occurs somewhere in any justly tuned diatonic scale), the minor triad built on scale degree VI contains intervals above the root of a Pythagorean minor third of 27:32, and a grave fifth of 27:40, each interval being smaller than pure by the syntonic comma 81:80. Hence, to secure a justly tuned triad on VI to serve as the tonic triad of the relative minor key, Holden raises scale degree V of A minor (E), D minor (A), and E minor (B) by the syntonic comma (shown in Figure 2.11 in rows 2, 7, and 12), allowing him to obtain A-, D-, and E-minor triads that reduce to the ratios 54:64:81, indicated in circles in Figure 2.12. However, as in the case of the two tunings of IV in a major key (63 and 64), there are situations in which V is conceived of variously. For example, a dominant seventh chord built on scale degree V of the relative minor calls on the grave V. Furthermore, as shown in rows 2, 7, and 12, scale degree IV of the relative minor must be tempered by the septimal comma 36:35 to conform to the proportion of 7:6 vis-à-vis II, the top interval of the justly tuned dominant seventh chord represented by the proportional series 4:5:6:7 and indicated in dashed squares in Figure 2.12.

Let us further examine the case of the tempered dominant seventh chord of D minor, the members of which are indicated in dashed squares in lines 2 and 5 of the chart. (In Figure 2.12 this corresponds to 35, 40, 50, 60; the seventh (G = 35), should be moved up an octave (to 70), resulting in a root position seventh chord of 4:5:6:7.) Here, the seventh (G) has been mentally adjusted by the septimal comma from 36 to 35. Holden concedes that conceptualizing the G as a product of the factors of 7 and 5 can help "reconcile this chord to

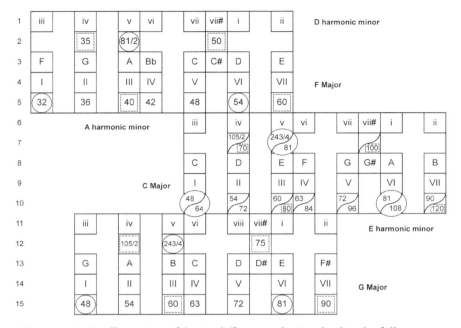

Figure 2.12 An illustration of the six different scales involved in the full chromatic gamut, with the temperings required for the minor mode. Roman numerals pertaining to minor scales are indicated by lower case letters. F major, D harmonic minor, and C major stand in proportion to a module of C taken as 48, while C major, A harmonic minor, G major, and E harmonic minor stand in proportion to a module of C taken as 64 (the raised leading tone G-sharp of A harmonic minor is shared with this module alone). Adjusted scale degrees in minor keys are indicated in rows 2, 7, and 12; all other pitches are shared with their relative major keys, and identical to the numbers in Figure 2.11. The fractions in these rows are an artifact of the low module values of this graphic representation (which range between 32 and 90 for the first module, and 48 and 120 for the second module), and reflect the aforementioned tempering of the minor triad and the dominant seventh chord built on V of minor. Members of the tonic triad of the minor keys are indicated in circles, and members of the dominant seventh chord of the minor keys are indicated in dashed squares.

the imagination."[85] Yet in view of the undeniable mental strain involved in comprehending a chord comprising only the higher prime factors 5 and 7, he concedes that we must "conclude that the key must effectually be changed, and the IV or the V, must be received as a new key, before either of their

---

[85] Ibid., part 2, §86, 368. He further posits that "the difficulty of conceiving such a parcel is the only reason why beginners in singing find it much harder to join a seventh to the leading chord of a flat [minor], than to that of a sharp [major] series." Ibid., part 2, §86, 368.

76 HEARING WITH THE MIND

substituted sixths [relative minors] can be introduced as such."[86] That is to say, according to this theory we cannot modulate to a minor key without first conceptualizing its relative major as a tonic. Most significantly, Holden here provides us with what is essentially a cognitive measure for knowing when we have modulated to a new key: if the mental effort of tempering the seventh of a leading-tone chord is excessive, we should consider ourselves in a new key and conceptualize the module accordingly.

Alternative tunings for scale degrees conceived according to different proportions or modules, Holden emphasizes, should never "be depressed or elevated in an unnatural or forced manner."[87] The designation of grave or acute describes a particular kind of hearing of a sound in relationship to a fundamental and a key. As such, it is intended only "to distinguish the different sounds of notes, which are apparently the same, on different occasions, according as one or another of the three scales takes place at the same time."[88]

It bears emphasizing that the proportions shown in Figure 2.11, which rely on two different modules, represent cognitive divisions of the module rather than a prescriptive system of tuning. However, Holden does offer a version of ⅕-comma meantone temperament at the very end of the *Essay* (§95–§96). In the Holden temperament, as Owen Jorgensen notes, the major thirds and fifths were

> the perfect complete compromises; that is, both the major thirds and the fifths were tempered by exactly the same amount—one-fifth syntonic comma. Therefore, Holden's major triads were not equal-beating, but his first inversion minor triads were equal-beating. . . . In the Holden temperament one needs to concentrate or strain to temper only the first major sixth with exact precision according to a specified known beat frequency. The remaining intervals are then tempered by means of the easiest equal-beating techniques ever discovered. The Holden temperament is thus easy to tune and even though equal beating techniques are used, the results are theoretically correct.[89]

---

[86] Ibid., part 2, §87, 368–369.

[87] Ibid., part 2, §84, 366. In the preface to his *Collection of Church Music*, Holden also emphasizes this point, noting that "an accent placed over any note does not intimate that the note is to be *distorted*, or put out of its natural place, but rather it shows what is its most natural place, in that occurrence." Holden, *Collection*, 5. Original italics.

[88] Holden, *Essay*, part 2, §84, 366.

[89] See Jorgensen, *Tuning*, 118–128, quotation at 118. I thank Danny Walden for this reference.

Double Fundamentals and Implied Sounds

As we have just seen, Holden regards attention as the mental faculty that determines the association of pitches within a given tonal context. In the first part of the *Essay* he proposes that our attention has a second function as well: it is the faculty by which we distinguish between consonance and dissonance, which he defines by writing, "When the several sounds mix and unite, in a manner agreeable to the hearer, it is called a *consonance*, or *consonant chord*; when they do not unite, but separately distract the attention of the ear, it is called a *dissonance*, or *dissonant chord*."[90]

This conception of consonance and dissonance is tightly linked with another distinction that Holden makes between concord and discord, namely, "two sounds are said to be *concord between themselves*, when both of them can be referred to one and the same fundamental perfect chord; and two sounds are called *discord*, when they cannot both be referred to one perfect chord."[91] Holden, following a long-established convention, defines a "perfect chord" as being composed of the first five partials of the harmonic series, in root position or inversion; in other words, a major triad.[92] Concord, then, for Holden is the common property of all of these intervals, while discord is the property of all other intervals.[93]

In characterizing concord, Holden notably foregrounds the criterion of perfect chord membership rather than any sonic character or aesthetic effect. Only subsequently does he invoke the traditional attribute of unified sound for concord, and this as a consequence of affiliation with the perfect chord, that is, with the first five partials of the harmonic series, by way of the mental phenomenon of attention:

> Allowing that the mind *naturally chuses* to conceive every sound in music as belonging to some perfect chord, it is plain, that two sounds will seem to unite, when both of them are included in the idea of one perfect chord, but

---

[90] Holden, *Essay*, part 1, §125, 113.

[91] Ibid., part 1, §153, 138.

[92] This chord, Holden claims, is ideally suited to be the final chord of a piece, as it does not generate expectations in its listeners for resolution or further motion. See ibid., part 1, §143, 127, and part 2, §76, 354. He further notes that in earlier times it frequently served as the final chord of music in a minor key by means of the raised Picardy third. See part 1, §269, 243, and part 2, §88, 372.

[93] According to James Tenney, the view that consonance is a result of triadic membership was an innovation of Rameau's that had been widely adopted in the eighteenth century. See James Tenney, *A History of "Consonance" and "Dissonance,"* part IV (New York: Excelsior, 1988), 65–85.

78 HEARING WITH THE MIND

separately *distract our attention*, when this cannot be done, or when they must necessarily be referred to two different fundamentals.[94]

Since discord is the subjective impression received when a lack of shared triadic membership causes a combination of notes to "distract our attention," it is plausible to infer that, conversely, in concord "the two sounds will seem to unite" because, as members of a single perfect chord, they are perceived as more unified. Unlike *consonance* and *dissonance*, therefore, which describe the impressions such sonorities make on the ear in terms of qualities such as unity, distinctness, agreeability, and their opposites, Holden's definitions of *concord* and *discord* speak to the cause of these impressions, namely, inclusion (or absence) in the perfect chord. Here consonance is implicitly reconfigured as a consequence of concord, as dissonance is of discord.[95] Moreover, Holden emphasizes, it is important to distinguish between "what only *divides our attention* and what *displeases*. The former of these is properly a *discord* in music, and the latter a *false, or inharmonical relation*."[96]

Holden regards consonance as a continuum bounded at one end by the five-note voicing of the perfect chord, the most unified chord possible, that exactly mimics the harmonic series of the first five partials in the proportion 1:2:3:4:5.[97] Any alteration in voicing, as well as doublings or omissions, detracts from the perfect chord's perceptual unity, as "the further we depart

---

[94] Holden, *Essay*, part 1, §153, 138.

[95] Thus the perfect chord for Holden is represented in our minds as a sort of categorial construct that determines our perception of intervals as concordant or discordant on the basis of membership or non-membership. This reverses the traditional understanding—still evident in the writings of Holden's Continental contemporaries—in which consonance is a primitive property of certain dyads and the triad, as the class of sonorities composed of such dyads, is consonant owing to the consonance of its components. See, e.g., Johann Philipp Kirnberger, *Die Kunst des reinen Satzes in der Musik* (Berlin: Voß, 1771), 1:23–26. I thank David E. Cohen for this idea, and Caleb Mutch for offering an important nuance. See also Caleb Mutch, "How the Triad Took (a) Root," *Journal of Music Theory* 66, no. 1 (2022): 43–62.

[96] Holden, *Essay*, part 1, §153, 139. Holden here tries to motivate a systematic distinction between consonance and dissonance on the one hand, and being in tune and being out of tune on the other hand. A harmonic relation may nevertheless be dissonant (e.g., 9:8, 9:5). Likewise, an inharmonic relation may be a consonance, such as a tempered third or fifth. For example, Holden notes that music performed on an organ, the pitches of which cannot be adjusted, is perceived much as it is when performed by violins (part 2, §94, 379). Furthermore, in the case of brass instruments such as the natural horn, "the interval between the II and IV is exactly 9 to 11: and yet this interval in compositions for a first and second horn, is very frequently used where the ear must inevitably conceive the sounds by the proportion of 6 to 7" (part 2, §94, 380). That is, although the second and fourth scale degrees function as the top two members of the dominant seventh chord built on the V of the scale, when they are performed by a natural horn whose fundamental is tuned to the tonic, Holden argues that we hear partials 9 and 11 of the tonic, yet understand them as if they were the partials 6 and 7 of the dominant.

[97] Ibid., part 2, §76, 354.

from the disposition of the most perfect of this or any other kind of chord, the greater imperfection is thereby introduced."[98] When we hear a triad voiced differently or with omitted sounds, he asserts, our minds nevertheless understand it as representing the proportions 1:2:3:4:5 in varying degrees of attenuation. Moreover, "not only one or two of the notes of this chord may be wanting, but they may all be wanting except one, and yet still the same idea conceived by the hearer, as if they were all joined together."[99]

Of course, in a real piece of music, many of the sounds that we hear will not be perfect chords in root position. Holden therefore makes some general recommendations for the sake of enhancing the unity of a sonority: the fundamental should be doubled where possible, as "the unity of the whole is destroyed, when the third or fifth attracts too much of our attention."[100] Chord voicing is also important because "our attention is carried either to the lowest or the highest sound which exists in a chord, rather than any of those which lie, as it were, concealed in the middle."[101] In the case of chord inversions, he proposes, it is best to reserve the highest position for an upper octave of the fundamental, in order "that the hearer, being disappointed of it below, may yet find it above."[102]

Relatedly, Holden's approach to dissonance develops from the assumption that such chords are perceptually less integrated. Here he invokes an idea proposed by the Swiss music theorist Jean-Adam Serre some two decades earlier in *Essais sur les principes de l'harmonie* (1753), namely, that dissonant chords have two fundamentals.[103] Holden asserts:

> No one sound can properly be called the *sole* fundamental of a dissonant chord, because of the divided attention which the discord creates, but notwithstanding we must be supposed to have two fundamentals partly in

---

[98] Ibid., part 1, §135, 120. Holden idiosyncratically argues that the most perfect configuration of the minor chord will omit its third entirely, corresponding to 1:2:3:4; see ibid., part 1, §140, 124–125.

[99] Ibid., part 1, §136, 120–121. Holden here and elsewhere posits something like Riemann's concept of *Klangvertretung*, in which a single note can represent a major or minor triad. See Hugo Riemann, "Die Natur der Harmonik," in *Sammlung musikalischer Vorträge*, ed. Paul Graf Waldersee (Leipzig: Breitkopf und Härtel, 1882), 4:184–185. For example, Holden writes, "It is necessary to conceive, not only the compound sounds of harmony, but also every single sound in melody, as belonging to some perfect chord, in order to account for their various effects, and the preference due to some successions of sounds rather than others" (part 1, §146, 131–132). I thank David E. Cohen for this observation.

[100] Ibid., part 1, §142, 125.

[101] Ibid., part 1, §144, 127.

[102] Ibid., part 1, §144, 128.

[103] For more on Serre's theory of dual fundamentals see Andrew Pau, "The Harmonic Theories of Jean-Adam Serre," *Intégral* 32 (2018): 3–5.

view, yet one of them may, for various reasons, claim the greater share of our regard; and therefore may be properly called *governing*.[104]

The fundamental of any well-formed chord is thus determined by the presence of an upper perfect fifth to that fundamental, which reinforces our perception of the fundamental as such.[105] Therefore, a chord containing several perfect fifths, such as a minor seventh chord, will have competing fundamentals.[106]

Holden observes: "The chord of the seventh, when the fundamental bears a less third, may be considered as a mixture of two perfect chords whose fundamentals are the two terms of the same less third."[107] That is, we can conceive of this either as a minor triad with an added seventh or as a major triad with an added sixth, the root of which is the first chord's minor third. To ascertain the governing fundamental, he proposes we rely on the voicing of the chord itself. **Figure 2.13** reproduces Holden's interpretation of a minor seventh chord of A–C–E–G as composed of members of two overlapping perfect chords A–C–E and C–E–G, which thus contains a double fundamental.

In providing guidelines for determining the fundamental of a dissonant chord, Holden's analysis in Figure 2.13 relies on his earlier claim that our attention tends to be drawn to the outer voices of a chord, that is, the bass

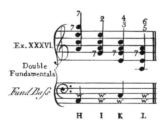

Figure 2.13 Seventh chord with double fundamental. Holden, *Essay,* part 1, plate 4.

---

[104] Holden, *Essay*, part 1, §160, 144.
[105] Holden writes, "Every note which has its own harmonics, and especially its fifth existing along with it, will in some degree attract our attention as a fundamental" (ibid., part 1, §176, 155–156).
[106] Two other intervals also possess the power to invoke fundamentals, namely, the diminished fifth, which has "a very peculiar property of referring the hearer to a fundamental note, at the distance of a greater third below its lower term," whether or not that note appears in the chord, and the tritone, which "refers to a fundamental, at the distance of a less third below its lower term" (especially when the perfect fifth above that fundamental is present), or it refers to the major third below the tritone's upper note (ibid., part 1, §179–180, 158).
[107] Ibid., part 1, §178, 156.

and soprano.[108] In root position (marked with the letter H in the figure), the A commands our attention through its presence as the lowest note in the chord, while the C is concealed within the inner voices; thus the A is indicated by the note on the fundamental bass stave, and the C is marked by a custodes. A similar situation occurs in the 4/3 inversion (letter K), where the C commands our attention, as it is the highest note of the chord and thus determines the fundamental bass, while the custodes indicates a conflicting fundamental on the A in the bass clef. In the case of the 6/5 (letter I) and 4/2 inversions (letter L), however, the attention is more or less equally divided between potential fundamentals, as shown by the custodes on both A and C. Here, Holden observes, the attention must be assisted through doubling or rendering the intended fundamental note conspicuous by other means:

> At L, the note A is uppermost, and C lowest; which causes a divided attention, or *double fundamental*, and the preference due to one or the other is to be determined from other considerations. The case is much the same at I, where both C and A are in the middle. In these ambiguous cases, if one of the notes which thus stand in competition be doubled, or be made more remarkable by falling in with the expectation of the hearer, the choice will be determined by either of these circumstances.[109]

The composer's choice of a specific chordal doubling, Holden here suggests, can fortify one or the other of the perfect fifths, strengthening its lower term's claim as fundamental and disambiguating the chord's quality (either a minor seventh chord or an added sixth chord). This, presumably, would be done in accord with the particular role that such a chord plays within its harmonic and melodic context. Much as in Rameau's *double emploi*, listeners can thus determine the chord's fundamental according to its fulfillment of melodic or harmonic expectation.[110]

---

[108] Holden, *Essay*, part 1, §144, 127. While Holden's belief that we can better perceive a chord's fundamental if it is placed in the soprano line may seem counterintuitive, it is likely that he arrived at this conclusion from observing that bass and soprano tend to attract most of our attention; see part 1, §199, 176, and §144, 128, respectively.

[109] Ibid., part 1, §178, 157.

[110] The criterion of expectation also applies to perfect chords with substituted sounds, or suspensions. To ascertain the fundamental of suspensions, Holden suggests that we search for notes that possess their own harmonics in the chord, particularly their fifth, as this "will in some degree attract our attention, [particularly if] such a note be rendered more distinguishable by its situation above or below the rest" (ibid., part 1, §176, 155–156). However, we must also factor in the syntactical context of the chords and "take into account the expectation, whether founded on nature or custom, of hearing such particular fundamentals in succession," as this may sometimes cause us to favor a different fundamental (ibid., part 1, §176, 155–156).

In the second, speculative part of the treatise, Holden formalizes these perceptual claims by introducing a new principle to explain various aspects of intervals, chords, and inversions, which he terms "implied sounds."[111] As we will see, this principle, an abstraction of the phenomenon of difference tones, allows him to provide a cognitive, rather than acoustic, account for the perceptual unity of consonance.

Holden asserts that, on hearing two different sounds, our imagination projects an implied sound, that is, a third sound equal to their frequency difference.[112] Say, for example, that we are in the key of C major, and hence measuring all sounds by a corresponding module of C (designated as f). If we hear G and E with the frequency $6f$ and $5f$, Holden predicts that our mind recognizes the difference of $6f - 5f = 1f$ and internally generates an implied sound of $1f$. Therefore, although the C is not present in the actual sonority (E, G), our imagination supplies the sound, as shown in **Figure 2.14**.

Figure 2.14 Implied difference tone generated by a minor third.

In the eighteenth century it was unclear whether difference tones were acoustical sounds or a psychophysical illusion.[113] Holden sidesteps this debate altogether by arguing that the implied sounds arise from the process of perception. He makes another analogy to visual experience:

> It may be inferred, that as the velocity with which one moving body approaches towards or recedes from another (which may be called its relative motion, and is equal to the difference of the two absolute motions) is a circumstance which always attracts part of our regard: so the difference

---

[111] Ibid., part 2, §76, 354.
[112] This can occur in different ways. The situation Holden describes involves two loud tones with similar frequencies f_1 and f_2. These generate a difference tone $(f_1 - f_2)$ and a summation tone $(f_1 + f_2)$, with beating.
[113] Difference tones, sometimes called Tartini tones, are now explained as being physical but intra-aural, arising from the non-linearity of the cochlear membrane; see Gerald D. Langner, *The Neural Code of Pitch and Harmony* (Cambridge: Cambridge University Press, 2015), 32.

of the vibrations, or the relative velocity of the pulses of one sound in comparison to those of the other, may some way be perceived, abstract from all considerations either of real or *sensible* coincident pulses.[114]

Holden applies his principle of implied sounds to explain a range of musical features. For example, he ascribes the power of "the most complete and natural arrangement" of the perfect chord (1:2:3:4:5) to the fact that the difference between any two pitches in this chord is always equal to one of the members of the chord itself.[115] He remarks, "The implied sounds, in this perfect chord, produce no other effect than that of fortifying, or doubling, all the real sounds of the chord, except the highest."[116] As shown in **Figure 2.15**, the root of the chord is thus always most strongly reinforced, as the difference of 1 appears four times in the sonority, the difference of 2 three times, the difference of 3 twice, and the difference 4 once, leaving only the highest note unfortified by implied sounds.

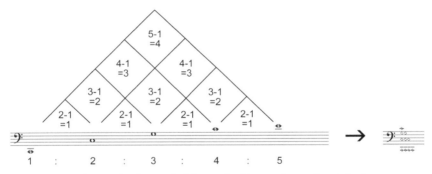

Figure 2.15 The implied sounds (shown in diamond note heads) of a perfect chord.

Holden maintains that implied sounds influence musical practice in a number of ways. The convention of enhancing an accompaniment with octave doublings serves to fortify the harmony, "because the manner of parcelling the vibrations of the key, by continual reduplication . . . essentially implies the coexistence of all its octaves, both above and below, not exceeding the limits of audible sound."[117]

---

[114] Holden, *Essay*, part 2, §75, 353–354.
[115] Ibid., part 2, §76, 354.
[116] Ibid., part 2, §76, 355.
[117] Ibid., part 2, §77, 356.

Implied sounds can also explain why we tend to prefer certain chord voicings. For example, the close-position first-inversion triad built on E4 (5:6:8, corresponding to E4, G4, and C5) comprises the implied sounds 6 − 5 = 1, 8 − 6 = 2, and 8 − 5 = 3, to which we project the implied sonority of C2, C3, and G3, all of which support our perception of the chord's root as C. However, if we change the voicing by raising the upper two notes by an octave to G5 and C6 (5:12:16), the resulting implied sounds are now 12 − 5 = 7, 16 − 12 = 4, and 16 − 5 = 11. Holden notes that even though these sounds stand in complex relations to the root, and are therefore barely perceivable, "still they contribute something toward the imperfection of the chord."[118]

Finally, implied sounds can help us understand why intervals tend to sound less consonant in very low registers: if an interval's implied sounds are lower than we can hear, then the interval is perceived as less consonant, as shown in **Figure 2.16**.[119]

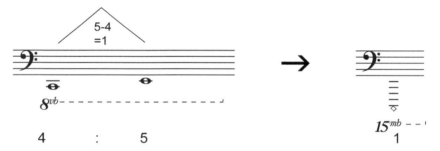

Figure 2.16 Inaudible implied sounds do not contribute to the consonance of dyads at the lower boundary of our hearing range.

To summarize, Holden understands consonance and dissonance as qualities occupying a continuous spectrum defined by greater and lesser degrees of perceptual unity. The degree to which a sonority will be classified as one or the other depends on two factors: (1) the degree of conformance of the specific voicing in question to the first five partials of the overtone series as embodied by the fundamental perfect chord 1:2:3:4:5, which also exhibits the most desirable configuration of implied sounds; and (2) the degree to which the sonority's implied sounds are (theoretically) perceptible, given the physiological boundaries of our hearing. Regardless of their ambiguous

---

[118] Ibid., part 2, §81, 360.
[119] Ibid., part 2, §91, 376–377.

status as acoustic or perceptual objects, the primary function of implied sounds is to supply the listener with additional information about the degree to which a chord is consonant or dissonant.

Conclusion

The *Essay towards a Rational System of Music* challenges our assumptions about what was possible for eighteenth-century music theory. Holden explicitly proposes a number of fundamental mechanisms of musical perception and then derives his entire theory as an orderly series of deductions therefrom. By approaching perception as composed of hierarchies of mental processes, Holden's methodology enables him to articulate a series of intuitions, many of which have since been rediscovered by the modern fields of music theory and music cognition.

For example, Holden's observation that our minds tend to automatically group a row of identical windows into units of twos and/or threes,[120] as well as his claim that our minds require "the number of measures in a strain [to] always be either two or three, or some product of these numbers"—are highly reminiscent of the psychological concept of chunking associated with George A. Miller.[121] Miller asserts that human thought is enabled by the ability of our minds to recode—to reorganize—a great deal of information into fewer, more informative units. This insight has been applied in the domain of music cognition by Jay Dowling, Diana Deutsch, and others to study how we recall melodies, hierarchically organized tonal sequences.[122]

Another way in which Holden's approach anticipates modern psychology lies in his conceptualization of perception as an active process in which the mind, in accordance with its innate cognitive preferences, extracts and organizes features of the musical input. Not unlike the approach of Fred Lerdahl and Ray Jackendoff's *Generative Theory of Tonal Music* (1983) and

[120] "When we cast our eyes on nine equidistant windows in a row, they are no sooner seen than subdivided into three times three: eight appears at first to be two fours, and each of these fours, two twos; seven we conceive as two threes disjointed, and one in the middle; six most naturally divides itself into two threes; but if seen along with nine, or immediately after it, we then trisect it, in conformity with nine, and it appears three twos: five becomes two twos disjoined and one in the middle; four becomes two twos, and single three or two need no subdivision." Holden, *Essay*, part 2, §11, 289.
[121] George A. Miller, "The Magical Number Seven, Plus or Minus Two: Some Limits on our Capacity for Processing Information," *Psychological Review* 63 (1956): 81–97.
[122] W. Jay Dowling, "Rhythmic Groups and Subjective Chunks in Memory for Melodies," *Perception & Psychophysics* 14 (1973): 37–40; Diana Deutsch, "The Processing of Structured and Unstructured Tonal Sequences," *Perception & Psychophysics* 28, no. 5 (1980): 381–389.

86 HEARING WITH THE MIND

other theories derived from the generative-syntax theories of Chomsky and his followers, Holden assumes that when we hear a sequence of musical sounds, our mind generates a parallel mental representation in the form of a path through divisions of the module.[123]

Holden's observation that anticipation is essential to our enjoyment of music has also been borne out by more contemporary research.[124] In a 2019 review, music psychologist David Huron traces the theorization of the link between prediction and musical pleasure to Leonard B. Meyer's *Emotion and Meaning in Music* (1956) and observes that this phenomenon has since been compellingly demonstrated across a number of musical parameters by scholars including Eugene Narmour, Elizabeth H. Margulis, and himself, and also, very recently, by psychologists using brain imaging techniques.[125]

Holden claims that our perception of groups of measures as phrases—what theorists of the eighteenth century commonly term "rhythm"—is dependent on tempo, and that it stalls beyond certain durational limits. At extremely slow tempi, he reports, "the memory has enough to do to retain the length of one single bar . . . here we cannot constitute any larger portions or strains, and one bar is all we regard at once; so that if the first line employ four bars, and the next line only three, we are not sensible of any impropriety."[126] This notion of definite cognitive constraints on the perception of duration has been corroborated by research on music and working memory. As music psychologist Justin London explains, given that our perception of beats as connected into groups tapers off at around two seconds per beat, "it makes sense that the absolute value for a measure might be from about 4–6 seconds (i.e., twice or three times the length of the 'slowest possible beat'),"[127] a hypothesis fully in line with Holden's introspectively-based assertions.

---

[123] Fred Lerdahl and Ray Jackendoff, *A Generative Theory of Tonal Music* (Cambridge, MA: MIT Press, 1983).

[124] This idea is succinctly stated in his *Collection of Church Music* as follows: "our enjoyment of music depends, in a great measure, on a faculty of retaining the ideas of former sounds, and anticipating those which are to follow." Holden, *Collection*, 4.

[125] See Leonard B. Meyer, *Emotion and Meaning in Music* (Chicago: University of Chicago Press, 1956); David Huron, "Musical Aesthetics: Uncertainty and Surprise Enhance Our Enjoyment of Music," *Current Biology* 29, no. 23 (2019): R1238–R1240; David Huron, *Sweet Anticipation: Music and the Psychology of Expectation* (Cambridge, MA: MIT Press, 2008); Eugene Narmour, *The Analysis and Cognition of Melodic Complexity: The Implication-Realization Model* (Chicago: University of Chicago Press, 1992); Elizabeth H. Margulis, "Melodic Expectation: A Discussion and Model" (PhD diss., Columbia University, 2003).

[126] Holden, *Essay*, part 2, §14, 290.

[127] Justin London, "Cognitive Constraints on Metric Systems: Some Observations and Hypotheses," *Music Perception* 19, no. 4 (2002): 536–537. London further provides a partial list of the many investigations of the mind's temporal (and attentional) limits since the 1930s.

A final way in which Holden's ideas anticipate later thought can be found in his observation that similarly small disparities go unperceived both in the pitch and temporal domains. "When the difference of two sounds is only a comma [81:80]," he writes, "we readily take one [of those sounds] instead of the other."[128] Analogously, "one bar may be, and always will be some very small matter longer than another, perhaps much more than one eightieth part of the whole bar, and yet we are sensible of no impropriety."[129] Holden here appears to intuit aspects of the phenomenon of "just noticeable difference" subsequently formalized under the name Weber's Law by Gustav Fechner in 1860, a phenomenon that lies at the heart of the modern field of psychophysics.[130]

Holden's explicit interest in the workings of the mental faculties invites a different kind of comparison with contemporary cognitive psychology. We can interpret the model of musical cognition the *Essay* implicitly suggests by reading it through the modern psychological lens of "hierarchical processing," as a mixture of "bottom-up" and "top-down processing." Bottom-up processing begins with sensory input, which is then "transformed and combined until we have formed a perception. The information is transmitted upwards from the bottom level (the sensory input) to higher, more cognitive levels."[131] In contrast, top-down processes require "our stored knowledge about the world in order to make sense of [sensory] input."[132] These concepts illuminate the originality and the complexity of Holden's approach, which, from a contemporary perspective, can be regarded as invoking both bottom-up and top-down processing at various stages of perception.

Figure 2.17 maps Holden's theory onto the tripartite hierarchy of mental processing widely accepted in modern psychology.[133] The low level comprises the level of the senses, and in the case of hearing, this entails the physiological transductions of the acoustic signal by the ear. As an example

---

[128] Holden, *Essay*, part 2, §42, 315.
[129] Ibid.
[130] The *Oxford Dictionary of Psychology* defines Weber's Law as "the proposition that the smallest detectable difference in the magnitude of a stimulus (the Weber fraction) is proportional to the magnitude of the lesser stimulus." "Weber's Law," in *A Dictionary of Psychology*, 4th ed., ed. Andrew M. Colman (Oxford: Oxford University Press, 2015), 818. See also Alexandra Hui, *The Psychophysical Ear: Musical Experiments, Experimental Sounds, 1840–1910* (Cambridge, MA: MIT Press, 2012).
[131] Paul Rookes and Jane Willson, *Perception: Theory, Development and Organisation* (London: Routledge, 2000), 13.
[132] Ibid.
[133] See Edward H. Adelson, "Layered Representations for Vision and Video," in *Proceedings of the IEEE Workshop on Representation of Visual Scenes* (Cambridge, MA, 1995): 3–9.

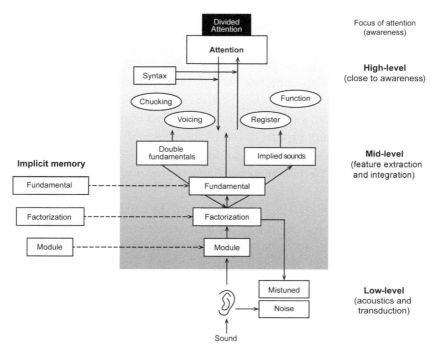

Figure 2.17 A flowchart of Holden's theory, mapped onto the modern tripartite division of the cognitive faculties involved in audition, shown at the right. The gray shading illustrates the gradual boundaries between low-, mid-, and high-level processing.

of low-level processing, for Holden it is the ear, not the mind, that evaluates the isochrony of successive pulses or vibrations.

With a few exceptions, the second, speculative part of Holden's *Essay* explores phenomena that we today ascribe to mid-level processing. According to the neuroscientist Daniel Pressnitzer, mid-level audition pertains to "processes that sit between an acoustical description of sound and the use of auditory information to guide behavior," in which a series of operations extract and integrate complex features, thereby distilling complex auditory stimuli into more hierarchically organized mental representations.[134] As an example of mid-level processing, consider Holden's claim that the

---

[134] Daniel Pressnitzer, "Mid-Level Audition," *Habilitation à Diriger des Recherches*, Université Paris Descartes et Département d'Etudes Cognitives, Ecole Normale Supérieure (2009), i.

fundamental bass arises as a consequence of the mind's acts of grouping. This instance demonstrates how the sounds conveyed by the ear are refined into more complex mental representations.

In contrast, the first, practical part of the *Essay* generally describes processes that we would today categorize as higher level, the stage at which auditory objects are perceived as coherent wholes participating in syntactical contexts. As an example of high-level processing, consider Holden's aforementioned account of the experience of hearing a minor seventh chord with two possible fundamentals, which divide our attention. These processes include the role of attention in selectively modulating our interpretation of sounds perceived, as well as the influence of stylistic conventions such as function, voicing, and register.

As Figure 2.17 illustrates, at the lowest level, the ear judges whether a sensed sound is regular. If not, it is discarded as noise; otherwise, the ear sends the sound up to the mid-level. At the mid-level, the mind performs a series of analyses: it decomposes the sound into small prime factors and, after first eliciting a module from them and dividing it, compares them to those previous divisions of the module, attempting to determine how the sound fits into the key. If the relationship is straightforward, the imagination projects a corresponding fundamental; if not, the sound is excluded as mistuned. If the sound is composed of more than one pitch, the mind ascertains its implied sounds and amends its fundamental as necessary. At successively higher levels, considerations of voicing and register, and subsequently function and chunking, affect the organizing of extracted features into coherent wholes, modulated by concerns of syntax and the faculty of attention.

Figure 2.17 thus interprets Holden's theory in light of the modern psychological concepts of bottom-up and top-down processing. Bottom-up processing is evident in the successive refinement of the internal representation of sounds, starting from the ear's low-level estimate of pitch and continuing through the acts of feature extraction and binding associated with mid-level processing and the resultant representations that are then further interpreted by high-level processes. However, this model of Holden's ideas also takes top-down processes into account. In tandem with bottom-up processing at the mid-level (such as grouping via a comparison with the module and associated factoring, as well as the assigning of implied sounds and projected fundamentals), an internal representation held in implicit memory—the previous module, factoring, and fundamental—directly affects the way in

90   HEARING WITH THE MIND

which these features are extracted and perceived at a given moment.[135] The influence of the immediate past thereby exerts a top-down influence on our interpretation of the current state.

At higher levels, these mid-level features are further organized by established musical conventions pertaining to register, voicing, and comparable qualities in order to chunk individual elements into larger units such as phrases or functions. Higher-level processing can also call on the attention to determine the function of the sounds perceived, particularly in ambiguous situations, such as the case we encountered at note 96, where a chord performed by natural horns can be functionally heard in terms that are considerably different from its acoustic expression. This, too, calls on top-down processing, as our interpretation relies on previously acquired syntactic knowledge to shape our experience of a current sound or event.

Perhaps the clearest example of the model's reliance on both top-down and bottom-up processes is seen in Holden's bifurcated approach to attention as both active and passive. On the one hand, we can place our attention on a certain sound as a tonic, exerting a top-down influence on the maintenance of the module and associated key relationships. On the other, our attention may be attracted by competing fundamentals or implied sounds and influenced from the bottom up. At the same time, considerations of syntax can affect both voluntary and involuntary forms of attention, as indicated with bidirectional arrows in Figure 2.17. Musical features that activate and divide our attention, located at the very top of the diagram, sometimes breach the focus of our conscious awareness, either by causing the sensation of distraction or by calling on our minds to actively intervene and interpret incoming stimuli (such as a modulation, a dissonance, or other rare or significant sonic objects).[136]

*

Considered within the context of eighteenth-century musical culture, Holden's *Essay* is exceptional by any measure. Like nothing else written in its day, it reveals a number of unexpected ways in which a rigorous engagement

---

[135] See Holden's derivation of the descending scale, and in particular his account of the decision to provide scale degree IV with a fundamental of V rather than II.

[136] My interpretation of Holden's account of divided attention as indicating a state of near or full awareness supports Natalie Phillips's theorization of late-eighteenth-century understandings of distraction as a productive cognitive state. See Natalie M. Phillips, *Distraction: Problems of Attention in Eighteenth-Century Literature* (Baltimore: Johns Hopkins University Press, 2016), 4.

with music theory can give rise to sophisticated insights about musical cognition. Much of the astonishing detail of Holden's theory arises from the fact that its innovative attempt to understand our perception of music relies upon concepts afforded by the faculty psychology of the time. Thus the *Essay* offers an unusually detailed and early account of how the mental faculties of attention, memory, and expectation shape our musical experience.

But beyond its deployment of available psychological theory, the most remarkable contribution of the *Essay* lies in the specific and highly original explanation Holden found of exactly *how* our minds go about perceiving pitches in relation to a key. To answer this difficult question, Holden invents the mental construct of the module. It is the module that allows him to make use of the familiar whole-number arithmetic of traditional music theory, with its proportions and string divisions. What the monochord and the module have in common—what makes this substitution possible—is that each is a construct comprising a quantity that can be divided and subdivided. Whether that quantity is a length of string or a length of time may have seemed, to a skilled practical mathematician such as Holden, irrelevant. Holden's innovation, regardless of whether he recognized it as such, was, first, to intuit the mathematical equivalence of string length and temporal duration, and then to see that a durational construct, situated within the listener's mind, could be used to *explain* the ear's mysterious ability to perceive music.

All this made it possible for the *Essay* to offer psychological interpretations of musical phenomena that were already thoroughly theorized in terms of a conceptual apparatus familiar from monochord divisions, namely, proportional arithmetical relations. Holden's greatest achievement thus lies in his detailed theorizing of our preference for simple integer-ratio rhythmic and pitch relations as emerging from the innate structure of our minds. (Indeed, from Holden's perspective, it was necessarily the case that the essential principles of his psychological theory and the music theory that described the musical practice of his day would be the same because he regarded these as reflecting the universal organizing principles of the human mind as such). More than any other factor, this insight allowed him to theorize the experience of the music of his time as arising from a set of foundational cognitive principles.

3

"Our Nurses' Tunes"

John Holden and Scottish Psalmody

On Sunday, Oct. 14 the college-chapel of Glasgow was opened. Principal Leechman preached in the forenoon, and Mr. John Hamilton, minister of the high church, in the afternoon. The music was performed in four parts, and the proper key for each tune pitched by a pitch pipe.[1]

Sunday, Oct. 14, 1764. The weather clear and cold. You have just taken your place in one of the pews of the new College Chapel in Glasgow. It is late afternoon, and the day's final rays of sunlight straggle in through the plain, narrow windows, illuminating the space and dappling the bare oakwood benches. The chapel is austere, as would be expected in any Presbyterian congregation, but the precisely cut masonry and carefully fitted furnishings speak of simple yet profound devotion. John Hamilton, minister of Glasgow Cathedral, stands beside the altar. Behind him is a man standing with his back towards you, facing the newly formed college band. This group of sixteen boys and men, some of whom you know, have been rehearsing in secret for the past few months, and from your seat in the pews you can see that they look nervous. Their director takes something out of his pocket, and you hear a shrill piping sound as the choir clears their throats. Musical instruments in the church! You have never heard of such an occurrence before in Scotland (although to be sure the decadent English tolerate such things), and feel rather scandalized. Moments later, however, the band burst into song, performing the psalm tune "St. Davids." And now your concern begins to resemble alarm, for they are singing the tune, which you and your fellow congregants have known all your lives, in an entirely new way. They are singing

---

[1] William Smellie, ed., "Affairs in Scotland (Oct. 1764)," *The Scots Magazine, and Edinburgh Literary Miscellany* (Edinburgh: W. Sands, 1764), 26:574.

Hearing with the Mind. Carmel Raz, Oxford University Press. © Oxford University Press 2025.
DOI: 10.1093/9780197786208.003.0004

in four-part harmony! From musical notation! You have never heard such music in a sacred context before; in every church you have attended all your life, the precentor would call out the words line by line while the congregation dutifully responded with the corresponding phrase of the psalm tune in an approximate lugubrious unison. And yet this novel harmonic music is strangely beautiful, with its changing (and sometimes pungently dissonant) sonorities cradling the words of Psalm 106. As you listen to the second verse, "God's mighty works who can express? or shew forth all his praise? Blessèd are they that judgment keep, and justly do always," you wonder whether perhaps those words might not also apply to this new style of singing, in spite of its controversial nature...

*

Sacred music in eighteenth-century Scotland was a highly idiosyncratic affair.[2] As a consequence of its fierce political and religious struggles against England, Scotland's churches had remained almost entirely devoid of polyphonic sacred music from 1643 to about 1748, and instrumental performance continued to be banned in establishment churches until 1863. The restoration of part singing in Scotland, which occurred as part of a movement known as the Monymusk revival starting in 1748, was controversial. This particular historical context profoundly shaped Holden's music-theoretical writings and, indeed, his approach to music more broadly.

Throughout his *Essay towards a Rational System of Music* (1770), Holden eschews any discussion of specific instrumental or concert repertoire, and at various junctures he asks his readers instead to consider familiar psalm tunes, often noting that any number of such tunes would serve his purpose.[3] Holden's decision to draw nearly all of his musical examples from English-language metrical psalmody reflects a unique triangulation between the specific context of music in Scotland, on the one hand, and his own conviction that he was identifying universal features of musical perception on the other.

---

[2] See, e.g., Henry G. Farmer, *A History of Music in Scotland* (London: Hinrichsen, 1947); Gordon J. Munro, "Scottish Church Music and Musicians, 1500–1700" (PhD diss., University of Glasgow, 1999); Timothy Duguid, *Metrical Psalmody in Print and Practice: English "Singing Psalms" and Scottish "Psalm Buiks," c. 1547–1640* (Farnham, Surrey: Ashgate, 2016); Timothy Duguid, "Early Modern Scottish Metrical Psalmody: Origins and Practice," *Yale Journal of Music and Religion* 7, no. 1 (2021): 1–23.

[3] In the second chapter of his *Essay*, Holden announces that he will, wherever possible, illustrate his theoretical observations with the "common Psalm tunes of the church, [which] are more universally known than any other pieces of music." He continues, "as we should first learn to spell and read our mother-tongue, so we should first learn to apply the scale, if I may use the expression, to our nurses [*sic*] tunes." Holden, *Essay*, part 1, §29, 23.

94   HEARING WITH THE MIND

The tunes he refers to formed the basis of worship practice for Presbyterian, Anglican, and Episcopal worship in Britain in his day.[4] By asking his readers to reflect on their own subjective experiences with melodies familiar to them since the nursery ("nurses tunes"), then, Holden could be sure that the insights he was describing would be available to all, regardless of their background and training.[5]

While theorists typically (and naturally) assume their readers' familiarity with the musical repertory to which their theories apply, in this case the repertoire to which Holden refers is highly unusual in several ways. Indeed, I can think of no other treatise that relies entirely on a tiny and simplified corpus of hymns in order to broadly generalize about features of sacred and secular music alike. It is made even more conspicuous by the fact that Holden *does* refer in general terms to various features of opera, instrumental music, and popular dance forms such as the minuet, but refrains almost entirely from making any concrete observations about this repertoire.

In what follows, I begin with the historically contingent, laying out the specific context to which Holden's writings respond, with a particular focus on the Monymusk revival. I then examine the account of contemporaneous psalm singing that we can glean from Holden's *A Collection of Church Music* (1766), as well as his *Essay*, and position his descriptions vis-à-vis his reception of the theories of Jean-Philippe Rameau.

The Monymusk Revival

Hymnody in Scotland in Holden's day had been profoundly shaped by the violent religious and political conflicts with England in the previous century. These upheavals, which culminated in the series of wars known as the

---

[4] The only examples that Holden includes which do not come from Scottish psalmody pertain to his example 30, which describes the "shake," a type of vocal ornamentation; to example 56, which features "Pasquali's first lesson" and demonstrates fundamental progressions; and to examples 71 and 72, which contain simple contrapuntal passages that demonstrate suspensions.

[5] Holden's writings thus starkly contrast with the most important theory of music penned in Scotland before 1750: Alexander Malcolm's ambitious *A Treatise on Musick, Speculative, Practical & Historical* (Edinburgh, 1721). Malcolm (1685–1763), a Scot who (unusually) shared Holden's Episcopalian faith and would later immigrate to Maryland, makes no mention of the practices of Presbyterian psalm singing whatsoever. Writing decades before the Monymusk revival, Malcolm may have seen no potential audience for music theory beyond an elite readership, as his references to Descartes, Locke, Vossius, and similar figures make clear. For more on Malcolm, see Paula J. Telesco, "Identifying the Unknown Source of a Pre-Rameau Harmonic Theorist: Who was Alexander Malcolm's Mysterious Ghostwriter?" *Eighteenth-Century Music* 17, no. 1 (2020): 37–52.

Wars of the Three Kingdoms (1638–1652), resulted in large part from the recurring attempts of King Charles I (1600–1649) of Scotland, England, and Ireland to unite his kingdoms into a single union characterized by some form of high-church Anglicanism. In Scotland this issue came to a head in 1637, when Charles attempted to impose a lightly revised form of the English Book of Common Prayer on Scotland.[6] Due to the efforts of the Scottish reformer John Knox (ca. 1514–1572), a disciple of John Calvin, the Scottish Kirk (church), had adhered to the Reformed tradition since its founding in 1560, and the Kirk's Presbyterian polity is enshrined in Knox's 1562 *Book of Common Order*, the Kirk's prayer and service book. Consequently, Scottish political and ecclesiastical leaders regarded Charles's initiative as a mortal affront, and in response drafted the so-called National Covenant of 1638. This document, which was eventually signed by nearly all of the Scottish nobility, rejected Episcopalian polity in favor of Knox's Presbyterianism and committed to austere Calvinist Presbyterian religious practices while rejecting any influence of Catholicism and High-Church Anglicanism.[7]

As a direct result of this religious turmoil, Scottish church officials purged their church services of features which were deemed insufficiently Reformed, such as instrumental performance.[8] Soon nearly all musicians were dismissed from prominent church positions in Scotland, and the band of the Chapel Royal was dispersed and the chapel organs dismantled and abandoned.[9] This was the second major transformation to the culture of Scottish sacred music; an earlier change had already occurred during the Scottish Reformation of the preceding century, which had seen the abandonment of polyphonic Catholic services in favor of metrical psalms sung in English and Gaelic, based largely on versions produced by John Knox, themselves modeled on Calvin's Geneva psalter. The "Scottish Psalter," as this

---

[6] In Scotland, King Charles I was already widely held in suspicion due to his marriage to the Roman Catholic princess Henrietta Maria of Bourbon and his patronage of Arminianism, most prominently through his appointment of the strongly anti-Calvinist William Laud as Archbishop of Canterbury.

[7] In 1560, Knox founded Presbyterianism, a Reformed denomination that eventually became the national church of Scotland. Its foundational creed is the Westminster Confession of Faith, drawn up in 1646 during the English Civil War by English religious reformers who were at the time allied with the Scots against Charles I. Presbyterian polity is not governed by a single head of the church, but rather by the nation's General Assembly, a body of elected church elders.

[8] See James Reid Baxter, "Ecclesiastical Music," in *The Oxford Companion to Scottish History*, ed. Michael Lynch (Oxford: Oxford University Press, 2001), in particular 432–433.

[9] Munro, "Scottish Church Music and Musicians," 70–71. Despite subsequent political instabilities, which included the brief restoration of the Catholic rite in the Chapel Royal by King James VII of Scotland/II of England in February 1688, sacred music-making continued to be widely condemned by Scottish citizens; even the refurbished Chapel was destroyed by an angry mob in December of that year.

96   HEARING WITH THE MIND

corpus of settings of the 150 biblical psalms came to be known, initially used 105 distinct psalm tunes in no fewer than thirty different meters, although the number of tunes fluctuated somewhat in subsequent editions. The 1635 edition of this collection, known as the "Great Psalter" and edited by Edward Millar, director of music at the Chapel Royal, while dominated by four-part settings in simple homophony, also contained a number of items featuring a more polyphonic texture.[10]

In the wake of the National Covenant, new, metrically simplified versions of the psalms were swiftly authorized by the leadership of the Presbyterian church, the General Assembly, in what became known as the "Scottish Metrical Psalter of 1650." This collection, and subsequent editions of it, included no music, and reduced the poetic meters from thirty to six.[11] By the second half of the seventeenth century, the canon of psalm tunes sung in Scottish churches had coalesced to twelve (or sometimes fourteen) tunes, first published in Aberdeen as *The Twelve Tunes for the Church of Scotland* (1666) and reissued five times until 1720. As Millar Patrick summarizes:

> By the end of the century these twelve [tunes] were canonized as embodying the accepted and inexpansible musical tradition of the Church of Scotland. At that point the canon was closed, and it remained fixed for a long time subsequently. All twelve . . . were common-metre tunes. Even the finest of the old tunes in other metres were forgotten.[12]

These successive transformations contributed greatly to the rapid transformation of devotional music making in Scottish churches. In order to support congregational participation in lieu of choral polyphony, churches widely adopted the English custom of "lining out," in which a precentor first sang out by himself a phrase comprising a line of the verse, followed by the congregation singing in a rough heterophonic unison.[13] Throughout the

---

[10] Psalm 18, set in five voices, probably composed by Andrew Blackhall, was included in the 1635 Great Psalter. See Munro, "Scottish Church Music and Musicians," 295.

[11] Ibid., 317. As an example of the changes, we can compare the Sternhold and Hopkins translation of Psalm 50 to that of the Scottish Metrical Psalter: Sternhold and Hopkins's translation employs the meter 10.10.10.10.11.11; the Scottish Psalter regularizes this into the ubiquitous Common Meter (8.6.8.6). Common Meter is typically set (in Scottish psalmody) to musical strophes comprising four phrases of four and three bars' length alternately (i.e., two pairs matching the pairs of verses); each syllable of text is set to one half note, with the first syllable of each strain set as an anacrusis.

[12] Millar Patrick, *Four Centuries of Scottish Psalmody* (Oxford: Oxford University Press, 1949), 111.

[13] This practice was intended to help illiterate churchgoers to understand the meaning of the text. See Nicholas Temperley, "The Old Way of Singing: Its Origins and Development," *Journal of*

second half of the seventeenth and well into the eighteenth century, lining out flourished as musical literacy and part singing receded from the popular memory. As the Edinburgh-based publisher and Monymusk revival leader Robert Bremner recalled in 1762, in this style, which he terms the "old way,"

> every Congregation, nay, every Individual, had different Graces to the same Note, which were dragged by many to such an immoderate Length, that one Corner of the Church, or the People in one Seat, had sung out the Line before another had half done; and from the whole there arose such a Mass of Confusion and Discord as quite debased this the noblest Part of divine Worship. This they called the old Way of singing, for which there were many Advocates.[14]

Bremner's comments echo those of other urbane overhearers of Reformed psalmody all over Europe.[15] By the late eighteenth century, Reformed psalmody had become notoriously sluggish, and, as the main notes of the psalm tune were performed at a very slow pace by untrained congregations, soon

---

the *American Musicological Society* 34, no. 3 (1981): 532–533. The 1644 Westminster Confession, to which most English Dissenters and all Scottish Presbyterians adhered, stated that while musical literacy was to be encouraged, "where many in the congregation cannot read, it is convenient that the minister, or some other fit person appointed by him and the other ruling officers, do read the psalm, line by line, before the singing thereof." *The Confession of Faith, and the Larger and Shorter Catechisms . . . appointed by the General Assembly of the Kirk of Scotland, to be a part of uniformity in religion, etc.,* 5th ed. (London: S. Cruttenden & T. Cox, 1717), 371.

[14] Robert Bremner, *The Rudiments of Music*, 2nd ed. (Edinburgh: Bremner, 1762), xiii. The first edition, published in 1756, does not mention the old way, but critiques "the present Irregularity and Confusion in performing that essential Part of Worship in our Churches," and blames it on the practice of singers "endeavouring to add more Graces than another, [so that] nothing else is now heard, by which the real Music is entirely lost: And what is still worse, each have different Graces to the same Note: So that those who have never been taught (who are by far the greater Number) hearing nothing regular to follow must sing at Random, and from the whole there arises such a Mass of Discord as makes this, the noblest Part of divine Worship, disagreeable and tiresome." Bremner, *The Rudiments of Music*, 1st ed. (Edinburgh: Bremner, 1756), vii–viii.

[15] Indeed, Charles Burney roundly mocks the Scottish style in his report on the singing of the Lutheran community in Bremen, where he reports that an organ gave out a "dismal" hymn tune "in the true dragging style of Sternhold and Hopkins" followed by the congregation lugubriously singing its verses a-capella. Burney further relates: "After hearing this tune, and these interludes, repeated ten or twelve times, I went to see the town, and returning to the cathedral, two hours after, I still found the people singing all in unison, and as loud as they could, the same tune, to the same accompaniment. I went to the post-office, to make dispositions for my departure; and, rather from curiosity than the love of such music, I returned once more to this church, and, to my great astonishment, still found them, vocally and organically performing the same ditty, the duration of which seems to have exceeded that of a Scots Hymn in the time of Charles I." Charles Burney, *The Present State of Music in Germany, the Netherlands, and United Provinces* (London: T. Becket, Strand, 1775), 2:280. I thank Matthew Hall for this reference.

98 HEARING WITH THE MIND

gave rise to heterophonous ornamentation and embellishments.[16] Music historian Nicholas Temperley surmises that the "old way's" lugubrious tempo was a result of the hesitancy of congregations left without musicians to lead or accompany them: over generations, tentative singers waiting for someone to move on to the next pitch would have given rise to what Temperley calls a "natural 'drag.'"[17]

Part singing was only gradually, and not without resistance, reintroduced into Scottish religious practice in the middle of the eighteenth century. This took place as part of a movement later known as the "Monymusk Revival," initiated by the Scottish laird Archibald Grant of Monymusk in 1748, a mere two years after the final suppression of the Jacobite Rebellion (and a crackdown on traditional Scottish practices, including the speaking of Gaelic and the wearing of tartan).[18] Grant (1696–1778), who had introduced English agricultural methods to improve his estate, was also eager to align Scottish sacred music more closely with English singing practices.[19] He sought out an English soldier stationed in nearby Aberdeen, one Thomas Channon, and hired him to teach part singing in his church and in surrounding communities. In his memorandum book of 1748, Grant reminds himself to:

> take measures for the church choir, and get all you can to join properly in praise, and to give books and premiums to encourage hopeful children; to read the line of Psalms plain and not drone the reading or singing . . . [get] Sc[h]ool Master [to] teach singing & all young people &

---

[16] As Bremner reports in 1762, "Endeavouring once to convince an old Man, who was Precentor in a Country-church, how absurd he rendered the Music, by allotting so many different Sounds to one Syllable, when there was only one intended; he replied, with a good deal of Briskness, that he did not value what any Man intended, and that he believed the People of the present Generation knew nothing of the Matter; for his Master was allowed to understand that Affair thoroughly, and he told him, there, ought to be eight Quavers in the first Note of the Elgin Tune." Bremner, *Rudiments*, xii.

[17] Temperley, "The Old Way of Singing," 523.

[18] A study of this movement can be found in Chapter 14 of Patrick, *Four Centuries of Scottish Psalmody*, 149–163. On the final suppression of the Jacobites, see Gregory Fremont-Barnes, *The Jacobite Rebellion 1745–46* (London: Bloomsbury Publishing, 2014).

[19] Grant, the representative for Aberdeenshire in the House of Commons from 1722–1732, had disgraced himself through involvement in the fraudulent speculations and subsequent collapse of the "Charitable Corporation" in 1732. Following his expulsion from the House of Commons, he returned to Scotland and tended to his estate. It seems likely, however, that his time in London had given him a taste for the kind of sacred music sung in England. For more on the Charitable Corporation scandal, see Peter Brealey, "The Charitable Corporation for the Relief of Industrious Poor: Philanthropy, Profit and Sleaze in London, 1707–1733," *History* 98, no. 333 (2013): 708–729; Grant's role in the affair is discussed in Ian J. Simpson, "Sir Archibald Grant and the Charitable Corporation," *The Scottish Historical Review* 44, no. 137 (1965): 52–62.

JOHN HOLDEN AND SCOTTISH PSALMODY   99

those with good voices when taught sit together in Church [are] to lead [the] rest.[20]

There was naturally some initial opposition to this bold move. An account of how this played out can be found in Thomas Ritchie's biography of Hume, published in 1807 but describing the year 1755. He writes:

> It was not uncommon [in 1755] for a congregation to divide themselves into two parties, one of which, in chaunting the psalms, followed the old, and the other the new mode of musical execution. . . . The moment the psalm was read from the pulpit, each side, in general chorus, commenced their operations; and as the pastor and clerk, or precentor, often differed in their sentiments, the church was immediately in an uproar. Blows and bruises were interchanged by the impassioned songsters, and in many parts of the country, the most serious disturbances took place.[21]

Ritchie's report is confirmed by the records of court proceedings brought against the Monymusk revival's most well-known antagonist, the weaver Gideon Duncan, that same year. Duncan, a gifted singer who had initially supported the reforms but then turned against them, lodged a vocal protest in this way and disturbed the precentor in Aberdeen, for which he was fined the very significant sum of £50.[22]

In spite of these controversies, the efforts of Grant and Channon to bring part singing to various rural parishes in Northeastern Scotland seem to have borne fruit.[23] This is evident in a report in the *Scots Magazine* of 1755, which praises "the progress of the poor illiterate country-people by the method of teaching now practiced among them."[24] The writer goes

---

[20] Archibald Grant, *Selections from the Monymusk Papers (1713–1755)*, ed. Henry Hamilton (Edinburgh: Scottish History Society, 1945), 171.

[21] Thomas E. Ritchie, *An Account of the Life and Writings of David Hume, Esq.* (London: T. Cadell and W. Davies, 1807), 57–58.

[22] See Gideon Duncan, *True Presbyterian: Or, a Brief Account of the New Singing, its Author and Progress in General . . . Together with the Authors* [sic] *Advice to the Ring-Leaders* (Glasgow: 1755). See also Duncan Campbell, "Notes on Church Music in Aberdeen," *Transactions of the Aberdeen Ecclesiological Society* 2 (1888): 22.

[23] On the rioting that took place after the 1755 choral performance, see ibid., 21.

[24] Anon., "An Account of a Late Improvement of Church-Music [April 1755]," *The Scots Magazine, and Edinburgh Literary Miscellany* 17 (Edinburgh: W. Sands, 1755), 190.

100   HEARING WITH THE MIND

on to note that the rural villagers had recently been taught to read music, observing that they

> not only perform with exactness some very difficult tunes which they have been taught, in three or four parts, but it is a certain fact, that several of them who have not the least skill of musical instruments and till lately knew not a single noise of vocal music, and of whom some were even judged incapable of learning, have, since their master left them, and without the least help from any other person, made out among themselves some tunes, which none of them ever heard sung, upon procuring proper sets of them.[25]

An Important landmark in the Monymusk Revival was a performance of four-part psalms by a choir composed of seventy parishioners from Monymusk and Fintray, which took place on January 2, 1755, in the Kirk of St. Nicholas in Aberdeen, led by Channon. (Indeed, the fact that the revival of part singing originated in this region may have had much to do with the fact that musical knowledge was better preserved in the north than in the south of the country due to the fact that Aberdeen was one of the few cities that had retained its "Sang School" after 1639 and could claim the seventeenth-century music publisher John Forbes as a native son.[26])

Aberdeen's example was rapidly imitated and the choir movement spread rapidly throughout Scotland. By November 1756, the city councils of Glasgow and Edinburgh had also appointed Englishmen (Thomas Moore and Cornforth Gilson respectively) to teach church music and part singing.[27]

---

[25] Ibid.

[26] There were some halting attempts to reintroduce part singing in Scotland before Grant and Channon's reforms, as is evident by the publication of a collection entitled *Twelve Tunes for the Church of Scotland, composed in four Parts, according to the Method used by the Master of the Musick School of Aberdeen* in 1714; however, forty years later, part singing had been largely forgotten throughout Scotland. As the aforementioned writer in *The Scots Magazine* lamented in 1755: "As to the harmony, or concord of different parts: This appears to have been used in church-music even in this country about the time of the reformation of religion, and for some considerable time afterwards; as is evident from sets of books published about that time; and continued in some measure till of late, as appears by a set published at Aberdeen in the year 1714. Yet through inattention it declined by degrees, till the harmony is now so entirely lost, that very few have any idea of this great effect in music, which is so proper to give it the grandeur and solemnity requisite in churches. In consequence of this, the tenor-part, on account of its peculiar melody, (it being the only part made entirely for the sake of air), engrossed any little regard which was shown to church-music; and people imagining they might perform it according to their various humours, the greatest confusion was soon introduced, and the melody too entirely lost in many places. This shows by experience, that if harmony is once disregarded, melody will likewise soon come to nothing[.]" Ibid. See also Campbell, "Notes on Church Music in Aberdeen," 19–21.

[27] Patrick, *Four Centuries of Scottish Psalmody*, 161.

JOHN HOLDEN AND SCOTTISH PSALMODY   101

Yet the process of learning to sing in parts left a considerable mark on con-temporaneous musical practice. As Moore, writing in Glasgow, notes in the preface to his *Psalm Singer's Pocket Companion* (1756):

> [Given that the] Method of Singing in several different Parts, does not ap-pear to have been much practiced in this Country [Scotland] for a long Time past, if ever it was, the best Way will be to sing only Tenor and Base, at first, in Time of public Worship, till the Congregation is grown pretty per-fect in such Tunes as they have not been used to sing formerly; and then to add more Parts, according as they have Voices that will suit them, and other Circumstances will admit.[28]

A situation such as the one Moore describes suggests a requirement for psalm settings to accommodate highly inexperienced choirs as well as poten-tial two, three, and four-voice performances.[29]

A 1764 satirical pamphlet purportedly by a professor at Glasgow University, actually authored by the Reverend William Thom—one of dozens of tirades the eccentric critic published against the university—sheds further light on the state of singing in that city at the time. Making a case for the construction of a new college chapel, the "professor" writes that in the new building, "[we will] do our utmost to improve our church music, which hitherto consists of little else but jarring and discordant sounds."[30] After fantasizing about the inauguration of a Gaelic hymnody based on Highland song, the author goes on to specify that while an organ would be ideal, it would be too expensive (and, more importantly, far too controver-sial). He writes:

> In the meantime (to save money) we will begin with a Pitch-pipe. Cheap music is good music. A pitch-pipe will strike the just tone: music in the

---

[28] Thomas Moore, *The Psalm Singer's Pocket Companion* (Glasgow, 1756), xi.
[29] A wry account of the challenges of setting up of a congregational choir from scratch can be found in the second edition of Bremner's *Rudiments* (1762), 43–64, advertised in the title as "A Plan for Teaching a Croud" on the title page, but entitled "A Plan for Teaching the Four Parts to any Number a House will conveniently hold, with as little Trouble, and as soon to four People" in the book itself. I thank David McGuiness for bringing this text to my attention.
[30] [William Thom], *The Motives which Have Determined the University of Glasgow to Desert the Blackfriar Church, and Betake Themselves to a Chapel In a Letter from Prof. — to H— M—, Esq; Airshire* (Glasgow: Printed and Sold by the Booksellers in Town and Country, 1764), 37. Thom here refers to a remarkably undisciplined musical culture, one where there was little agreement as to the method of singing the psalm tunes, much less pitching them. The larger context of the pamphlet does not allow us to ascertain whether this depiction is satirical as well.

102   HEARING WITH THE MIND

four parts will be sung in our chapel with admirable harmony, to the great
delight and devotion of the worshippers, and the great wonder and enter-
tainment of the whole city.[31]

While Thom's main goal in the pamphlet is to lampoon the pretensions of
members of the college, his reference to the musical performance turned
out to be accurate: we learn from the *Scots Magazine* that "On Sunday, Oct.
14, [1764] the college-chapel of Glasgow was opened. . . . The music was
performed in four parts, and the key for each tune pitched by a pitch-pipe."[32]
   The pitch pipe, a loud stopped wooden pipe typically thirteen inches long
with a slider to move the stopper (akin to the modern-day slide whistle) and
a gauge showing the position of the slider for the various notes of the scale,
was considered necessary in order for the precentor to be able to pitch each
psalm appropriately. It first became widely used in the early eighteenth century: in 1746, the English composer and writer William Tans'ur described it
as having been

> but little in vogue with us, till within these thirty Years; for, I remember,
> I went several miles to see the first I heard talked of; which Instrument is
> greatly improved to what it was in former days, and is of singular use in
> all kinds of Music, i.e. for setting of many unfixed Instruments in Tune,
> as well as in Vocal Music; we having it now so as to carry in a Pocket,
> and on whose Register or Stop, is marked the several Letters of the Scale
> of Music; which Tones, either Flat, or Sharp, or Natural, being given by
> drawing the Register, which enlarges the Tube, or Cavity, so as to con-
> tain such a quantity of Air, as will produce any degree of Sound, whether
> grave or acute, &c.[33]

The pitch pipe swiftly became a concrete representative of the new way of
singing in Scotland. The aforementioned fervent opponent of the Monymusk
Revival, Gideon Duncan, published his dissent in the form of a fifteen-page
pamphlet consisting of rhyming couplets, concluding with the peroration: "Come let us all with solemn Voice unite / Stop those proud Singers,

---

[31] Ibid., 41.
[32] *The Scots Magazine* 26 (October 1764): 574. The earliest record I have found of Holden
"instructing the band" at the College chapel dates from May 15, 1765. Minutes of the University of
Glasgow Senate for May 15, 1765, Glasgow University Archive, vol. 26643:41.
[33] William Tans'ur, *A New Musical Grammar: Or, the Harmonical Spectator, Containing All the
Technical Parts of Musick, etc.* (London: Robinson, 1746), 55.

and their graceless Pipe."[34] As late as 1816, Sir Walter Scott used the pitch pipe to symbolize the struggle between the old way and the new, depicting Robert Paterson ("Old Mortality"), an eighteenth-century stonemason and admirer of the Covenanters, as one whose "spirit had been sorely vexed by hearing, in a certain Aberdonian kirk, the psalmody directed by a pitch-pipe, or some similar instrument, which was to Old Mortality the abomination of abominations."[35]

The revival of religious part singing throughout Scotland thus seems to have been exceptionally swift and successful. Yet in the 1750s and 1760s, Scottish cities remained largely dependent on foreigners for musical instruction. In the domain of sacred music, these were typically Englishmen such as Channon, Moore, Gilson, and Holden, while cities that had a more robust scene of secular music, such as Edinburgh or Aberdeen, attracted iterant Italian musicians, some of whom would later settle there.[36] By 1766, the year Holden published his *Collection*, it was clear that music in Scotland more generally was ascendant. Holden's ambitious writings seem to reflect a confidence that his intended readers, at least, would be interested not only in learning to sing in parts, but also in learning about music theory.

Psalmody in the *Collection* and the *Essay*

Holden was the director of the chapel choir at Glasgow University from 1766 (if not earlier) to 1772, a position in which he would have taught singers and led part singing in weekly church services. He published his own four-part arrangements for this particular group—twenty-two psalms and two anthems—as *A Collection of Church Music* in 1766. Moreover, in his *Essay*, which came out four years later, Holden turns to psalmody at a number of

---

[34] Duncan, *True Presbyterian,* 14.

[35] "Old Mortality" here refers to the Covenanter stonemason Robert Paterson, the main character of Sir Walter Scott's novel of the same name. See Sir Walter Scott, *Old Mortality*, ed. Jane Stevenson and Peter Davidson (Oxford: Oxford University Press, 2009), 72. See also George Leitch, "Notice of a Mahogany Pitchpipe Formerly Used in Cults Parish Church, Fife," *Proceedings of the Society of Antiquaries of Scotland* 40 (December 11, 1905): 43–46.

[36] Some of the most notable of these figures include the prominent Italian composer Niccolò Pasquali, whom Holden cites in his *Essay*, and Francesco Barsanti, who published one of the earliest collections of Scottish folk songs in 1742. In the 1740s, Scottish composers such as James Oswald and William McGibbon established a style of secular music there that drew inspiration from traditional Scottish songs.

104   HEARING WITH THE MIND

points in order to illustrate theoretical issues—ranging from ascertaining the key of a piece to the phenomena of musical expectation and harmonic motion, as well as the perception of phrase length at different tempi—that would have been particularly salient in unaccompanied congregational singing by relatively untrained singers.

Holden's *Collection*, the first of his works to raise these issues, bears the sort of floridly descriptive title characteristic of books published in the eighteenth century, namely:

> *A Collection of Church Music: Consisting of New Setts of the Common Psalm-Tunes, with some other Pieces; Adapted to the Several Metres in the Version Authorized by the General Assembly, Composed with a View to Render the Just Performance of Each Part More Easy to Learners; and the United Effect of the Whole More Full and Pathetic: And also to Exemplify New Discoveries in the Scale of Music, Principally Designed for the Use of the University of Glasgow.*

As its title explicitly indicates, the *Collection* was conceived mainly for Holden's charges in the university chapel choir. Unlike other eighteenth-century British psalm books, which occasionally include an introduction that conveys basic musical rudiments, Holden's ten-page preface is clearly aimed at musically literate readers with a serious interest in theoretical concerns. Moreover, Holden refers to theoretical concepts—such as Rameau's fundamental progression—which could have been known to only a small number of music lovers at the time. Radically different from other sets of hymns published in Britain at the time, it is a hybrid document that employs the relatively simple form of four-part psalm settings to argue for advanced speculative ideas about musical harmony and tuning.

In the preface of the *Collection* Holden justifies his publication of new harmonizations of tunes set to the metrical psalm translations authorized by the General Assembly.[37] He begins by justifying the publication of yet another set of arrangements of Scottish metrical psalms: his versions, he says, have a narrower range than usual, comprising the two octaves and a fourth between the G on the lowest line of the bass stave and the C above middle C—a range that permits choirs to eliminate their dependency on female

---

[37] On the challenge of keeping the range of Scottish church hymns within easily singable boundaries, see Timothy Duguid, "Early Modern Scottish Metrical Psalmody," 1–23.

JOHN HOLDEN AND SCOTTISH PSALMODY   105

singers entirely. The Glasgow college chapel choir would have consisted only of boys and men, but Holden suggests that this would be advantageous for parish choirs as well, as well-trained boys would subsequently grow into competent adult singers.

It soon becomes clear, however, that neither the range of tunes nor the composition of parish choirs is the preface's main concern. Rather, Holden asserts, his psalm settings depart from convention through a number of alterations designed "either to render the harmonies of some chords more full and solemn or the successions of chords more natural and better connected, or to give the parts a more graceful and easy-singing melody."[38] These enhancements, he tells us, result primarily from his inclusion of discords, which "are as carefully avoided in our [i.e., contemporaneous] setts of psalm tunes, as if they had been destructive of all the purposes of music."[39]

Holden's claim regarding the lack of dissonances in other psalm tune settings is no rhetorical exaggeration: these intervals are, in fact, almost entirely absent from the two collections that Holden briefly mentions by name in the preface (Caleb Ashworth, *A Collection of Tunes*, 1762 and Robert Bremner, *Rudiments of Music*, 1762). The Monymusk revival had indeed restored part singing, but almost entirely in root-position consonant triads, reviving the simple homophonic arrangements characteristic of the period following the Scottish Reformation.[40]

Beyond the music-theoretical topics discussed in the preface, to which we will shortly return, the text also indicates that Holden possessed considerable familiarity with practical exigencies of choral singing. Thus, for example, he admonishes his readers to be vigilant with regard to various bad habits— ranging from "the custom of singing as loud as possible" which he regards as "the most universal and the most intolerable fault among such as would be esteemed the best performers" to the "false notion [that] has prevailed among some singers, and even some teachers, that the bass ought, to keep a

---

[38] Holden, *Collection*, 4.

[39] Ibid.

[40] Even before the National Covenant, moreover, various editions of Psalteries reveal the progressive simplification of the harmonies involved in psalm tunes. According to historian of Scottish church music Gordon Munro, in the settings found in the 1625 Aberdeen Psaltery: "The majority of chords are in root position; all the settings are homophonic and homorhythmic, except for occasional passing notes (including accented passing notes and passing sevenths); and quaver harmony notes are cleverly used to avoid consecutive fifths or octaves. . . . There are relatively few dissonant suspensions: all are 4–3." In contrast, Munro continues, the arrangements in the 1633 edition boast "far fewer first inversion chords, no six-four chords, no six-five chords, no quaver motion of any kind, and only one 4–3 suspension." Munro, *Scottish Church Music*, 307–308.

106 HEARING WITH THE MIND

little behind, and, as it were, drag after the other parts."[41] He also offers pragmatic advice, such as recommending that "the contra [alto] singers be placed near the bass, both because the concords are generally more perfect between the bass and contra, than between any of the other parts," and that "it is much better to have no contra at all than to have it too loud."[42]

Yet while Holden evidently had experience with singers, he appears to have had far less facility with four-part writing than one might expect. Nearly every one of his psalm settings contains idiosyncratic—if not downright erroneous—treatments of voice leading. As an example, we can compare his setting of the tune "St. Davids" with the setting by Ashworth, whose collection he cites in his Preface (**Figure 3.1** and **3.2**; the melodies of both settings can be found in the tenor as is customary in British psalmody). Holden's setting clearly differs from standard common-practice part writing, in that it frequently omits chord tones (in particular thirds), while overusing unison and octave doublings and including parallel fifths and octaves. Moreover, his handling of dissonance is at times unorthodox, to say the least (see, e.g., measures 4 and 13 of the tune "St. Davids," in which there is a dissonant passing chord on the downbeat, and the chord it resolves to, instead of being simply a tonic triad, has its fifth replaced by the sixth above the bass). Holden claims in the Preface that his innovations produce "more full and solemn" harmonies and "more natural and better" chord progressions.[43] Yet the *Collection's* frequent eccentricities—which often contradict the practical precepts for part writing that Holden himself provides in his treatise of 1770—indicate that he may not have had much (if any) training in four-part writing.[44]

Holden returns to the example of psalm tunes in the second chapter of the first (practical) part of his *Essay*, when he introduces the difference between music in major and minor keys. Having established that each scale degree has a different quality, Holden remarks that "we are supposed to keep the key note [i.e., the tonic] constantly in view, during the whole course of a tune; and to consider all the other notes of the scale chiefly with regard to their

---

[41] Holden, *Collection*, 10–11.

[42] Ibid., 12.

[43] Holden, *Collection*, 4. Voice-leading idiosyncrasies are also characteristic of the settings found in the 1635 and 1666 Scottish Psalter, indicating that the conventions of concerted part-writing may not have been highly prized in Scottish psalm settings. On this point see e.g., Munro, *Scottish Church Music and Musicians*, 313, 317–318.

[44] See Munro, *Scottish Church Music*, 317–318.

JOHN HOLDEN AND SCOTTISH PSALMODY   107

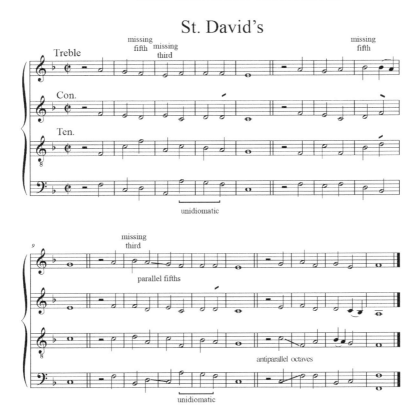

Figure 3.1 Holden's setting of St. David's tune, with voice-leading infelicities indicated. Note that the melody is in the tenor. The diacritics indicate tuning adjustments and will be explained shortly. Holden, *Collection*, 4.

several relations to the key."[45] In order to demonstrate the resulting perceptual effect of this phenomenon, he suggests that the reader:

> sing, or play over the well known Psalm tune, called *York* or *Stilt*, as represented, Ex. IX and if he sings, let him apply the syllable *la* to each note, rather than the words of any verse; the hearer will find, all along, a sort of expectation of something still to follow, until the performer arrives at the last note of the fourth line; at which note, the expectation ceases, and the hearer is perfectly satisfied with a conclusion on that note: whereas, if the

---

[45] Holden, *Essay*, Part 1, §28, 23.

Figure 3.2 Ashworth's setting of the same tune (note the antiparallel octaves between the tenor and bass in measures 6 and 7 as well as measures 15 and 16). Caleb Ashworth, *A Collection of Tunes Suited to the Several Metres Commonly Used in Public Worship, with an Introduction to the Art of Singing and Plain Composition* (London: J. Buckland, 1762), 27.

performer were to stop at any other place of the tune such an abrupt ending would be quite unsatisfactory, and a plain disappointment to the hearer.[46]

This is a very early instance of tonal implications being attributed simply to the notes of a melody (i.e., the scale degrees of a key), independently of any harmonic properties attending a contrapuntal or harmonic setting of that

---

[46] Ibid., Part 1, §29, 23–24. Holden concludes, "This natural expectation, which never ceases till some proper succession of notes occur, leading to, and terminating upon one certain sound, is not peculiar to a few tunes only, but is common to all." Ibid.

melody.[47] Moreover, the action of expectation or desire is explicitly assigned to the listener, rather than to the intervals or notes themselves.[48] As we saw in the previous chapter, this move of locating musical experience in the subject is a hallmark of Holden's method.[49] The fact that Holden refers to a familiar tune to make this example should not obscure the point that the hearing he attributes to the listener is dependent not on the listener's familiarity with the tune itself but rather on the acquisition of tonal competence.

Holden revisits psalm singing in the fourth chapter of the first part of the *Essay*, when discussing the tricky question of determining the most successful tempo, in absolute terms, for a musical performance.[50] After conceding that it is impossible to find a single tempo that would work for all possible listeners, he proposes that the ideal tempo for a given congregation's practice of psalm singing depends not only on factors including the size and ability of the choir but also their age and intelligence:

> [As] those who have a brisker flow of spirits, a more ready conception, and a quicker succession of ideas, require quicker music, for the same expression, and *vice versa*, we may conclude that the same Psalm ought to be sung quicker, when the congregation consists mostly of young people; and slower, when the greater part are old; quicker, in general in a town, than in a country church; quicker, in places where music is more generally practised; and slower, where it is less in use.[51]

Holden takes up this notion of tempo obliquely in the second, speculative part of the treatise, where he points out that slow tempi hinder us from perceiving irregular phrase lengths. As an example, he turns to "the slow

---

[47] For a slightly earlier example see "Solfier," in *Dictionnaire de musique*, ed. Jean-Jacques Rousseau (Paris: chez la veuve Duchesne, 1768), 446. See also David E. Cohen, "Rousseau as Music Theorist: Harmony, Mode, and (*L'Unité de) Mélodie*" in "Colloquy on "Rousseau in 2013: Afterthoughts on a Tercentenary," *Journal of the American Musicological Society* 66, no. 1 (2013): 275–280; "Gunn, John" in *A Dictionary of Music and Musicians* 1, ed. George Grove (Boston: Ditson & Co., 1879), 641.

[48] See, e.g., David E. Cohen, "'The Imperfect Seeks its Perfection': Harmonic Progression, Directed Motion, and Aristotelian Physics," *Music Theory Spectrum* 23, no. 2 (2001): 139–169; Caleb Mutch, "Studies in the History of the Cadence" (PhD diss., Columbia University, 2015), 192–199.

[49] This notion can be traced to the music theory of René Descartes and is already strongly present in Rameau; see Jairo Moreno, *Musical Representations, Subjects, and Objects: The Construction of Musical Thought in Zarlino, Descartes, Rameau, and Weber* (Bloomington: Indiana University Press, 2004); Cohen, "The 'Gift of Nature.'"

[50] Holden describes this challenge as "the most undetermined matter that we meet with in the whole science of music." Holden, *Essay*, Part 1, §106, 91.

[51] Ibid. "Flow of spirits" here seems intended in the metaphorical sense of high spirits or active disposition, rather than in a technically medical sense, as Holden never returns to the notion of physiological spirits in the treatise.

110 HEARING WITH THE MIND

tedious way of singing psalms, which is in use among us," a likely reference
to the old way of singing, which persisted in some parts of Scotland well into
the twentieth century, and which would have still been widespread in the
1760s.[52] Here, he continues:

> The memory has enough to do to retain the length of one single bar; as the
> number of psalm singers, who do not keep true time, plainly evidences: here
> we cannot constitute any larger portions or strains, and one bar is all we re-
> gard at once; so that if the first line employ four bars, and the next line only
> three, we are not sensible of any impropriety.[53]

Holden's reference to the first line employing four bars and the next three
appears to describe Common Meter (8.6.8.6.), the most ubiquitous meter
for the psalms. This meter was traditionally cast in British psalmody in
two paired phrases, in which the rhythm of the first phrase of each pair
was usually one semibreve followed by six minims and ending with a semibreve (equaling five semibreves) and that of the second phrase was usually
one semibreve followed by four minims and a semibreve (equaling four
semibreves).[54]

Holden would have likely taken the example of the different ways in
which the irregular phrases of psalm tunes tended to be performed at
different tempi from his own experience leading the Glasgow College
choir.[55] Indeed, this practical engagement with psalmody may have
helped shape his approach to the relationship between tempo and the
perception of phrase symmetry, as well as his insistence on the crucial
importance of our mind's preference for isochronous grouping in perception.[56] (Isochrony is, as we saw in the previous chapter, the fundamental principle of his theory of music.)

---

[52] Temperley, "The Old Way of Singing," 511–544.

[53] Holden, *Essay*, part 2, §14, 290.

[54] The semibreves presumably may have helped all members of the congregation to sing together
by orienting them to the first and last note of each phrase.

[55] In his *Collection*, moreover, Holden also rejects the idea that the bass should drag behind the
other voices (see Holden, *Collection*, 11), quoted on page 108; this may be related to Temperley's conjecture (see page 101). Given that he makes this point in a preface to a collection of four-part psalm
settings, it seems possible that such dilatory habits might have spilled over from the old way into
the new.

[56] Indeed, part singing could be very slow in the eighteenth century: as late as 1787, the Scottish
church musician William Taas defined a semibreve, the basic unit used in the notation of psalm
tunes, as having the duration of "four seconds," or the length of time "one can conveniently sing
without breathing." William Taas, *Elements of Music* (Aberdeen, 1787), 34.

Figure 3.3 St. David's tune as it appears in its original notation by Ravenscroft. Thomas Ravenscroft, *The Whole Booke of Psalmes: With The Humnes Evangelicall, and Songs Spiritual* (London, The Company of Stationers, 1621), 85.

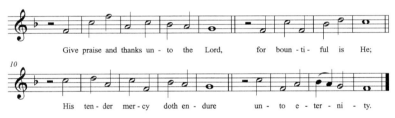

Figure 3.4 St. David's tune as it appears in settings by Holden and by Ashworth. Holden, *Collection*, 4; Ashworth, *Collection*, 27.

Figure 3.3 shows the psalm tune "St. David" as it originally appears in the collection authored by the composer of the tune, Thomas Ravenscroft, in 1621. Eighteenth-century sources, including Holden's and Ashworth's versions shown in **Figure 3.4**, largely preserve this traditional notation while altering the semibreves that open and conclude the phrases to a minim rest followed by a minim, which serves as an anacrusis to the next measure.[57]

---

[57] It seems evident to me from the context of Holden's discussion, however, that he regards the notation of Common Measure hymns (a five-measure phrase followed by a four–measure phrase) as mere convention, as the first measure in each case is a rest followed by an upbeat. Holden says explicitly, moreover, that "five bars in one strain is not used, and seven proves much more intolerable." He then goes on immediately to exemplify this point ("For an instance of this it may be observed…") with the case of a Common Measure psalm tune, which he describes as "the first line employ[s] four bars, and the next only three." Holden, *Essay*, part 2, §14, 290–291.

112 HEARING WITH THE MIND

When such lines are sung at a slow tempo, Holden suggests, our minds do not really register the discrepancies between the length of the different phrases, which is to say, the lines seem to be approximately of the same length.

However, Holden continues, were we to perform the same tune at a fast tempo, we would unconsciously regularize it as follows:

> we will find ourselves under a necessity either of holding on the last notes of the second and fourth lines, or of resting at the ends of these lines, till the time of the first and third lines be elapsed: because in this way of singing the memory retains the whole length of the line, and therefore we must make these lengths all equal; they must be isochronous strains.[58]

Holden here reveals that he essentially regards the psalm phrases (when performed quickly) as differing still further from their notation by conforming to phrases of equal length. He ascribes this change to our mental preference for isochrony at the level of phrase groups, which he regards as a cognitive imperative, noting that "the number of measures in a strain must always be either two or three, or some product of these numbers . . . five bars in one strain is not used."[59] In practice, Holden thus maintains, all four phrases would be altered to conform to the four-measure phrases characteristic of the eighteenth century, as illustrated in my interpretation of his account in **Figure 3.5**.

Figure 3.5 My illustration of Holden's claim that we unconsciously regularize common-meter psalm tunes into phrases composed of four measures.

---

[58] Ibid.
[59] Holden, *Essay*, part 2, §14, 290. Note also his phrasing in the passage noted above: "We *must* make these lengths all equal; they *must* be isochronous strains." Ibid., 291.

JOHN HOLDEN AND SCOTTISH PSALMODY   113

Holden also turns to psalm singing to describe the psychological tendency to regard the first sound we hear as the tonic note of a major scale. He illustrates this experience by recounting how

> the church-clerk, or precentor, who begins the psalm tune immediately after the sounding of a pitch pipe, will be abundantly sensible of this [psychological phenomenon]: for if the intended tune be one of a flat series [i.e., minor key], which has the sixth of the natural scale for its key; or if it be one of a sharp series [i.e., major key], and yet the first note be not the principal key; he will, in either of these cases, find that it requires some care and attention to begin his tune right: whereas, if the first note be the key note of a sharp series, he cannot easily go wrong.[60]

Here Holden provides insight into the subjective experience of a precentor tasked with setting the pitch of a tune and intoning it. He continues by observing that "when the tune is ushered in by an organ prelude, or any other symphony, these difficulties are entirely evaded by the impressions of the preceeding [*sic*] notes."[61] Holden is clearly drawing on musical experiences he must have had in England, as there were no organs whatsoever in churches in Scotland at the time. In Glasgow, the Cathedral's organ loft, defaced in the Reformation, remained in that state throughout the eighteenth century, and while an organ was installed there in 1802, it was forbidden to be used for worship.[62] Even in Episcopal churches in Scotland, such as St. Andrew's in Glasgow, which Holden attended, the use of organs in worship was banned until the 1790s.[63] Yet Holden clearly assumes that his readers would have been agnostic on this matter, and indeed such liberal sentiments seem to have been held by members of the Glasgow community, as reflected in the words of Thom's aforementioned satirical "Professor," who admits that "an organ is what we would intend" but registers his concern "lest the captious people hereabout should be too much disgusted."[64] Intriguingly, however, Holden

---

[60] Holden, *Essay*, part 2, §24, 301.

[61] Ibid.

[62] In 1777, the historian John Gibson observed that "on the west end [of the cathedral] stood the organ-loft, ornamented by a variety of figures, now defaced." John Gibson, *History of Glasgow: From the Earliest Accounts to the Present Time* (Glasgow: Chapman, 1777), 134. This text was included without change in the *Encyclopedia Britannica* as late as 1823.

[63] See Geoffrey B. Payzant, "The Organ Controversy in Scotland," *The Dalhousie Review* 32, no. 4 (1953): 44–48; James Inglis, "The Scottish Churches and the Organ in the Nineteenth Century" (PhD diss., University of Glasgow, 1987).

[64] [Thom], *The Motives*, 41.

114 HEARING WITH THE MIND

supplies figured bass signatures below the bass line of his psalm arrangements, suggesting that instrumental accompaniment might have played a role of some kind in rehearsal or individual study, at least in private.

Holden also turns to psalm tunes at least twice to illustrate more sophisticated music-theoretical issues which pertain to the ideas of Rameau. Before turning to these, let us quickly survey the ways in which Holden imported many of the French theorist's main ideas and terms into his theory. Holden begins his chapter "Of Fundamental Progressions" by comparing musical cadences to "points in writing, or pauses in speaking."[65] These, he continues,

> serve in the same manner to distinguish the ending of every smaller portion, or phrase, as well as of the whole piece: and as there are different sorts of pauses necessary to mark out the larger, and the less divisions of a sentence, so there are different sorts of cadences, some more and some less satisfactory and conclusive.[66]

Holden here regards cadences from a formal, rather than harmonic, point of view, by comparing them to grammatical *clausulae*, making no mention of harmonic considerations. He then goes on to define the various types of cadences as comprising two successive chords, the second of which, he says, always arrives on the accented part of the measure. In offering a linguistic analogy and in his insistence that a cadence conclude on the downbeat of a measure, Holden diverges from Rameau, who defines cadences in terms of their harmonic content and is slightly more tolerant about their possible metrical positions, and Holden aligns himself with older traditions pertaining to cadential theorizing.[67]

A *perfect regular cadence*, for Holden, is a succession in which the fundamental bass, bearing a dominant seventh chord, rises a fourth or falls a fifth

---

[65] Holden, *Essay*, part 1, §189, 166.
[66] Ibid.
[67] Compare Holden's prescriptive attitude with Rameau's more expansive notion in *Génération Harmonique;* namely: "Toute Cadence, ou tout repos se forme par le passage de la Dominante-tonique, ou de la Soudominante à la Tonique, qui doit se faire entendre pour lors dans le premier moment de la mesure; excepté dans de certains cas de fantaisie où cette régle générale peut être transgressée, surtout dans une Mesure à trois Tems, où la Cadence peut se terminer sur le dernier Tems." Rameau, *Génération harmonique,* 173–174. "This tonic must make itself heard on the first beat of the measure, except in certain cases in which you are acting on whim, in which this general rule can be broken, especially in a three-beat measure, where the cadence can end on the last beat." Deborah Hayes and Jean-Philippe Rameau, "Rameau's Theory of Harmonic Generation: An Annotated Translation and Commentary of *Génération harmonique, ou Traité de musique théorique et pratique*" (PhD diss., Stanford University, 1968), 199. On the history of the cadence see Mutch, "Studies in the History of the Cadence."

JOHN HOLDEN AND SCOTTISH PSALMODY   115

to the fundamental of the "chord of the key," that is, the root-position major or minor triad on the tonic:

> The perfect regular cadence upon the principal key must always conclude the piece, like the full stop in writing; and there is no other way in nature of coming into the chord of the key, without leaving an expectation of something more to follow, but by the chord of the 5*f* [i.e., V], bearing a greater third and seventh, as a leading chord.[68]

This passage, as we saw in the previous chapter, exemplifies Holden's methodology by explicitly foregrounding the notion of a listener's expectation, rather than simply inter-chordal relationships. Here, too, Holden departs from Rameau in that the French theorist regards virtually any major or minor triad (*accord parfait*) as a potential tonic, and thus as a point of repose.[69] For Holden, however, a tonic chord can cause the listener to expect something to follow, depending on the manner by which it is arrived at, as well as its voicing, implying the notion that different V-I cadences have different levels of finality.

Holden defines the fundamental bass as a reference point "from whence every consonance is derived, and [showing] to what note each of the sounds ought to be referred." This definition is basically identical to Rameau's approach in the *Traité*, though written in a far more compact and legible style.[70] He next goes on to introduce and explain the Frenchman's concept

---

[68] Holden, *Essay*, part 1, §189, 167.

[69] See, e.g., "Il faut remarquer avant toute chose, que nous ne donnons le nom de Tonique qu'aux Nottes qui portent l'Accord parfait, et celuy de Dominante qu'à celles qui portent l'accord de la Septiéme; que la Notte tonique ne peut paroître qu'aprés une Dominante dont la Tierce est majeure, et dont cette Tierce fait la fausse-Quinte avec sa Septiéme, que si la Tierce de cette Dominante n'est point majeure, et que les intervales de la fausse-Quinte ou du Triton n'y ayent point lieu entre sa Tierce et sa Septiéme, elle ne peut être suvie que d'une autre Dominante; qu'ainsi il est à propos de distinguer ces Dominantes en appellant Dominantes toniques, celles qui contiendront dans leur accord de septiéme un intervale de fausse-Quinte ou de Triton, et simplement Dominantes, celles où ces intervales ne paroîtront point." Jean-Philippe Rameau, *Traité de l'harmonie reduite à ses principes naturels* (Paris: Ballard, 1722), 68. "Notice particularly that we give the name tonic only to those notes which bear a perfect chord, while we give the name dominant only to those which bear a seventh chord; that the tonic note may appear only after a dominant whose third is major and forms a false fifth with the seventh of the dominant; that unless the third of this dominant is major and either the interval of the false fifth or the tritone occurs between its third and its seventh, it can be followed only by another dominant; and that we should therefore differentiate dominants by calling those whose seventh chords contain either a false fifth or a tritone dominant-tonics and those in whose seventh chords these intervals do not appear simply dominants." Jean-Philippe Rameau, *Treatise on Harmony*, trans. Philip Gossett (New York: Dover Publications, 1971), 83.

[70] In the *Traité*, Rameau defines the fundamental bass as "the source (*le Principe*) of harmony" and notes that it "determines all rules concerning consonances" (*La Basse-Fondamentale détermine toutes les Règles qui concernent les Consonances*). Rameau, *Treatise on Harmony*, iii; Rameau, *Traité*, xvii, xix.

116 HEARING WITH THE MIND

and taxonomy of cadences (*regular*, i.e., authentic, fundamental bass motion down by fifth or up by fourth; *irregular*, i.e., half, motion down by a fourth or up by a fifth; and *rompue*, i.e., deceptive (which Holden calls *false*), motion up by step from V to VI).[71] He likewise adopts Rameau's idea that these progressions serve as templates for harmonic progression more generally, remarking that "every fundamental passage from one chord to another, may be an *imitation* of some of the above described cadences; especially the passages from an unaccented to an accented note, as from the end of one bar or measure to the beginning of the next."[72]

Holden next goes on to observe that

> in the common psalm tunes, for instance, there is ordinarily a cadence imitated at every bar [fundamental bass motion up or down by fifth]; and these imitations are often introduced, without altering any of the degrees of the scale, although some occasional alteration would be necessary to make the cadence perfect in its kind. Thus the passage from the 3d *f* [i.e., III] to the 6th [VI] accented, or from the 6th to the 3d accented [i.e., VI to III], is often admitted without altering the 5th [of the scale]; though the former imitates a regular [i.e., authentic], and the last an irregular [i.e., half] cadence, in both of which the sharp 5th is an essential note.[73]

Within the course of the phrase, Holden clarifies, the cadence need not be strictly imitated, and the progression can remain diatonic (that is, the chromatic pitch that would tonicize the target chord is not necessarily required). While this point is small, it serves to exemplify Holden's familiarity with the theories of Rameau, as well as his attempt to extend them to the psalm tunes of his own day and place.

Holden's interest in explaining his experiences with congregational psalm singing by means of Rameau's theories is made more explicit in another passage in the *Essay*, in which he addresses the dependency of melodic errors on the regularity of the implied fundamental succession. He observes that

---

[71] I use Roman numerals here purely for the sake of convenience. They were, of course, part of neither Rameau's nor Holden's analytic language.

[72] Holden, *Essay*, part 1, §214, 192. Original italics.

[73] Ibid., 192–193. Holden here describes the movement from III to VI or VI to III, that is, fifth motion between two chords imitating either a regular or an irregular (i.e., authentic or half) cadence such that one of the chords is in a relationship to the other as V to I. Since the I chord is minor in this context, its V would need to have its third raised if the intention is to produce a correct authentic or half cadence.

untrained singers often pick up tunes by ear, and in doing so learn only parts of the tune correctly. Fixing such a singer's mistakes is easy when the fundamental succession that supports the erroneous note and the right note is the same, but much harder when the implied fundamental succession which underpins their error is different. It is still more difficult, Holden continues, if the untaught singer's erroneous notes

> belong to a more easy fundamental succession, than the true ones; which often happens. For instance, a whole congregation will very frequently conclude a line of a psalm-tune, by ascending from the 7th to the K, when the real tune descends from the K to the 7th; and in this case the wrong notes suggest a regular cadence on the K, and the right notes belong to a cadence on the 7th; and it is next to an impossibility to correct this fault; for the regular cadence is so natural in itself, and so intimately connected with that part of the tune, according to their way, that nothing else will satisfy them in its place.[74]

With this example, Holden describes a plausible situation in which the "true" melody might pause at VII (over a half cadence), but the untaught singer rather paused at VIII over an authentic cadence, as in **Figures 3.6a** and **3.6b**.

Figure 3.6 Compare (a), the "true" melody with a less intuitive fundamental bass with (b), which features the wrong melody with a more intuitive fundamental bass.

---

[74] Holden, *Essay*, part 1, §223, 204.

118 HEARING WITH THE MIND

Holden's example is interesting because it implicitly assigns a cognitive value to the "ease," that is, the regularity, of the fundamental progression—the prototype of which, of course, is the regular cadence—by linking deviations from it to increased mental effort: given insufficient or unreliable information, such as the incorrect impression of a tune retained by beginning singers, our minds tend to prefer melodies compatible with the easiest fundamental progression, that is, the regular cadence. (They will literally regularize them.) Changing the mental representation of erroneous melodies, then, is easy if to do so entails no loss in regularity of the fundamental progression, but a melody that implies a less regular fundamental progression is more difficult to learn. More generally, the point is that Holden here acknowledges that some fundamental bass progressions are more regular than others, and that this degree of regularity affects our ability to grasp and retain their associated melodies.

Rameau's profound influence is evident throughout the *Collection* and the *Essay*, and Holden's attempt to understand practices of Scottish psalm singing through the prism of the Frenchman's thought reflects his own universalizing worldview. In both works, Holden repeatedly emphasizes the importance of fundamental progressions in shaping the way in which we sing and hear music. Thus, in the preface to the *Collection*, he writes:

The influence of the fundamental progression is not confined to such singers as are complete judges of harmony, but extends to every one who has any taste for music; and any enjoyment of what he performs, although he be totally ignorant of the fundamental principles, and seem to sing by natural instinct only. It is as natural to conform our sounds to certain fundamental progressions, in singing to conform our [bodily] attitudes to certain laws of gravity and motion, in walking.[75]

This passage reappears almost verbatim in the *Essay* in two locations, with Holden comparing the influence of the fundamental progression on our perception of music to any number of "customary operations, both of mind and body, which we perform, without ever studying the philosophical principles of the matter."[76] This idea that the fundamental bass exerts a determinative influence on our musical behavior—and indeed, does so without our even

---

[75] Holden, *Collection*, 5.
[76] Holden, *Essay*, part 1, §220, 202.

being conscious of it, as Holden makes clear in his *Essay*[77]—is obviously a restatement of the Frenchman's famous claim that the fundamental bass is "the ear's unique compass, the musician's invisible guide, which has always conducted him in all his musical works without his having yet become aware of it."[78]

Although book III of Rameau's *Traité* had been translated into English in 1737 and a handful of British authors prior to 1766 had invoked Rameau's concepts in their writings, the Frenchman's ideas were not widely understood, much less accepted, in the British Isles at the time.[79] Holden's understanding and espousal of Rameau's theoretical constructs is unprecedented in British music-theoretical thought, as is his adoption of a deductive approach to theorizing about music, a point which is explored at greater length in the previous chapter.

The Preface also reflects Rameau's ideas in other notable ways. Consider Holden's justification of his editorial decision to precede nearly all chromatically raised notes with a grace note a semitone higher, which he explains as follows:

> The sharpened notes which occasionally enter, in the course of a tune, are difficult to beginners; and in order to come at the proper sounds of such notes, it is necessary to introduce them by touching slightly upon the degree next above, and from thence descending by the step a semitone, upon the sharpened note: and therefore a small *appoggiatura* is generally placed upon the note which ought to be so touched for the learners direction, in the following setts.[80]

---

[77] The passage from the Preface quoted above appears verbatim in Holden, *Essay*, part 1, §277, 248.

[78] "L'unique Boussole de l'Oreille, ce guide invisible du Musicien, qui l'a toujours conduit dans toutes ses productions, sans qu'il s'en soit encore apperçû." Rameau, *Génération harmonique, Préface*, fᵉiii; translated in Hayes, "Rameau's Theory of Harmonic Generation," 16. As David E. Cohen has demonstrated, Rameau assumes that the *corps sonore*, and consequently the fundamental bass, enable the ear to elicit specifically musical experiences from certain combinations of sounds, regardless of whether the listener is actively (that is, intellectually) aware of these music-theoretical constructs. Cohen, "The 'Gift of Nature,'" 68–92.

[79] Burney, for example, states: "After frequent perusals and consultations of Rameau's theoretical works, and a long acquaintance with the writings of his learned commentator D'Alembert, and panegyrists the Abbé Roussier, M. De la Borde, &c. if any one were to ask me to point out what was the discovery or invention upon which his system was founded, I should find it a difficult task." Charles Burney, *A General History of Music: From the Earliest Ages to the Present Period* (London: The Author, 1789), 4:612. Erwin Jacobi discusses some aspects of the British reception of Rameau after Holden's time; see Erwin R. Jacobi, "Harmonic Theory in England after the Time of Rameau," *Journal of Music Theory* 1, no. 2 (1957): 126–146; another valuable source is the index of Kassler's *The Science of Music*, which, upon perusal, reveals the paucity of Rameau's reception in the British Isles.

[80] Holden, *Collection*, 9.

120 HEARING WITH THE MIND

In *Génération harmonique* (1737), Rameau similarly suggests tackling the interval of a minor (that is, a chromatic) semitone by first singing the note a whole step above the previous note as an appoggiatura.[81] He justifies this practice by asserting, "Nature operates in this way on your ear, without your thinking about it, to make you reach a less natural interval through a more natural one."[82] While Holden makes no mention of different sized semitones in his preface, he follows Rameau by inserting appoggiaturas before nearly every chromatic semitone in the set.[83]

Rameau's influence can also be found in Holden's elucidation of the *accents aigus* and *graves* that are printed above various notes in his arrangements. These diacritics are intended to indicate ad hoc tuning adjustments, what he terms "occasional temperament," that is, "those tiny elevations and depressions . . . [which] take place whenever the sounds are entirely under the direction of the ear, as with the voice or violin."[84] Here we learn that Holden does not assume equal temperament (and indeed, such an assumption would have been very unusual for a theorist or musician of his time and place).[85] The alterations he proposes, he maintains, do "not intimate that the note is to be distorted, or put out of its natural place, but it rather shews what is its most natural place, in that occurrence."[86] The accent marks are thus intended to guide singers in making the adjustments that would naturally occur when singing unaccompanied, while also ensuring that all participants in the musical activity are, in fact, assuming the same fundamental progression and hence the same occasional temperament.[87] In the *Collection*, he accordingly employs an acute accent to mark a slight rise in pitch and a grave accent to indicate a slight lowering of the pitch—changes, Holden emphasizes, that

---

[81] Rameau, "*la petit Note breve*," in *Génération*, 85; translated in Hayes, "Rameau's Theory," 109.

[82] "La nature opérant ainsi sur votre Oreille, sans que vous y pensiez, pour vous faire arriver à un intervale peu naturel par ceux qui le sont." Rameau, *Génération*, 85; translated in Hayes, "Rameau's Theory," 109.

[83] *Génération* had not yet been translated into English by 1766, indicating that Holden must have read French. A related observation regarding the appoggiatura is offered by Pier Francesco Tosi in his *Opinioni de' cantori antichi, e moderni o sieno osservazioni sopra il canto figurato* (Bologna: Lelio dalla Volpe, 1723), 22, in the paragraph beginning "Giacche non è possibile." The English translation by John Ernest Galliard, *Observations on the Florid Song; or Sentiments on the Ancient and Modern Singer* (London: J. Wilcox, 1743), may have been known to Holden; the relevant passage there occurs in Chapter 2, §15, 38. It is thus possible that Holden was not directly influenced by Rameau on this point, but simply reporting on a well-known performance strategy. I thank Matthew Hall and David E. Cohen for discussing this point with me.

[84] Holden, *Collection*, 5.

[85] The temperaments in "common use" in Britain at the time were in all probability varieties of meantone; see note 71 in Chapter 6.

[86] Holden, *Collection*, 6.

[87] Ibid.

indicate a far smaller alteration than would result from adding an accidental before the note.

In the *Essay*, as we saw in the previous chapter, Holden proposes a version of ⅕-comma meantone temperament as the closest practical approximation to the tuning implied by the psychologically informed theory of tonal perception that he advocates (aspects of which a variable pitch ensemble, such as a-cappella singers, can enact with relative ease). In the preface, he simply sketches out his system by explaining that (under certain circumstances that he does not enumerate in the *Collection*), in a piece in a single (major) key, two diatonic scale degrees should be altered: IV should be lowered (by the ratio of 64:63), and VI should be raised (by the ratio of 81:80, or the syntonic comma). As major-mode music at this time virtually always tonicizes or modulates to the dominant, he also includes the corresponding fourth and sixth scale degrees of the key of V, which is to say that, in certain contexts, I should be lowered by 64:63 and III should be raised by 81:80. Therefore, no fewer than four scale degrees—I, III, IV, and VI—can and indeed must at times be altered in a piece of (major-key) music that is "connected," which he defines as transpiring "when the principal key is never so far relinquished, but that the ear expects to return to it again, and conclude with a cadence upon it."[88] The same holds true, of course, in minor, which for Holden is built on the VI of the major scale, so that a connected (minor-key) piece will feature a III and a VI each lowered by 64:63, a V raised by 81:80, and a tempered I, which is *lowered* by 81:80.[89]

There is a final aspect of the *Collection* which is worth considering, namely, Holden's attitude to dissonance. As mentioned earlier, Holden asserts that by the judicious use of discord in his psalm settings, he can create fuller sonorities, smoother chordal connections, and more singable vocal parts in his arrangements.[90] He then goes on to make the point that, when properly contextualized, dissonances are an acceptable, indeed a desirable, feature of musical discourse:

Both the theory and practice of the moderns sufficiently prove, that although a dissonant chord is not agreeable nor satisfactory, when considered

---

[88] Ibid., 6. In contrast, of course, Holden allows that "a piece of music may pass into different keys, quite foreign to the principal key, and incompatible with it; in which cases, the hearer's expectation may be said to wander from one key to another, and the occasional vary, according to the various keys which succeed each other." Ibid.

[89] Ibid.

[90] Ibid., 4. Whether Holden succeeds in this goal is a different matter.

122 HEARING WITH THE MIND

singly, that is, without regard to what has preceded, or is to follow; yet when such chords are introduced in proper successions they become not barely tolerable, but confessedly preferable to consonant chords, in their respective places.[91]

Dissonances, that is, which when properly integrated within the flow of consonant sonorities, acquire an unexpected quality that we might call contextual euphony.[92] Such an explicit acknowledgement of the positive potential of dissonance is highly unusual within music theoretical thought in the eighteenth century. A far more traditional eighteenth-century approach to dissonance is articulated, for example, in Ashworth's aforementioned collection, which describes concords as "several sounds which[,] when they are made at the same time, afford the ear a sensible pleasure," whereas discords, "if made together, would give the ear pain."[93]

Holden's more positive view of dissonance, however, is not merely the result of his understanding that its use can aid in the accomplishment of the three desiderata with which he begins. Rather, his argument for the inclusion of discord in psalm settings is strategic: it flows seamlessly into a discussion of the important role of dissonant chords in connecting consonant chords, a role predicated on the mind's ability to synthesize temporally distinct events. He writes:

> Our enjoyment of music depends, in a great measure, on a faculty of retaining the ideas of former sounds, and anticipating those which are to follow; and we have equal reason to exclude, out of speech, all auxiliary words which have no determinate *meaning*, unless joined with others, as to exclude dissonances, out of harmony, because they *express* nothing when alone.[94]

Holden here first calls our attention to the fact of the temporality of music and, equally important, the existence of a "faculty" of our minds equal to the

---

[91] Ibid.

[92] On the reciprocal relationship between consonance and dissonance see David E. Cohen, "Boethius and the Enchiriadis Theory: The Metaphysics of Consonance and the Concept of Organum" (PhD diss., Brandeis University, 1993).

[93] Caleb Ashworth, *A Collection of Tunes Suited to the Several Metres Commonly Used in Public Worship, with an Introduction to the Art of Singing and Plain Composition* (London: J. Buckland, 1762), 7. Discords, Ashworth later explains, are included only "for the sake of variety, . . . [as] an echo to the sense when the sentiment is harsh, and to make the following concords the more pleasing to the ear." Ibid., 21.

[94] Holden, *Collection*, 4. Original italics.

JOHN HOLDEN AND SCOTTISH PSALMODY   123

demands imposed by that temporality, the mental capacity to link together musical events occurring at disparate times—such as a dissonance, its preparation, and its resolution—into a coherent synthesis.[95]

Turning to one of the oldest and most widely used tropes of music theory, Holden goes on to compare the organization of music to that of discourse, asserting that, as dissonances serve a connecting function linking successive harmonies, they can be regarded as analogous to syncategorematic words. He then concludes:

> To shew how the several chords which are admitted in harmony, are connected one with another, in every strain which gives us pleasure, may not improperly be called the Syntax of music, and is a subject of which no competent idea can be conveyed within the limits of a preface.[96]

Syntax, in eighteenth-century English grammar texts, is the part of the discipline concerned with the "right placing or joining Words together in a Sentence."[97] In keeping with the generally prescriptive style of grammatical texts at this time, discussions of syntax typically state various rules pertaining to relations between parts of speech.[98] A modern reader may be disappointed by the paltriness of the eighteenth century's version of the concept given what syntax has come to signify in linguistics and, subsequently, in music theory. And indeed, Holden never returns to this evocative notion of a musical "syntax," suggesting that the term may not have been a particularly apt characterization of a theoretical project with aims as profound as his.[99]

That project, as we saw in the previous chapter, is intimately connected with the temporality of music, in that the theory Holden advances to explain

---

[95] This depiction of our perception of music in time is similar to an account articulated by the Scottish writer on aesthetics Alexander Gerard in 1759: "Whenever our pleasure arises from a succession of sounds, it is a perception of a complicated nature; made up of a sensation of the present sound or note, and an idea or remembrance of the foregoing, which, by their mixture and concurrence, produce such a mysterious delight, as neither could have produced alone. It is often heightened, likewise, by an anticipation of the succeeding notes." Alexander Gerard, *An Essay on Taste* (London: A. Millar, 1759), 61. Despite the evident similarities in their approaches, however, Gerard, unlike Holden, next likens our perception of beauty in music to that of beauty in visual forms, proposing that similar laws govern the modalities of hearing and sight.

[96] Holden, *Collection*, 4.

[97] James Greenwood, *An Essay Towards a Practical English Grammar: Describing the Genius and Nature of the English Tongue* (London: R. Tookey, 1711), 35. According to historian of grammar Ian Michael, "The set definition of syntax remains virtually unchanged throughout the whole period [i.e., the eighteenth century]." Ian Michael, *English Grammatical Categories: And the Tradition to 1800* (Cambridge: Cambridge University Press, 2010), 466.

[98] Ibid., 468.

[99] Holden is the only eighteenth-century theorist who I know to have described the rules governing the connection of chords using the term "syntax," a fact which is nonetheless of interest.

124 HEARING WITH THE MIND

"how the several chords which are admitted in harmony, are connected one with another" turns out to be, in fact, a theory of how our minds cognitively process music—an activity, like music itself, that is inherently temporal. The phenomenon of dissonance makes it evident that what we hear *as* music is a function of temporal expectation: in the experience of a prepared and re-solved dissonance, events that by themselves would be unpleasant become pleasant because they occur in a sequence governed by certain laws. What is more, the idea of temporal sequence entails the necessity that there be a mental capacity to integrate events separated in time. It is this ambitious vision that Holden pursues in the forthcoming treatise that "he hopes shortly to finish."[100]

As this chapter has demonstrated, Holden's ideas seem to have flourished not in spite but rather because of the idiosyncratic context of Scottish music in the 1760s. His work leading choirs in a culture in which part singing had long been suppressed may have not only inspired his psychological observations but also encouraged him to notice the music-psychological aspects and potential of Rameau's writings, traits which went largely un-remarked upon in the eighteenth century.[101] At the same time, the provin-ciality of his circumstances and training, as well as his isolation from the music-theoretical centers of Europe, would have paradoxically given him the courage to attempt to establish the theory of music, as he puts it, "upon prin-ciples in a great measure new, which seem to be more and more satisfactory, than any which have formerly been pursued; and from whence several of the most remarkable and hitherto unintelligible circumstances in the practice of music are accounted for, with the greatest ease and perspicuity."[102]

---

[100] Holden's attitude toward dissonance as functioning as the connective tissue that supports har-monic progression is relatively new in the history of Western music theory at this point. A possible antecedent is Jean-Philippe Rameau, who regards dissonances as being internally compelled to move to more stable consonances. In the 1737 English translation of book III of Rameau's *Traité*, which Holden cites in his *Essay*, we find the following passage: "A discord instead of being troublesome to a Composer, on the contrary, it gives him a greater Liberty, for in all Progressions of a Bass as-cending a Second, a Fourth, or a Sixth, there will always be found one Note in the upper Parts, which having made a Consonant Interval with the first Note of the Bass, may, without altering it, make the Seventh to the second Note of that Bass, which ought to be practised as often as possible." Jean-Philippe Rameau, *A Treatise of Musick: Containing the Principles of Composition* (London: Robert Brown, 1737), 20. Still, the distance from statements of this kind to Holden's more general approach is substantial.

[101] Cohen, "The 'Gift of Nature,'" 68–92.

[102] Holden, *Collection*, 4.

4

To "Fill Up, Completely, the Whole Capacity of the Mind"

Listening with Attention in Late Eighteenth-Century Scotland

In his essay, "Of the Nature of that Imitation which Takes Place in What Are Called the Imitative Arts" (1795), the Scottish philosopher and economist Adam Smith famously opines that

> [instrumental] music seldom means to tell any particular story, or to imitate any particular event, or in general to suggest any particular object, distinct from that combination of sounds of which itself is composed. Its meaning, therefore, may be said to be complete in itself, and to require no interpreters to explain it.[1]

The enjoyment afforded by instrumental music, Smith holds, is similar to the intellectual gratification we derive from the contemplation of a "great system in any other science";[2] the pleasure thereby generated—a pleasure both sensual and intellectual—is quite distinct from the expressive, plot-driven delights engendered by the imitative arts.

Smith's account of attending to instrumental music will likely seem familiar to contemporary readers trained in the post-Hanslickian tradition of listening with its focus on structures and relations intrinsic to "the music itself."[3] Indeed, over the past thirty years, music scholars have generally

---

[1] Adam Smith, *Of the Nature of that Imitation which Takes Place in What Are Called the Imitative Arts* (Dublin: Wogan and Byrne, 1795), 173–174.

[2] Ibid.

[3] Wayne Slawson observes that Smith's "aesthetic defense of instrumental music seems very close to the view taken for granted by musicians of the present day." Wayne Slawson, "Forked Tongues: Structural Illusions in Music," in *Reflecting Senses: Perception and Appearance in Literature, Culture and the Arts*, ed. Walter Pape and Frederick Burwick (Berlin: de Gruyter, 2011), 250.

Hearing with the Mind. Carmel Raz, Oxford University Press. © Oxford University Press 2025.
DOI: 10.1093/9780197786208.003.0005

126   HEARING WITH THE MIND

discerned in this account a harbinger of the listening practices associated with so-called absolute music.[4] In point of fact, Smith's attitude radically departs from other eighteenth-century theories about musical meaning: as Wilhelm Seidel emphasizes, "Adam Smith's essay is wholly exceptional; there is nothing from the period, and for a long time afterwards, to which it can be compared."[5]

Seidel has ascribed the stark divergences between Smith's approach to instrumental music and those of his contemporaries solely to the philosopher's formidable intellectual abilities.[6] Here I offer a complementary perspective on Smith's innovations by contextualizing his ideas as part of a broader development in British conceptions of musical listening, and of attention itself, in the decades around 1760 and 1770, when the philosophical psychology expounded by Smith's compatriot, Thomas Reid, made a profound impact upon conceptualizations of attention. (Indeed, as a number of scholars have noted, the origins of our current understandings of attention—and our ongoing fascination with its often-capricious nature—are largely traceable to the eighteenth century: to novel approaches to the mental faculties native to German philosophy and British empiricist psychology, to innovations

---

[4] For example, Lydia Goehr regards Smith's writings on instrumental music as anticipating the Romantic notion that "the lack of intermediary, concrete, literary or visual content made it possible for instrumental music to rise above the status of a medium to actually embody and become a higher truth." Lydia Goehr, *The Imaginary Museum of Musical Works: An Essay in the Philosophy of Music: An Essay in the Philosophy of Music* (Oxford: Clarendon Press, 1992), 154. According to Mark Evan Bonds, "Smith's valorization of the 'intellectual pleasure' afforded by music emphasizes the act of contemplation in ways that prefigure nineteenth-century debates about the aesthetic perception of the art." Mark Evan Bonds, *Absolute Music: The History of an Idea* (Oxford: Oxford University Press, 2014), 106. See also Birgit Klose in "Die erste Ästhetik der Absoluten Musik: Adam Smith und sein Essay 'über die sogenannten Imitativen Künste'" (PhD diss., Philipps-Universität Marburg, 1996) and Alexander Wilfing, *Re-Reading Hanslick's Aesthetics: Die Rezeption Eduard Hanslicks im englischen Sprachraum und ihre diskursiven Grundlagen* (Vienna: Hollitzer, 2019). On absolute music see Bonds, *Absolute Music,* Carl Dahlhaus, *Die Idee der absoluten Musik* (Kassel: Barenreiter: 1978), and Daniel Chua, *Absolute Music and the Construction of Meaning* (Cambridge: Cambridge University Press, 1999).

[5] "Der Essay von Adam Smith ist einzigartig; es gibt nichts was seinerzeit und lange danach zur Seite hätte gestellt werden können." Wilhelm Seidel, "Der Essay von Adam Smith über die Musik— Eine Einführung," *Musiktheorie* 3 (2000): 197.

[6] Seidel regards Smith's essay as a "unique testimonial of the fascination the process [of hearing instrumental music] has exerted on a clear and free mind" [Der Essay ist ein einzigartiger Beleg für die Faszination die der Vorgang auf einen klaren und freien Kopf ausgeübt hat]. Wilhelm Seidel, "Die Sprache der Musik," in *Festschrift Klaus Wolfgang Niemöller zum 60. Geburtstag,* ed. Jobst P. Fricke (Regensburg: Bosse Verlag, 1989), 495. On the relationship between Smith's approach and other theories of British aesthetics see Anselm Gerhard, *London und der Klassizismus in der Musik: Die Idee der "absoluten Musik" und Muzio Clementis Klavierwerke* (Stuttgart and Weimar: J.B. Metzler, 2002), 122–150; Maria Semi, *Music as a Science of Mankind in Eighteenth-Century Britain,* trans. Timothy Keates (Farnham: Ashgate, 2012), 93–103.

in the methods characteristic of the natural sciences, and to new modes of conceptualizing the essence of aesthetic experiences.)[7]

This chapter thus seeks to clarify the connection between Smith's view of instrumental music and Reid's psychological theories by identifying a probably missing link between them, namely, John Holden. More specifically: I argue that Smith's original characterization of instrumental music prefigures attitudes that (re)emerged toward the end of the nineteenth century precisely because of the indebtedness of that characterization to the ideas of Reid, which continued to exert a substantial influence on philosophies of language and perception in the nineteenth and twentieth centuries.[8] I further demonstrate that the crucial link between the two thinkers lies in the music-theoretical approach advanced by Holden, an acquaintance of Reid, who adapted many of Reid's precepts into a theory of musical cognition that anticipates later experimental work on perception in significant ways, as discussed in Chapter 2.

Reid is regarded as a founding figure of the Scottish School of Common Sense or alternately, Common Sense Realism, the philosophical school associated with the Scottish Enlightenment. Rejecting both the "Way of Ideas" endorsed by Descartes and Locke, and the skepticism advocated by Hume, Reid maintained that the intuitive judgements common to humanity at a species level form a sufficient basis for philosophical inquiry. That is to say, his philosophical method rests explicitly upon first principles that are at once common to all persons and immediately accessible, possessing "the consent of ages and nations, of the learned and unlearned, [which] ought to have great authority with regard to first principles, where every man is a competent judge."[9] Reid's ideas were hugely influential in their day, both in

---

[7] A spate of recent literary and historical studies have sought to investigate the various contexts from which our modern notions of attentiveness emerged; works that examine the late eighteenth century include Phillips's *Distraction*, Lily Gurton-Wachter's *Watchwords: Romanticism and the Poetics of Attention* (Stanford: Stanford University Press, 2016); Nicholas Mathew, "Interesting Haydn: On Attention's Materials," *Journal of the American Musicological Society* 71, no. 3 (2018): 655–701; Francesca Brittan and Carmel Raz, eds., "Colloquy: Attention, Anxiety, and Audition's Histories," *Journal of the American Musicological Society* 72, no. 2 (2019): 541–580.

[8] Holly K. Andersen and Rick Grush maintain that, rather than arising solely as a synthesis of experimental work conducted in Germany in the last three decades of the nineteenth century, much of it influenced by the renowned psychologist Wilhelm Wundt, William James's "specious present doctrine" drew on a tradition of Anglophone theorizing about time and memory that began with Reid, in precisely those passages that we will turn to in the current chapter in due course. This tradition, they claim, also influenced Husserl. Holly K. Andersen and Rick Grush, "A Brief History of Time-Consciousness: Historical Precursors to James and Husserl," *Journal of the History of Philosophy* 47, no. 2 (2009): 277–278.

[9] Thomas Reid, *Essays on the Intellectual Powers of Man* (Edinburgh: J. Bell, 1785), VI, 570.

128 HEARING WITH THE MIND

Europe and across the Atlantic, and the legacy of his thought still resonates in contemporary theories of perception.[10]

My purpose here is to recover the core psychological assumptions pertaining to memory and attention in Reid's psychology, to examine how these conceptualizations undergird Holden's theory of musical perception, and to explore how traces of these ideas may have filtered through Holden via another minor Scottish music theorist, Thomas Robertson, to Smith, who, in his treatment of instrumental music, transmuted them into a psychologically-informed account of structural listening unalloyed by any mimetic or expressive imperative. Offering a conception of attentive listening that differs starkly from contemporaneous attitudes toward attention as largely dependent on wonder or astonishment, as well as from subsequent early Romantic notions of the transport afforded by instrumental music,[11] both Holden's proto-cognitive theory of the perception of tonal relationships and Smith's proto-Hanslickian depiction of listening to instrumental music suggest that Reid's innovative psychology facilitated the development of new accounts of musical experience.

Reid's Approaches to Attention and Memory

Thomas Reid's approach to the faculties of attention and memory and their relationship to perception is exceptional in the context of eighteenth-century British philosophy.[12] In proposing a model in which the mind actively directs its attention to different objects, Reid discarded the doctrine

---

[10] See, e.g., Nicholas Wolterstorff, *Thomas Reid and the Story of Epistemology* (Cambridge: Cambridge University Press, 2004); Terence Cuneo and René van Woudenberg, eds., *The Cambridge Companion to Thomas Reid* (Cambridge: Cambridge University Press, 2004); Ryan Nichols, *Thomas Reid's Theory of Perception* (Oxford: Oxford University Press, 2007); Ryan Nichols and Gideon Yaffe, "Thomas Reid," in *The Stanford Encyclopedia of Philosophy* (Winter 2016 Edition), ed. Edward N. Zalta, https://plato. stanford.edu/archives/win2016/entries/reid/; Charles Bradford Bow, ed., *Common Sense in the Scottish Enlightenment* (Oxford: Oxford University Press, 2018).

[11] See, e.g., the understandings of attention described in Matthew Riley, *Musical Listening in the German Enlightenment: Attention, Wonder and Astonishment* (Burlington, VT: Ashgate, 2004), and the account of intense listening depicted in Wilhelm Wackenroder, *Phantasien über die Kunst von einem kunstliebenden Klosterbruder* (Berlin: Johann F. Unger, 1797). On Smith's use of "transport" as a metaphor for the conveyance of interpersonal emotions and on the influence of this on subsequent notions of emotional experience, see Miranda Burgess, "Transport: Mobility, Anxiety, and the Romantic Poetics of Feeling," *Studies in Romanticism* 49, no. 2 (2010): 229–260.

[12] See Gideon Yaffe, "Thomas Reid on Consciousness and Attention," *Canadian Journal of Philosophy* 39, no. 2 (2009), 166; Rebecca Copenhaver, "Reid on Memory and Personal Identity," *The Stanford Encyclopedia of Philosophy* (Winter 2018 Edition), ed. Edward N. Zalta, https://plato.stanf ord.edu/archives/win2018/entries/reid-memory-identity.

of the association of ideas, one of the key presuppositions motivating the work of his contemporaries, the Associationist philosophers. This school of thought, the leading lights of which included David Hartley, David Hume, and Joseph Priestley, posited a passive mind that acquires complex concepts and behavior by learning through repetition, or by identifying relational determinations such as cause and effect. Unlike the Associationists, Reid held that the mind actively participates in processes of perception through operations of the faculties of attention and memory.

The mind's ability to selectively direct its attention is central to Reid's conception of conscious experience. In his *Essays on the Intellectual Powers of Man* (1785; henceforth *EIP*), he describes the mind as constantly thronged with ideas and thoughts, some of which are unnoticed and immediately forgotten, while others are deemed deserving of attention, and intentionally "stopped, examined, and arranged."[13] In perception, he writes,

> we seem to treat the thoughts that present themselves to the fancy in crowds, as a great man treats those that attend his levee. They are all ambitious of his attention; he goes round the circle, bestowing a bow upon one, a smile upon another; asks a short question of a third; while a fourth is honored with a particular conference; and the greater part have no particular mark of attention, but go as they came. It is true, he can give no mark of his attention to those who were not there, but he has a sufficient number for making a choice and distinction.[14]

By comparing the mind to an aristocrat judiciously bestowing the favor of his regard, Reid emphasizes the significant degree of agency we have in deciding upon what to attend. Attention here is selective: the mind, using its own criteria, chooses to notice some ideas more than others.[15]

The originality of Reid's theory emerges when we compare it to Associationist accounts of auditory attention. Let us consider the familiar situation of attending to a specific part of a musical texture. Reid observes in *An Inquiry Into the Human Mind* (1764; henceforth *IHM*), "I have been assured, by persons of the best skill in music, that, in hearing a tune upon

---

[13] Reid, *EIP*, IV, 408.
[14] Ibid.
[15] While Reid concedes that there are cases of involuntary attention, in *An Inquiry Into the Human Mind* (henceforth *IHM*), he notes that it may also be possible to "withhold attention . . . by strong resolution and practice." Reid, *An Inquiry Into the Human Mind*, 2nd ed. (Edinburgh: A Millar, 1765), II, 61.

130   HEARING WITH THE MIND

the harpsichord, when they give attention to the treble, they do not hear the bass; and when they attend to the bass, they do not perceive the air of the treble."[16] In contrast, in *Observations on Man, His Frame, His Duty, and His Expectations* (1749), Hartley notes, "in a concert, some one instrument generally strikes the ear more than the rest, so of the complex vibrations which exist in the medullary substance, some one part will prevail over the rest, and present the corresponding idea to the mind."[17] In this latter, materialist scenario, a dominant stimulus impresses the corresponding idea, that is, mental representation, upon the neural substance and thus the passive mind, without the mediation of selective attention on the part of the listener. For Reid, however, the mind plays an active role in selecting objects of attention for itself.

While Reid generally regards attention as a voluntary operation of the mind,[18] he also admits the opposite scenario. Sometimes, when our minds are "so vacant of all project and design as to let our thoughts take their own course, without the least check or direction: . . . some object will present itself, which is too interesting not to engage the attention, and rouse the active or contemplative powers that were at rest."[19] In practice, he concludes, the majority of our everyday experiences are composed of mixtures of these two modes of mental activity, at least if we are awake and in full possession of our faculties.[20]

---

[16] Reid, *IHM*, VI, 228. This is the only reference to musical perception in *IHM*, where Reid clarifies that he will not treat the question of a "musical ear," such as would afford "the perceptions of harmony and melody, and of all the charms of music," which he regarded as of a "higher order" than bare sensory perception. Ibid., IV, 91. Reid's attitude toward music as articulated in his *Lecture Notes on the Fine Arts*, written in 1774, suggests that he did not regard himself as an authority on music. He writes: "There are undoubtedly certain relations of sounds which are pleasant and which we call harmonious. They please from causes which we cannot explain. All their vibrations bear a certain ratio, and the more simple the ratio the more harmonious. There is evidently likewise a key in music, so that all notes must be within it or else they become harsh and disagreeable. We cannot explain whence arises this pleasure either." Peter Kivy, ed., *Thomas Reid's Lectures on the Fine Arts, Transcribed from the Original Manuscript, with an Introduction and Notes* (The Hague: Martinus Nijhoff, 1973), 49. An examination of Reid's "common sense" approach to aesthetics and sense perception can be found in Leslie E. Brown, "Thomas Reid and the Perception of Music: Sense vs. Reason," *International Review of the Aesthetics and Sociology of Music* 20, no. 2 (1989): 121–140.

[17] David Hartley, *Observations on Man, his Frame, his Duty, and his Expectations*, part 1 (London: S. Richardson, 1749), 64.

[18] These passages occur in a discussion of the nature of the succession of thoughts in our minds, in which Reid suggests that there are two kinds of trains of thought: they are either "such as flow spontaneously, like water from a fountain, without any exertion of a governing principle to arrange them; or they are regulated and directed by an active effort of the mind, with some view and intention." Reid, *EIP*, 407.

[19] Ibid.

[20] Ibid.

Reid's approach to the faculty of memory also differed significantly from those of his contemporaries. He rejects the Aristotelian model of the faculty of memory as a repository of images that are recalled and reapprehended, a tradition which, via Locke, informed the Associationist conception of the act of memory as entailing a present encounter with an image of the original object of perception.[21] Rather, he argued, the memory "produces a continuance or renewal of a former acquaintance with the things remembered."[22] That is, the "former acquaintance," or the previous apprehension, is not replaced by a new, present apprehension of an image of the imagination (a simulacrum of the original object), but rather the first apprehension of the object is protracted or "renewed." As Rebecca Copenhaver specifies, for Reid, "memory is an act that preserves a past apprehension."[23]

Reid regards the faculty of memory, like the faculties of consciousness and perception, as an "original faculty," which "make[s] a part of the constitution of the mind."[24] Memory, he maintains, "implies a conception and belief of past duration" and thus serves to extend our consciousness beyond the specious present; it is intimately bound up with the experience of sensation and perception in prolonging the availability of an object for attention and evaluation.[25] Our consciousness (awareness of the operations of our minds) and perception (awareness of external objects) are thus made up of a succession of infinitesimally brief events, which are unified by memory.[26] As Reid maintains in *EIP*:

---

[21] Through Locke, the traditional "store-house" metaphor directly informed the approaches of Associationists such as Hartley, who thought that memories were preserved in material form as vibratory traces or pathways within the brain, as well as Hume, who claimed that the memory of previous impressions was distinct from present impressions in reappearing with diminished vivacity before the mind's eye. John Locke, *An Essay Concerning Human Understanding* (London: Holt, 1690), bk. 2, chap. 10, §2, 65–67; Hartley, *Observations on Man*, 374–382; David Hume, *Treatise of Human Nature* (London: John Noon, 1739), part 1, chap. 1, §3, 24–25.

[22] Reid, *EIP*, III, 305.

[23] Copenhaver, "Reid on Memory and Personal Identity," 177.

[24] Reid, *EIP*, I, 15.

[25] Reid, *EIP*, III, 305; Reid writes: "our notions of duration, as well as our belief of it, is got by the faculty of memory." Ibid., 310. However, he also remarks that "our consciousness, our memory, and every operation of the mind, are still flowing like the water of a river, or like time itself....Consciousness, and every kind of thought, is transient and momentary, and has no continued existence." Ibid., 336. As Marina Folescu has demonstrated, for Reid, "some of the operations of our minds should themselves be construed as events, with duration . . . [enabling us] to be conscious of operations while they're occurring." Marina Folescu, "Remembering Events: A Reidean Account of (Episodic) Memory," *Philosophy and Phenomenological Research* 97, no. 2 (2018): 316.

[26] I adopt this distinction between consciousness and awareness from Folescu, who notes: "Consciousness is thought by Reid to function like perception, except that it takes as objects occurrent operations of our minds, and not things in the outside world." Ibid., 309.

132 HEARING WITH THE MIND

> Our senses and our consciousness are continually shifting from one object to another; their operations are transient and momentary, and leave no distinct notion of their objects, until they are recalled by memory, examined with attention, and compared with other things.[27]

I shall pass over further details of Reid's model of perception and the unusual role of memory therein to examine the repeated appeals to auditory experience in his argument.[28] Sounds, which are durational, provided Reid with a compelling example of the duration of a present perception.[29]

In *IHM,* Reid invokes the classic example of unheard and unnoticed sounds, variations of which had been treated by Locke and Leibniz, among others, to argue that perceiving a sensation involves not only recognition in the moment, but some kind of memory of the experience as well.[30] To illustrate his point, he invokes the chimes of a clock, perhaps one of the most paradigmatically durational sounds in that each single chime itself lasts a specific length of time, while the chiming as a whole signals the demarcation of longer time units:

> In proportion as the attention is more or less turned to a sensation, or diverted from it, that sensation is more or less perceived and remembered. . . . When we are engaged in earnest conversation, the clock may strike by us without being heard; at least we remember not the next moment that we did hear it. The noise and tumult of a great trading city, is not heard by them who have lived in it all their days; but it stuns those strangers who have lived in the peaceful retirement of the country.[31]

Reid returns to this example in *Essays on the Active Powers of Man* (1788; henceforth *EAP),* where he writes,

---

[27] Reid, *EIP,* III, 324.

[28] See, e.g., Copenhaver, "Reid on Memory and Personal Identity"; Marina Folescu, "Thomas Reid's View of Memorial Conception," *The Journal of Scottish Philosophy* 16, no. 3 (2018): 211–26; Folescu, "Remembering Events"; and Nichols, *Thomas Reid's Theory of Perception.*

[29] Reid emphatically ruled out the possibility of durationless events, writing "Has not every Sensation, every Perception, every thought some duration? If so, memory only gives it a longer duration, and it is difficult to say where one ends & the other begins." Reid, *Correspondence,* 314.

[30] Locke, *Essay,* bk. 2, §4; and Gottfried Leibniz, *Nouveaux essais sur l'entendement humain* (1704), chap. 9, §1–7. For earlier treatments of unheard and hence unnoticed sounds, see Aristotle, *De sensu* 447a12–b21; Augustine, *De musica,* bk. 6, chap. 21; and Lactantius, *De opificio Dei,* cap. xvi.6–8. I thank David E. Cohen for these last three citations.

[31] Reid, *IHM,* 61.

While two persons are engaged in interesting discourse, the clock strikes within their hearing, to which they give no attention. What is the consequence? The next minute they know not whether the clock struck or not. Yet their ears were not shut. The usual impression was made upon the organ of hearing, and upon the auditory nerve and brain; but from inattention the sound either was not perceived, or passed in the twinkling of an eye, without leaving the least vestige in the memory.[32]

Both of these passages make the same argument. We may disregard the sounds of a clock because our attention is diverted by competing distractions, or we may hear it chime and immediately forget. In both cases, we physically sense the chime but the sound does not attain the status of a perception in our awareness, and hence we do not notice it. The first passage also brings up an additional scenario, in which we become inured to a certain sound through long exposure and so cease to be aware of hearing it: the sound still affects our auditory organs, but we do not notice or remember it, so no perception takes place.

In a letter to his student, the philosopher Dugald Stewart, Reid spells out how our apprehension of an auditory stimulus requires that a sensation become a perception via operations of attention and memory:

When we hear the clock strike, there are three things that concur, first, the impression upon the bodily organ secondly, the Sensation of Sound, thirdly the belief that the clock strikes. It is probable that the first of these takes place when we are most inattentive & even when we sleep, because it seems to be the necessary Effect of a mechanical Cause. But whether the Sensation and Perception take place when I cannot in the least remember them, next Moment I know not, and I think it impossible I should know it.[33]

Reid here observes that no attention or consciousness is necessary for sound to physiologically affect the auditory organ; indeed, this can even occur in sleep. A sonic impression thus gives rise to a sensation, but perception, properly speaking, only occurs once we register the sound, and this awareness is dependent on the mental faculties of memory and of attention.

---

[32] Thomas Reid, *Essays on the Active Powers of Man* (Edinburgh: J. Bell, 1788), 79.
[33] Reid, *Correspondence*, 213–214.

134   HEARING WITH THE MIND

In the same letter, Reid returns to the case in which we fail to hear the clock strike when engrossed in conversation, and suggests that either the sound is not noticed, and thus never forms part of a perception, or it is noticed but immediately forgotten.[34] Reid inclines to the second option but declines to adjudicate between the two. For our purposes, however, this distinction makes no difference: the main point is that perception depends on the operations of attention and memory.

To conclude our discussion of Reid's attitude toward attention and memory, it is enlightening to consider an experiment pertaining to the perception of time that he describes in *EIP*, the goal of which is to ascertain how accurately we can reproduce brief temporal intervals:

> It may be observed, that one who has given attention to the motion of a second pendulum, will be able to beat seconds for a minute with a very small error. When he continues this exercise long, as for five or ten minutes, he is apt to err, more even than in proportion to the time.[35]

This phenomenon, Reid avers, supports his hypothesis that memory and attention play a crucial part in perception. As he goes on to conclude: "It is difficult to attend long to the moments as they pass, without wandering after some other object of thought."[36] I take his experiment to illustrate how, for the duration of one minute, the original pendulum is still fresh in the memory, and the mind can still attend to the task at hand by measuring out each interval with reference to the original duration. The deterioration of the task performance is caused by the attenuation of the impression of the pendulum's motion in our memory and/or a relaxation of concentration, as

---

[34] Reid summarizes his distinction between sensation and perception in a letter to Lord Kames of December 20, 1778, where he writes that these "two different Operations of the Senses, which almost always go together, and on that Account are confounded both by the Vulgar and by Philosophers; but which may be distinguished in thought. The first is *Perception*; by which I mean an Information of something about us by our Senses, and nothing more. The second is Sensation; by which I mean the feelings we have by our Senses whether pleasant painfull or indifferent." Reid, *Correspondence,* 112. In his comprehensive monograph on Reid's philosophy of perception, Ryan Nichols remarks on the fact that Reid is widely recognized as the first [early modern] philosopher to make this distinction with such a high degree of clarity, and further demonstrates that Reid intentionally marginalized the role of sensations in the process of perception. He interprets Reid as claiming that sensations "manifest our bodies' reactions to the world" and thereby serve the teleological requirement of prolonging our corporeal existence. Nichols, *Thomas Reid's Theory of Perception*, 152–53; see also 143–60 more generally.

[35] Reid, *EIP*, III, 331.

[36] Ibid.

the mind begins to wander, bringing about the compounding of the reproduction error.[37]

To summarize, Reid's psychology makes an original attempt to isolate and emphasize the constitutive roles played by attention and memory in our acts of perception. In endeavoring to ascertain exactly how these different aspects of perception interact, he calls upon the case of listening to different lines of a musical texture, the phenomenon of disregarding sounds by dint of distraction or familiarity, and our capacity to retain and reproduce measured intervals of time. Taken together, these examples outline a theoretical approach to the experience of listening that relies upon introspection, self-reports, and experiments, in an effort to isolate the crucial importance of the faculty of attention, in particular, in our ability to make sense of the sounds around us.

Holden's Proto-Cognitive Music Theory

Thomas Reid and John Holden were contemporaries at the University of Glasgow, where Reid succeeded Smith in the chair of moral philosophy. As discussed in Chapter 1, we have firm evidence that Reid and Holden were acquainted and that the philosopher held the musician in high regard. Although Holden does not cite Reid in his *Essay*, these thinkers share certain distinctive conceptual strategies with regard to the working of our perception.[38]

---

[37] Reid's experiment is strikingly similar to a contemporary psychological measure known as a "continuous performance task," the goal of which is "to measure individual differences in sustained attention." Hettie Roebuck, Claudia Freigang, and Johanna G. Barry, "Continuous Performance Tasks: Not Just About Sustaining Attention," *The Journal of Speech, Language, and Hearing Research* 59, no. 3 (2016): 501. Reid goes on to describe a second experiment involving the beating of durational units of a second, although its purpose is different. He writes: "I have found by some experiments, that a man may beat seconds for one minute, without erring above one second in the whole sixty; and I doubt not but by long practice he might do it still more accurately. From this I think it follows, that the sixtieth part of a second of time is discernible by the human mind." Reid, *EIP*, III, 331. Reid's estimate of a deviation of one sixtieth of a second, i.e., of 1.67%, in timing accuracy is remarkably close to the empirical results found in contemporary psychology: Bruno H. Repp reports that trained musicians tapping along to a metronome produce deviations ranging from .05% to 2%; however, untrained participants display larger deviations of around 4%. Bruno H. Repp, "Sensorimotor Synchronization: A Review of the Tapping Literature," *Psychonomic Bulletin & Review* 1, no. 6 (2005): 975.

[38] Jamie C. Kassler has linked the importance of language in Holden's theory (and his desire to explicate the "syntax" of music), to a similar focus on language in the philosophy of the Scottish School of Common Sense. Kassler, "British Writings on Music," 84–87. In subsequent writings, she describes his *Essay* as "a curious mixture of French music-theoretic doctrines transmuted into the logic of the Scottish school of common-sense philosophy," Kassler, *The Science of Music*, 1:527–529. Leslie Brown has also related Holden's reliance on intuition, his method of generalization from particular

136   HEARING WITH THE MIND

As we saw in Chapter 2, Holden uses the faculty of attention to theorize our experience of certain musical phenomena, such as tonicity ("key note") and tonality (sense of key),[39] seventh and ninth chords, deceptive cadences, and the selection of a harmonic "fundamental" (identification of a root in ambiguous cases). He maintains that our sense of key results from an active engagement of the attention selectively interpreting incoming sounds, and he further proposes that selective acts of attention allow the listener to determine the fundamental of a chord and endow it with syntactic significance within the context of the harmonic progression in which it occurs. This substantial treatment of attention—and more particularly, the operations of a selective mode of attending as a necessary but not sufficient condition for musical perception—appears to me to respond to ideas raised by Reid.

Holden appears to implicitly distinguish between—and invoke—both voluntary and involuntary forms of attention throughout the treatise.[40] The former is typically described using active verbs, by which the listener actively "turns," "fixes," or "places his attention" on a certain sound as a key note, whereas in the case of the latter, the listener's attention is passively "attracted," "distracted," or "diverted" by prominent sonic features, such as a particular line, sound, or instrument. Thus, for example, the bass line "attract[s] the hearer's attention at the ends of several strains, or portions of the tune, and more particularly at the final conclusion."[41] Our attention can be "carried away by an imitation,"[42] and in a concerto, the solo line should "engage the hearer[']s attention more particularly, while the others serve only to fill up the harmony."[43] The same holds true within a chord: "our attention is carried either to the lowest or the highest sound which exists in a chord, rather than

---

instances, and the use of analogy, to the methodologies laid out in the *IHM*. Brown, "The Common Sense School and the Science of Music," 122–132.

[39] Holden means by "key note" essentially what we mean by "tonic" in the sense of tonic note or pitch class, although he also uses the same term to denote the home key of a whole piece (or of a section or even a phrase).

[40] This bipartite division of attention is reminiscent of contemporaneous approaches to the subject, which typically held that there are two kinds of attention: one spontaneous, the other voluntary. For example, as Matthew Riley has shown, Georg Friedrich Meier speaks of "natural" and "arbitrary" attention, while Sulzer distinguishes "compulsive" and "voluntary" attention. Both of these writers link the first, involuntary form of attention to affective states such as transport or wonder. Riley, *Musical Listening*, 29–30. See also Michael Hagner, "Toward a History of Attention in Culture and Science," *MLN* 118, no. 3 (2003): 672. Holden, it should be emphasized, does not discuss experiences of transport or wonder.

[41] Holden, *Essay*, part 1, §61, 57.

[42] Ibid., part 2, §320, 277.

[43] Ibid., part 1, §124, 110.

any of those which lie, as it were, concealed in the middle,"[44] and the resolution of the leading tone "inevitably attracts the chief attention of the hearer."[45] All of these cases depict the involuntary instinct of our minds to attend to prominent features in the environment.

Voluntary attention, on the other hand, for Holden typically pertains to those cases in which the mind undertakes an act of reinterpretation in response to an ambiguity in the musical discourse. It is typically activated after an equivocal or unpleasant sound of some kind has "divided" or "distracted" the attention as a consequence of two apparently equivalent options attracting our involuntary attention in equal measure.

An unusual appeal to the selective nature of voluntary attention can be found in Holden's rather remarkable new definition of consonance and dissonance. He proposes that consonances are intervals whose notes can be referred to a single "perfect chord," which he defines as the first five partials of the harmonic series, and thus a single, shared fundamental.[46] Holden then offers an original explanation based in part on the success or failure of involuntary attention:

> Allowing that the mind *naturally chuses* to conceive every sound in music as belonging to some perfect chord, it is plain, that two sounds will seem to unite, when both of them are included in the idea of one perfect chord, but separately *distract our attention,* when this cannot be done, or when they must necessarily be referred to two different fundamentals.[47]

Here Holden claims that the mind instinctively attempts to interpret every note as a member of a perfect chord, and when confronted by multiple notes in harmony, tries to assign as many of them as possible to a single perfect chord. When the mind is unable to do, as is the case when there appears to be

---

[44] Ibid., part 1, §144, 127.

[45] Ibid., part 1, §193, 169.

[46] A contemporary reviewer hails this as "a new definition of concord and discord, which appears to us very clear and conclusive, and entirely settles all those disputes and cavils, which this point has occasioned among former writers on this subject" (Anon., "Holden's Essay towards a Rational System of Music," *The Critical Review* 33 (June 1772): 324). Note that this definition is familiar to us from the works of Riemann, who defines the qualities of consonance and dissonance as dependent on an interval's triadic membership, or the lack thereof; see, e.g., Hugo Riemann, "Konsonanz," *Musik-Lexikon* (Leipzig: Verlag des Bibliographischen Instituts, 1882), 477; these ideas are already evident in his *Musikalische Syntaxis: Grundriß einer harmonischen Satzbildungslehre* (Leipzig: Breitkopf & Härtel 1877), 5. I have been unable to locate this way of defining consonance or dissonance in any theorist prior to Holden thus far. I thank David E. Cohen for his help on this point.

[47] Holden, *Essay*, part 1, §153, 138. Original italics.

138   HEARING WITH THE MIND

more than one fundamental for a given chord, voluntary attention can help make a determination based on syntactical expectation.

As an example, let us return to Holden's invocation of attention to theorize dissonant chords such as seventh chords, discussed in Chapter 2. In this situation, Holden argues, our attention is divided when a seventh chord contains two perfect fifths, as is the case with both the major seventh and the minor seventh chords.[48] Here, he observes, "no one sound can properly be called the *sole* fundamental of a dissonant chord, because of the divided attention which the discord creates."[49] The psychological experience of hearing these dissonant chords entails divided attention: the state of having "two fundamentals partly in view."[50] In some of these cases, however, we can prioritize one of the two fundamentals because "one of them may, for various reasons, claim the greater share of our regard; and therefore may be properly called *governing*."[51]

Holden also appeals to the phenomenon of divided attention to explain cases of syntactical substitutions such as deceptive cadences. He proposes that in such cases, the attention is divided between an actual event and the memory of the anticipation of a different projected event: "The attention may at any time be supposed to be divided between the fundamental, which would naturally have taken place, and that which is introduced, partly unexpected, by the false [deceptive] cadence."[52] Holden here suggests that we retroactively determine the correct fundamental in a passage following contextual cues. While this occurs nearly instantaneously, hearing a deceptive

---

[48] Holden maintains that the reason the dominant seventh chord does not divide the attention is due to the presence of the diminished fifth, which "inevitably attracts the chief attention of the hearer," so that "it immediately becomes our principal concern to hear its terms reduced to a better agreement." Holden, *Essay*, part 1, §193, 169. Indeed, he regards the resolution of this chord as a model for harmonic movement. He writes: "[It] is the most satisfactory of all passages, and the only one which puts an end to the expectation of the hearer; and by examining what way this expectation may be satisfied, and the attention wholly fixed upon one chord, we come easily to understand how it may be disappointed or suspended, and the attention carried to different chords." Ibid., part 1, §193, 168–169.

[49] Ibid., part 1, §160, 144.

[50] Ibid.

[51] Ibid. Original italics. Holden credits the assignment of two fundamentals to these seventh chords to the Swiss music theorist Jean Adam Serre, who proposed this idea in *Essais sur les principes de l'harmonie* (1753). See page 82 in Chapter 2. Holden's own innovations include invoking the concept of attention to provide a psychological account of the chord's effects and aligning double fundamentals with divided attention, as well as proposing that the problem of determining which of the two competing fundamentals is the primary one can be resolved by appealing to syntactical considerations, taking "into account the expectation, whether founded on nature or custom, of hearing such particular fundamentals in succession." Holden, *Essay*, part 1, §176, 156.

[52] Ibid., §160, 144; and §217, 197–198. Holden uses the term "cadence" to describe an (expected) dominant-to-tonic harmonic progression, regardless of its location in the phrase.

LISTENING WITH ATTENTION 139

cadence as deceptive entails first the mental projection of the expected tonic and then the retention of that projection, at least momentarily, in spite of the onset of the different bass note that the music substitutes for it.

Let us now turn to the role of memory in Holden's approach to the determination of a key. "In the practice of music," he writes, "the key note is constantly kept in mind; and all other notes which are admitted, are some way compared with the key [note]"—this retention of the tonic being the *sine qua non* that enables us to feel closure at the end of a piece.[53] It is "the uninterrupted remembrance of one certain sound, as a principal key, which connects together all the successive parts of one piece of music."[54] Holden thus theorizes the experience of being in a key as a memory of a tonic, across a continuous span of time, that functions as a touchstone to which all incoming sounds are referred. "These comparisons," he continues, with their "consequent perceptions, are, indeed, the very essence of music."[55]

As a familiar analogy to the experience of being in a key—that is, of retaining as a reference point the key note—Holden offers the sensation of hearing bagpipe music, with its tonic drone: "The key might be held on in the bass from first to last, as in the musette and bagpipe music; and although it should not actually be so held on, yet it is undoubtedly always kept in mind in all connected pieces."[56] This notion of keeping the key note in mind demonstrates that Holden regards the perception of the key as dependent upon the memory in prolonging the apparent duration of a percept even in its physical absence. It thereby recalls Reid's approach to the crucial role of the operations of memory and attention in extending the duration of perception, discussed earlier.

Just how the mind accomplishes this retention of the key note is explained, in the second part of the *Essay*, by Holden's music-psychological construct of the module. As we saw in Chapter 2, Holden's module is a brief time span, a unit of measure precisely long enough to comprise two or three vibrational cycles of the "key note" (the tonic pitch in a specific octave). It is the module—and not an impression of the key note itself as a pitch—that our memory unconsciously retains as we hear a piece of music, where it functions as the referential element against which the frequencies of all other pitches, such as the other scale degrees, are proportionally measured as interval ratios.[57]

[53] Ibid., §29, 25.
[54] Ibid., §31, 27.
[55] Ibid., §29, 25.
[56] Ibid., §321, 278.
[57] An explanation of the module can be found in Chapter 2 of this book.

140 HEARING WITH THE MIND

He thereby reveals the mental mechanism that enables sounds to be heard as having specific intervallic, music-syntactical, and qualitative relations to the key note. The result is a psychologistic account of purely musical phenomena as tonal functions, chord progressions, and the qualities of scale degrees. Holden's module—and its retention in memory—is thus the indispensable basis of our subjective experience of music as being "in a key."

Holden's psychological construct could have been, I suggest, a response to Reid's claim that the operation of perception always includes the operations of memory and attention as well. Specifically, the module ingeniously meets the theoretical requirement that the tonic be somehow preserved in memory as a point of comparison without actually being phenomenally present, in order for incoming sounds to be grasped in relation to it. Hearing in a key, for Holden, obliges the mind to evaluate all pitches according to an infinitesimally small measure that represents the tonic pitch; it does so nearly instantaneously upon our hearing the sounds in relationship not only to their immediate context but also to the key note itself as represented by the module. The involuntary retention of the module, and its use as a basis for further mental operations, *is*, in effect, attending to music.

Of course, music frequently modulates, and these changes require the demarcation of new keys (and correspondingly, new modules). Holden describes this experience as follows:

> When a person has fixed his attention on one sound as a key note, if other sounds accidentally intrude upon his ears, which belong to a different key, they have often a most disagreeable effect; although these other sounds, when compared with their proper key be ever so agreeable. . . .Before he [the listener] can hear such incompatible sounds with satisfaction, he must quit his own former key, and turn his attention altogether on that to which they properly belong.[58]

Attention here is depicted as a mental strategy whereby the listener can perceive, and move between, different keys. The listener initially "fixes" a single key note—represented by the module—in their mind as a yardstick against which to measure sounds extraneous to the key.[59] Upon

---

[58] Ibid., part 1, §31, 27.
[59] This excerpt comes from the first part of the treatise; Holden introduces the module only in the second part.

LISTENING WITH ATTENTION 141

hearing a significant number of non-diatonic tones, an unpleasant, distracted feeling alerts the listeners' minds that the music has modulated, prompting our voluntary attention to seek out and establish a new key note (and module) according to which the incoming sounds can be heard as pleasing. Redirecting the attention to select and retain a new tonal reference point compatible with these new sounds enables the listener to render pleasant sounds that would be discordant when measured against the initial key note.

Finally, I would like to examine some of Holden's empirical reports on the perception of rhythm, which seem to me to be very much in the vein of Reid's investigation of the perception of time. When we listen to the ticks of a watch, he writes,

> although the alternate stronger pulses . . . be undoubtedly all equal, yet when we count one, and pass over the next, and count the next, and pass over the next, and so on; we imagine the pulses which we count, to be really stronger than the intermediate ones, which we pass over. The superior regard which we bestow on the counted pulses, is, here, the sole cause of these imaginary accents.[60]

Holden here reports, in what might may well be the earliest depiction of the important psychological phenomenon of subjective rhythmicization, that the apparent difference in the volume of watch ticks is caused by our attention and imagination, rather than by any acoustic discrepancy.[61] He goes on to suggest that the reader attempt to group beats into units of threes and fours (composed of two groups of twos). Our ability to impose such accents, he continues, indicates that

> the ideas of the length of the whole measure, and of its proper divisions and subdivisions, are as constantly kept in mind, through the course of a tune, as the key note was observed to be . . . and that the constant remembrance of the measure and its divisions regulates the time and accent of each sound which we use; much in the same manner as the key note determines the tune or pitch of all those which follow it.[62]

---

[60] Ibid., §95, 83; see also part 2, §14, 294.
[61] Holden predates Kirnberger by six years; see Kirnberger, *Die Kunst des reinen Satzes* 2, 114–115.
[62] Holden, *Essay*, part 1, §96, 84–85.

142   HEARING WITH THE MIND

This passage sets out the background for Holden's speculative theory of pitch perception, which is modeled on rhythm, as we saw in Chapter 2. Central to our interests here is the important role he ascribes to the faculty of memory in retaining the key and the measure in our perception of music.

Holden also spends considerable time discussing the constraints on our faculty of memory, specifically with regard to integrating rhythmical events into larger groups. For example, he suggests that the restricted capacity of our memory leads our minds to group music into measures, and measures into phrases made up of groups of twos and threes and their multiples:

> Besides this distribution of music into measures, the mind extends its view, and as far as the memory can be supposed distinctly to retain, goes on to constitute some number of measures into isochronous phrases, or strains of a tune; and these strains may contain a greater number of measures in quick time than in slow, because of the inability of the memory; but here, as before, the number of measures in a strain must always be either two or three, or some product of these numbers: for here five bars in one strain is not used, and seven proves much more intolerable.[63]

To summarize, Holden claims that our experience of hearing music requires the mind to constantly retain the key note (as represented by the module) and the measure as a reference against which we compare all subsequent sounds. The faculty of memory is essential to our ability to perceive music, as is the faculty of attention, which he understands as taking both voluntary and involuntary forms, frequently manifesting a combination of both. Holden describes a mode of attending to the relationships between sounds: a strategy of hearing in the present while retaining and comparing with past sounds. His account of listening thus has much in common with that outlined by Reid, even if his main focus is on musical cognition.

Thomas Robertson: An Interlude

While Adam Smith's activity at the University of Glasgow overlapped with Holden's at various times during the 1760s, there is no evidence that they

---

[63] Ibid., part 2, §14, 290.

ever met.[64] Smith does not appear to have owned a copy of Holden's *Essay*.[65] Holden's *Collection* was issued in 1766, three years after Smith had left the University of Glasgow to tutor Henry Scott, Duke of Buccleuch. Despite these chronological discrepancies, the similarities between Holden's and Smith's musical approaches are striking, suggesting that they may have exchanged ideas in some form.

One possible intermediary between Holden and Smith may have been the Scottish music theorist Thomas Robertson (ca. 1745–1799).[66] Smith and Robertson moved in similar circles and were acquainted with each other (possibly as fellow founding members of the Royal Society of Edinburgh).[67] In fact, Smith was sufficiently close to Robertson that, at some point during the course of the year 1783, he read a manuscript version of Robertson's *Inquiry into the Fine Arts: Of the Fine Arts which Refer to the Ear*.[68] Later that same year he wrote to his publisher, Thomas Cadell, on Robertson's behalf, recommending the book and noting that he had read "the theory (not the History) and was much instructed."[69] Robertson, as we shall see, clearly borrows some of Holden's ideas; his *Inquiry* could thus have served as an avenue by which Holden's thinking exercised an influence on that of Smith.

Robertson's *Inquiry* expounds far-reaching theoretical speculations about the complementary roles of the mind and the senses in perceiving music in a uniquely chatty and, for the time, pugnacious tone. Generally speaking, the *Inquiry* holds little interest for the modern-day reader, as its author's understanding of music theory is slight, and his accounts of other sources seem largely paraphrased (often inaccurately) from the work of contemporary

---

[64] Anselm Gerhard cursorily mentions a similarity between the writings of Holden and Smith, and posits that the latter may have had access to the former's work ("Holden, dessen Schrift dem zur gleichen Zeit an derselben Universität lehrenden Adam Smith mit Sicherheit bekannt gewesen sein dürfte"). Gerhard makes a few minor errors in his brief treatment of Holden: his chronology is faulty (Smith left Glasgow in 1763, and Holden's *Essay* came out in 1770), and Holden's work does not reflect Associationism, as Gerhard states. Gerhard, *London und der Klassizismus in der Musik*, 137.

[65] Hiroshi Mizuta, *Adam Smith's Library: A Catalogue* (Oxford: Clarendon Press, 2000).

[66] Robertson was the minister of the parish of Dalmeny near Edinburgh. He was a member of the prestigious "Edinburgh Speculative Society and a recipient of an honorary doctorate of divinity from the University of Edinburgh. In addition to his musical writings, he published an "Essay on the Character of Hamlet" (1790) in the *Transactions of the Royal Society of Edinburgh* (incidentally, it appeared in the same issue as Walter Young's "Essay on Rhythmical Measures"), and a well-regarded *History of Mary Queen of Scots* (1793). See "Alphabetical List of Members," in *History of the Speculative Society of Edinburgh* (Edinburgh: Speculative Society of Edinburgh, 1845), 465–477.

[67] See the *Former Fellows of the Royal Society of Edinburgh: Biographical Index Part 2* (Edinburgh: The Royal Society of Edinburgh, 2006), 794, 855.

[68] Letter 231 to William Strahan, Edinburgh, October 6, 1783, in *The Glasgow Edition of the Works and Correspondence of Adam Smith VI: Correspondence*, ed. Ernest Campbell Mossner and Ian Simpson Ross (Oxford: Oxford University Press, 1987), 269.

[69] Ibid.

144 HEARING WITH THE MIND

writers such as Holden or Charles Burney. Robertson's prose style is un-wittingly entertaining, however, if mostly for its bombast and lack of self-awareness.[70] Initially conceived as a multivolume work dealing with the fine arts in turn, the *Inquiry* was the subject of scathing reviews, including a critique (anonymously penned by Burney) in *The Monthly Review* of 1786, which characterized it as "affected, obscure, superficial, and abounding with provincial barbarisms."[71] This evaluation may have led Robertson to abandon his project, and no further volumes of the work were published.

The *Inquiry* is divided into seven chapters and a postscript: the first two deal with the theory of ancient and modern music respectively, while the third engages in what the author terms "Speculations in Music." The subsequent three chapters present a history of music from antiquity until the middle of the eighteenth century, while the final section attempts to draw speculative conclusions from reports on the music of the South Sea Islanders by members of Captain Cook's expedition.[72]

Robertson mentions Holden by name in his first chapter, alongside the Scottish music theorist Alexander Malcolm and the French mathematician Jean le Rond d'Alembert. While he praises Malcolm to the skies, he is more circumspect about d'Alembert.[73] When it comes to Holden, however, Robertson simply writes that "Holden, the latest writer, has communicated several illustrations."[74] He continues:

---

[70] Consider, for example, the following justification Robertson provides for his theory: "Upon what is a System of Music to be founded? It would be pleasant to find that we are not under the dominion of an arbitrary rule; but can see a reason, why every thing is as it is. But how seldom have we that pleasure! A dogmatism reigns among Musicians; and we are tempted, in revenge, to look behind the scene, to see the springs that permit their puppet to move only in a certain way." Thomas Robertson, *An Inquiry into the Fine Arts* (Edinburgh: T. Cadell, 1784), 224–225.

[71] Anon., "Robertson's Inquiry into the Fine Arts," *The Monthly Review, or, Literary Journal* 74 (April 1786): 248. See also Anon., "Robertson's Inquiry into the Fine Arts," *The English Review* 5 (June 1785): 401–410. Even a more generous account in *The New Review* noted the work was "deficient in style" even if "many of [Robertson's] observations are no less conspicuous for their good sense, than for the strength of the expression in which he has clothed them." The author of the review, however, admitted that he did "not understand the subject, which however seems to be treated with great perspicuity and genius." Anon., "Robertson's Inquiry into the Fine Arts," *The New Review* (May 1785): 339–340, 345.

[72] Here Robertson mostly paraphrases and quotes from the passages on music and dance that appear in James Cook, *A Voyage to the Pacific Ocean: Undertaken by the Command of His Majesty, for Making Discoveries in the Northern Hemisphere* (London: Strahan, 1784), 1:247–255, 288, 298, 396–397.

[73] Robertson writes: "As [d'Alembert's] system is built entirely upon physics, I have never been able to look upon it in any other light than as narrow and delusive. Yet, in justice to the abilities of D'Alembert it is to be confessed, that although his plan be awkwardly contrived, addressing one reader in the text and another in the notes, and consequently exceptionable to both, yet his attempt is the first that has been made among the moderns, to introduce perspicuity and analysis into the subject." Robertson, *Inquiry*, 25–26.

[74] Ibid., 26.

These authors [that is, Malcolm, d'Alembert, and Holden] are mentioned, however, solely upon this account, that it has been their intention to write more intelligibly than others, and to carry the literary Public, in some degree, along with them. Notwithstanding of this, there is no doubt that they will prove by far too difficult to be soon understood, and quite unequal to the wishes of any reader, who, without studying music minutely, desires to obtain some accurate, though general, knowledge of the science.[75]

As this passage makes clear, Robertson was familiar with the aforementioned authors but thought their work was too difficult for the general public, for whom his own treatise was intended.

Robertson's laconic reference to Holden's work is peculiar given his own substantial debt to the *Essay*. In fact, a considerable number of many of Robertson's definitions appear to be rewordings of passages in the *Essay*, and many of his theoretical speculations seem in fact to be literal interpretations of Holden's metaphors. Take, for example, the analogy upon which Holden bases his theory, namely, his hypothesis that "the pleasure which results from the hearing *well-tuned sounds* owe[s] its rise to the same principle which influences us to be delighted with *well-timed notes*."[76] This becomes, in Robertson's telling, the simple assertion that "Tune is nothing else but Time."[77] He subsequently develops this claim into an idiosyncratic variation on coincidence theory, presenting the "rhythms" of various integer ratios associated with familiar musical intervals, as shown in **Figure 4.1**.[78]

Robertson thus appropriates the rhythm-pitch analogy that is the impetus for Holden's theory while omitting most of that theory itself in favor of a much

---

[75] Ibid.

[76] Holden, *Essay*, part 2, §18, 296. Original italics.

[77] Robertson, *Inquiry*, 49.

[78] "As Tune is, strictly speaking, nothing else than Time, let the two Sounds be considered as two Rythmi, flowing forward at one and the same time. The pulses will either coincide with one another constantly, as in the case of the Unison; where the two Rythmi move with an equal pace: or they will only coincide at certain points; sooner, as in the case of Octave; later, as in the case of a Fourth, and others; the Rythmi moving there with unequal speed." Ibid., 228. Other examples of such borrowing include Holden's claim that "there are certain due proportions of time which must be adhered to, with the greatest care and exactness, or otherwise the music will appear wild and incoherent: and in order to discuss this matter fully, we must first take notice of the natural division of all regular music into certain equal-timed parcels, called measures" with Robertson's assertion that "unless [the mind] perceive a strict equality, she abandons entirely her employment, finding the Time to be wild and unintelligible. The fundamental condition required, therefore, in that Music, is, that Sounds be divided into precisely equal portions of Time." Holden, *Essay*, part 1, §68, 64; Robertson, *Inquiry*, 52. Other striking similarities can be ascertained between Holden, *Essay*, part 2, §46, 321 and Robertson, *Inquiry*, 234–235.

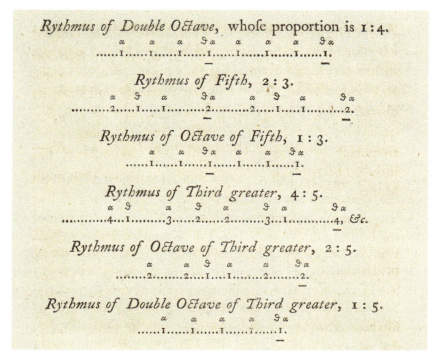

Figure 4.1 Some of Robertson's "Various Rhythms of Intervals." The Greek letter α marks the rhythms of the faster term; an ϑ marks the rhythms of the slower term. The Arabic numerals indicate the proportional size of the distances to their immediate left.

more traditional explanatory model.[79] But far more importantly, he seems to miss the point of the analogy in a crucial way: Holden maintains that the prevalence of duple and triple rhythmic groupings in music indicates that

---

[79] Related analogies between rhythm and pitch would independently resurface over the nineteenth and twentieth centuries, notably in the writings of Hector Berlioz, Moritz Hauptmann, Friedrich Opelt, Henry Cowell, and Karlheinz Stockhausen. More recently, the music psychologist Guy Madison has proposed that our "perception of rhythm is built on similar principles as the perception of pitch, since both seem to utilise small integer subdivision of intervals. The metrical structure is in this sense analogous to the partials (overtones) that characterise complex auditory tones." See Hector Berlioz, "Strauss. Son Orchestre, Ses Walses. — De L'avenir Du Rhythme." *Journal des Debates* (November 10, 1837): 1; Moritz Hauptmann, *Die Natur der Harmonik und Metrik* (Leipzig: Breitkopf und Härtel, 1853); Alexander Rehding, "Three Music-Theory Lessons," *Journal of the Royal Musical Association* 14, no. 2 (2016): 273–275; Guy Madison, "Sensori-motor Synchronisation Variability Decreases as the Number of Metrical Levels in the Stimulus Signal Increases," *Acta Psychologica* 147 (2014): 15. I thank Rick Cohn for bringing Madison to my attention.

LISTENING WITH ATTENTION   147

our minds have an innate preference for simple integer ratios in the domains of rhythm and pitch. Robertson eschews any kind of psychological claim and describes highly complex rhythms, including those of the diesis 44:45 and the tempered fifth 667:1000, proportions which are by definition excluded from Holden's theorization of our mental predilections.

More specifically to the point under discussion in the present chapter, Robertson also appropriates Holden's account of subjective rhythmicization when he claims that

> the mind is found to watch over the whole of [the Measure]; its beginning, its middle parts, and its end. As the end, however, of one, and the beginning of the following Measure, may be held to be the same thing, unless there be any considerable interruption, the end is neglected, as coinciding with the beginning. The beginning, and the divisions in the middle, therefore, occupy the attention. Weight and stress are mentally laid upon them. The imagination comes in, and fancies she hears those instants sounding louder than the rest, and the performer, to aid the imagination actually strikes them louder. The fancy, working powerfully at these moments, has made the voice to swell, the hand, or the foot, or even the whole body, to move.[80]

In Robertson's account, our minds attend to the beginnings and middles of measures, and it is this act of attention that causes accents ("weight and stress") to be imposed upon these metric locations. So far, this aligns with Holden. However, in characteristic fashion, Robertson attempts to expand the original, and in doing so, he misses the point by proposing that this mental act then leads the performer to in fact play them louder. Robertson's explanation thus borrows Holden's neat account of the mental character of metric accents, only to revert immediately after to an earlier (or perhaps simpler) attitude with regard to meter.[81] Most important for our purposes,

---

[80] Robertson, *Inquiry,* 54.

[81] Robertson's account is thus similar to that of Descartes in the *Compendium Musicae* (1618/ 1650) who writes: "Few are aware how in music with diminution, employing many voices, this time division is brought to the listener's attention without the use of measures; this, I say, is accomplished in vocal music by stronger breathing and on instruments by stronger pressure, so that at the beginning of each measure the sound is produced more distinctly; singers and instrumentalists observe this instinctively, especially in connection with tunes to which we are accustomed to dance and sway. Here we accompany each beat of the music by a corresponding motion of our body; we are quite naturally impelled to do this by the music." (Pauci autem advertunt, quo pacto haec mensura sive battuta, in musicâ valde diminutâ et multarum vocum, auribus exhibeatur. Quod dico fieri tantùm quâdam spiritûs intensione in vocali musicâ, vel tactûs in instrumentis, ita vt initio

148   HEARING WITH THE MIND

however, is the fact that Robertson's account contains a key element that he doubtlessly learned from Holden, namely, the origin of metric accents in mental attention.[82]

There is one further passage in Robertson's work worth noting before we turn to Smith. In a truculent rant, Robertson attacks "unthinking men . . . [who] throw contempt upon Modern Music [that is, instrumental music]."[83] He continues:

> It is mere Sound, without words, say they; but even admitting this to be so, Is mere Sound a thing of nought? Does it not place an image before you? Do not you see it passing by? *its very speechlessness filling you with attention? Numeris & Modis*, say the ancients themselves, less acquainted with it, *inest quœdam tacita vis* [In rhythms and harmonies there is a certain inaudible power]. Rail at Modern Music; but expect not to rail, without feeling her power, as your punishment. Smiling at your scorn, in place of answer, she will call for experiment; and *be you ready to attend her*. You hear her Melodies, and you are melted with feeling: you hear her Harmonies, and you are overcome with reverence. Your frame is overset. Look, next, to whole audiences enraptured; tears in every eye; thrillings through ever'y nerve. Where have I fallen? will she then ask of you. Have I not excited all the feelings of the breast? Have I not *overpowered the mind;* and the audience, unable to withstand my charms, interrupted me with an universal applause?[84]

In this passage Robertson links the attention to instrumental music in a way that seems to closely anticipate aspects of Smith's account of the power of instrumental music, as we will see in the next section.

---

cuiusque battutae distinctiùs sonus emittatur. Quod naturaliter observant cantores, et qui ludunt instrumentis, praecipue in cantilenis ad quarum numeros solemus saltare et tripudiare: haec enim regula ibi servatur, vt singulis corporis motibus singulas Musicae battutas distinguamus. Ad quod agendum etiam naturaliter impellimur à Musicâ). René Descartes, *Compendium of Music* trans. Walter Robert (Bloomington, IN: American Institute of Musicology, 1961), 14–15; René Descartes, *Musicae Compendium* (Utrecht: Trajectum ad Rhenum, 1650), 9–10.

[82] It seems highly unlikely that Robertson was aware of Kirnberger's independent reports on the mental nature of metrical accents in *Die Kunst des reinen Satzes in der Musik* (1776), which does not appear in the list of works Robertson claims to have consulted (indeed, no German-language source appears in this list; see Robertson, *Inquiry*, 264–265).

[83] Robertson, *Inquiry*, 446–447.

[84] Ibid. My italics. I thank David E. Cohen for this translation.

Smith and Holden

Following this short detour, let us now return to Smith and Holden. Both of them take complementary approaches to the same question, namely, the relationship between the workings and preferences of our psychological faculties and the features of musical works that generate pleasure.[85] Holden seeks to study the mental faculties involved in the perception and enjoyment of music, whereas Smith attempts to characterize the features of music that aid our minds in perceiving and enjoying it.

Smith describes the experience of listening to instrumental music as follows:

> Instrumental Music, by a proper arrangement, by a quicker or slower succession of acute and grave, of resembling and contrasted sounds, can not only accommodate itself to the gay, the sedate, or the melancholy mood; but if *the mind is so far vacant* as not to be disturbed by any disorderly passion, it can, at least for the moment, and to a certain degree, produce every possible modification of each of those moods or dispositions....In a concert of instrumental Music the *attention* is engaged, with pleasure and delight, to listen to a combination of the most agreeable and melodious sounds, which follow one another....The mind being thus successively occupied by a train of objects, of which the nature, succession, and connection correspond, sometimes to the gay, sometimes to the tranquil, and sometimes to the melancholy mood or disposition, it is itself successively led into each of those moods or dispositions; and is thus brought into a sort of harmony or concord with the Music which so agreeably engages its attention.[86]

Within this fairly standard account of music's ability to move the passions of a receptive listener, Smith makes it clear that our enjoyment of instrumental

---

[85] Smith's interest in musical perception is also evident in an anecdote related by the French geologist Barthélemy Faujas de Saint-Fond, who was invited by Smith to listen to a competition of highland bagpipers. Saint-Fond relates that Smith observed, "I shall put you to a proof which will be very interesting for me; for I shall take you to hear a kind of music of which it is impossible that you can have formed any idea, and it will afford me great pleasure to know the impression it makes upon you," and that Smith subsequently "requested me to pay my whole attention to the music, and to explain to him afterwards the impression it made upon me." Barthélemy Faujas de Saint-Fond, *Travels in England, Scotland, and the Hebrides: Undertaken for the Purpose of Examining the State of the Arts, the Sciences, Natural History and Manners, in Great Britain* (London: James Ridgeway, 1799), 2:242–243.

[86] Smith, *Philosophical Essays*, 163. My italics.

150   HEARING WITH THE MIND

music requires the attention to be unburdened by competing distractions, so that the appropriately "vacant" or receptive mind can attend to the succession of sounds and attune itself to various suggested moods correspondingly.[87] As this passage discloses, Smith, like Holden and Reid, regards the faculty of attention as essential to the perception of music.

A few pages later, Smith fleshes out the way in which instrumental music engages the attention, namely by virtue of the relationship between successive sounds,

> by the sweetness of its sounds it awakens agreeably, and calls upon the attention; by their connection and affinity it naturally detains that attention, which follows easily a series of agreeable sounds, which have all a certain relation both to a common, fundamental, or leading note, called the key note; and to a certain succession or combination of notes, called the song or composition.[88]

This conception of musical pleasure is remarkably similar to the approach described in Holden's *Essay* seven years earlier, whereby "[a] great part of our enjoyment of music is owing to the perception of certain due relations among the several successive sounds, when compared with one determinate sound as a key."[89] Holden underscores this point repeatedly, noting that "so long as a hearer remains in suspense, which note to fix upon for a key, he has no enjoyment of the music."[90] Both Holden and Smith thus emphasize that musical pleasure arises from perceiving the relationship between sounds heard in the moment and a key note retained in the mind.

To demonstrate the unusual nature of this approach, let us consider the account of the pleasures of music provided by the aforementioned Hartley, which is in many ways far more typical of eighteenth-century (English) thought:

---

[87] This account is reminiscent of Smith's notion of sympathy as presented throughout *The Theory of Moral Sentiments* (1795), whereby the impartial spectator must feel a simulacrum of the passion or emotion undergone by the sufferer in order to experience compassion. On the relationship between the *Theory of Moral Sentiments* and the *Essays* see Boris Voigt, "Musikästhetik für den Homo oeconomicus: Adam Smith über Gefühle, Markt und Musik," *Zeitschrift für Ästhetik und allgemeine Kunstwissenschaft* 58, no. 1 (2013): 97–120.

[88] Smith, *Philosophical Essays*, 171.

[89] Holden, *Essay*, part 1, §25, 18.

[90] Ibid., part 1, §30, 26.

> The pleasures of music are composed . . . partly of the original, corporeal pleasures of sound, and partly of associated ones. When these pleasures are arrived at tolerable perfection, and the several compounding parts cemented sufficiently by association, they are transferred back again upon a great variety of objects and ideas, and diffuse joy, good-will, anger, compassion, sorrow, melancholy, [etc.] upon the various scenes and events.[91]

For Hartley, pleasure arises in the first place from the sensuous nature of the sounds and then from the memories or associations they evoke. Here, the mind is truly passive, responding in a quasi-mechanical way to purely physical stimuli and their specific associations. Hartley thus offers a fundamentally different portrayal of the role of the mind and of listening from that of Smith, who continues:

> By means of this relationship [to the key note], each foregoing sound seems to introduce, and as it were prepare the mind for the following: by its rhythmus, by its time and measure, it disposes that succession of sounds into a certain arrangement, which renders the whole more easy to be comprehended and remembered.[92]

This passage explains how syntactical features—such as the specific arrangement of tones in relation to a tonic and the organization of sounds into rhythmic units—assist the mind in understanding and retaining the music. This subject comes up frequently in Holden's *Essay*, and indeed is the driving force behind the psychological speculations in the second part of the work. Moreover, Holden explicitly conceives of his theory in these terms, writing that "to shew how the several chords which are admitted in harmony, are connected with another, in every strain which gives us pleasure, may not improperly be called the *Syntax* of music."[93]

Next, Smith observes:

> Time and measure are to instrumental Music what order and method are to discourse; they break it into proper parts and divisions, by which we are enabled both to remember better what is gone before, and frequently

---

[91] Hartley, *Observations on Man*, 233.
[92] Smith, *Philosophical Essays*, 171.
[93] Holden, *Collection*, 4. Original italics.

152   HEARING WITH THE MIND

to foresee somewhat of what is to come after; we frequently foresee the return of a period which we know must correspond to another which we remember to have gone before; and, according to the saying of an ancient philosopher and musician, the enjoyment of Music arises partly from memory and partly from foresight.[94]

In this passage, Smith compares rhythm and meter to grammatical divisions, which assist in both retention and anticipation. He explicitly links the pleasure of music to our ability to remember what has already occurred while predicting what is to come. A compact formulation of this idea can be found in Holden's preface to his 1766 collection of hymns, where he asserts, "Our enjoyment of music depends, in a great measure, on a faculty of retaining the ideas of former sounds, and anticipating those which are to follow."[95]

Smith further maintains:

Without this order and method we could remember very little of what had gone before, and we could foresee still less of what was to come after; and the whole enjoyment of Music would be equal to little more than the effect of the particular sounds which rung in our ears at every particular instant.[96]

While our enjoyment depends to some considerable degree on memory and anticipation, it is the fact that the sounds are already highly organized that makes it possible for those sources of pleasure to operate in us when we hear music. Holden makes a similar observation in the quite different context of his own treatment of scale-degree qualities: "If we were only to consider musical sounds singly, without any regard to their relations to, or dependencies upon each other; no such properties as these [that is, scale degree

---

[94] Smith, *Philosophical Essays*, 171. The "ancient philosopher" cited by Smith may be Aristoxenus, who writes: "Comprehension of music comes from two things, perception and memory: for we have to perceive what is coming to be and remember what has come to be." Andrew Barker, *Greek Musical Writings* 2: *Harmonic and Acoustic Theory* (Cambridge: Cambridge University Press, 1989), 155. Alternately, Birgit Klose maintains that Smith here refers to Augustine; see Klose, *Die erste Ästhetik der absoluten Musik*, 117–118. Similar ideas about the pleasures of anticipation were also articulated by the Aberdeen philosopher Alexander Gerard in his *Essay on Taste* (1756), namely: "Whenever our pleasure arises from a succession of sounds, it is a perception of a complicated nature; made up of a *sensation* of the present sound or note, and an *idea* or remembrance of the foregoing, which, by their mixture and concurrence, produce such a mysterious delight, as neither could have produced alone. It is often heightened, likewise, by an *anticipation* of the succeeding notes. Hence it proceeds in part that we are in general best pleased with pieces of music which we are acquainted with." Gerard, *Essay on Taste*, 61–62. Original italics.

[95] Holden, *Collection*, 4.

[96] Smith, *Philosophical Essays*, 172.

qualities] could be attributed to any one sound more than another."[97] Reid, we recall, claims that the operations of our senses are ephemeral without the functioning of memory and attention to lend them continuity. Here, similarly, both Smith and Holden envision counterexamples to normal musical perception, in which without the aid of memory and attention, sounds impact us merely as disconnected instantaneous stimuli and hence are not truly *heard*.[98]

Smith then encapsulates the manner in which the various aspects of instrumental music come together to affect the mind.

> A well-composed concerto of instrumental Music, by the number and variety of the instruments, by the variety of the parts which are performed by them, and the perfect concord or correspondence of all these different parts; by the exact harmony or coincidence of all the different sounds which are heard at the same time, and by that happy variety of measure which regulates the succession of those which are heard at different times, presents an object so agreeable, so great, so various, and so interesting, *that alone, and without suggesting any other object, either by imitation or otherwise, it can occupy, and as it were fill up, completely the whole capacity of the mind, so as to leave no part of its attention vacant for thinking of any thing else.*[99]

A sonorous object, Smith suggests, can completely occupy the attention without requiring any kind of programmatic supplement, and indeed, he later observes, any additional stimuli would only disturb us by distracting our attention.[100] There is no equivalent to this claim in Holden's *Essay*. While Holden acknowledges the possibility that instrumental music can be devoid of a particular expression, he does so ruefully, maintaining that "there may be pieces of music, as there are pictures, which are, simply, fine and pretty, without aiming directly at any particular expression[,] and among this class, it is to be feared, too many of our instrumental pieces may justly be ranked."[101] Although the few words Holden dedicates to expression elsewhere in his

---

[97] Holden, *Essay,* part 1, §21, 15.

[98] Reid, *EIP*, 324.

[99] Smith, *Philosophical Essays,* 172. My italics. This passage is remarkably similar to Robertson's aforementioned paean to instrumental music. Robertson, *Inquiry*, 446–447.

[100] "Such instrumental Music, not only does not require, but it does not admit of any accompaniment. A song or a dance, by demanding an attention which we have not to spare, would disturb, instead of heightening, the effect of the Music." Smith, *Philosophical Essays,* 172.

[101] Holden, *Essay,* part 1, §123, 108.

154 HEARING WITH THE MIND

treatise mostly recapitulate standard tropes, his entire project is a bold attempt at providing a theoretical account of syntactical ways in which we perceive abstract relationships between sounds, regardless of programmatic content.[102]

Listening with Attention

The similarities in the thinking of Reid, Holden, and Smith constitute a compact and suggestive case study in the circulation of concrete ideas about perception and listening between philosophy and music. Taken together, their innovative works suggest that the Scottish Enlightenment attitude to psychology, in particular as exemplified by Reid, enabled a new kind of theorizing about the musical experience, one that foregrounded the importance of the faculty of attention in the process of perceiving sound and music—whether in the context of Smith's introspective analysis of the experience of hearing instrumental music, or in that of Holden's proto-cognitive account of our ability to grasp and enjoy tonal relations.

To see Smith thus, not as an outlier but as a participant in a broader conversation about listening and attention, has enabled us better to understand the context from which his original ideas emerged. The same widened perspective also allows us to trace the legacy of Reid's psychology and Holden's proto-cognitive concepts in the writings of subsequent generations of musical and philosophical thinkers.[103] Finally, the hypothesis I have presented here—that Reid's ideas on attention, reconfigured into a theory of musical

---

[102] Ibid., part 1, §104, 89–90.

[103] To give a concrete example from each domain: the Scottish clergyman and musician Walter Young, to whom we will turn in the next chapter, applies the faculty of attention, alongside other mental operations, to theorize musical and poetic rhythm, while Reid's student, the philosopher Dugald Stewart, addressed both the case of the unheard clock chime and the experience of listening to different parts of a musical texture in his own investigation of attention. See Walter Young, "Essay on Rythmical Measures," *Transactions of The Royal Society of Edinburgh* 2, no. 2 (1790): 55–110. Stewart writes: "It is commonly understood, I believe, that in a concert of music a good ear can attend to the different parts of the music separately, or can attend to them all at once, and feel the full effect of the harmony. If the doctrine, however, which I have endeavoured to establish, be admitted, it will follow, that in the latter case the mind is constantly varying its attention from the one part of the music to the other, and that its operations are so rapid, as to give us no perception of an interval of time." Dugald Stewart, "Of Attention," in *Elements of the Philosophy of the Human Mind* (London: A. Strahan and T. Cadell, 1792), 1:130. In his treatment of the case of the unheard clock, Stewart dissents from Reid by arguing that we can at times be conscious of a perception without recollecting it. He includes an amusing counter example: if a preacher were suddenly to stop his sermon, all the sleeping congregation members would wake up with a start, but they would be unable to remember what they had heard. Ibid., 104.

cognition by Holden, influenced Smith's thinking about music—suggests that the interchange between eighteenth-century Scottish philosophy and music theory were both reciprocal and remarkably rich.

Studying the advent of new attitudes toward the perception of music in late eighteenth-century Scotland thus allows us to historicize a key episode in the history of audition, namely the emergence of a discourse concerned with focused listening to instrumental music, by showing how Reid's innovations in theorizing perception enabled both Holden and Smith to arrive at conclusions that would only surface again many decades later, in different cultures and philosophical contexts. The originality of their ideas in the late eighteenth-century landscape, as well as the ways in which Smith, in particular, anticipates subsequent notions about musical autonomy in the writings of figures such as Eduard Hanslick, testifies to the fact that these novel approaches to the faculty of attention enabled the articulation of a conception of music that still resonates today.[104]

---

[104] Thomas Grey, review of *Absolute Music: The History of an Idea*, by Mark Evan Bonds, *Music Theory Spectrum* 38, no. 1 (2016): 128.

5

Rhythm as a Universal "Science of Man"

Walter Young's "Essay on Rhythmical Measures" (1790)

"The rhythmical constitution of man, which, being a part of his nature, must be fundamentally the same, in all ages and amongst all nations."[1] This quintessentially eighteenth-century claim is made by the Scottish cleric and musician Walter Young (1745–1814) in his "Essay on Rhythmical Measures," one of the most remarkable music-theoretical documents of the Scottish Enlightenment. By "rhythmical constitution," Walter (I use his first name henceforth to distinguish him from his sister, Anne, the subject of the next chapter) means mental capacities and operations of the kind that are today studied in the field of music cognition. His goal in the "Essay" is to reveal exactly how our innate "rhythmical constitution" determines the rhythmic properties of works of music and poetry and contributes to our appreciation of them. Although the bulk of the "Essay" deals with the perception of rhythm in the domain of music, it provides an account of poetic rhythm as well. It also includes original conjectures as to the development of rhythmic preferences across time and culture, and it tries to reconcile reports of ancient Greek music and poetry with the universal cognitive principles which, in Walter's view, hold across all times and places.

First read to the Royal Society of Edinburgh in 1786, the "Essay" attracted a moderate amount of attention upon publication in the Society's *Transactions* four years later, garnering praise in *The Critical Review* (which declared the work "highly worthy [of] the attention"); *The Monthly Review* (which lauded it as "ingenious and well-written"); and even overseas, with the *Göttingische Anzeigen von gelehrten Sachen* commenting on its "very fine psychological and aesthetic observations."[2] In his entry on the "Essay" in

---

[1] Young, "An Essay on Rythmical Measures," 99. Henceforth in quoting from Young, I silently modernize his spelling of "rythm" as "rhythm."

[2] *The Critical Review or Annals of Literature* 1 (February 1791): 126; *The Monthly Review or Literary Journal* 5 (June 1791): 193; *Göttingische Anzeigen von gelehrten Sachen* 90 (June 4, 1791): 903. The latter's original text is "enthält sehr feine Bemerkungen psychologischer und ästhetischer Art über den Tonfall, und seine Grundregeln."

Hearing with the Mind. Carmel Raz, Oxford University Press. © Oxford University Press 2025.
DOI: 10.1093/9780197786208.003.0006

RHYTHM AS A UNIVERSAL "SCIENCE OF MAN"   157

the *Allgemeine Literatur der Musik*, Johann Nikolaus Forkel reproduced the *Göttingische Anzeigen*'s favorable remarks on the work's psychological and aesthetic observations,[3] while Forkel's student, the British polymath Thomas Young (of no relation), deemed it "excellent."[4]

Among modern scholars, however, Walter's "Essay" has thus far been of interest almost exclusively to students of prosody, in which field there has been a steady trickle of scholarly engagement with the work since the 1900s,[5] mostly in studies concerning the groundswell of mid- and late-eighteenth century applications of musical concepts to poetic meter.[6] In the domain of music, in contrast, the "Essay" has received little comment apart from Jamie C. Kassler, who summarizes some of the work's key findings and comments on its philosophical orientation in *The Science of Music in Britain, 1714–1830*.[7] This reception is surprising given that the innovations of the "Essay" seem largely, if not entirely, confined to its musical part, which takes up the bulk of the treatise.[8]

---

[3] Forkel writes, "Eine Abhandlung . . . welche sehr feine psychologische und ästhetische Bemerkungen enthält" and cites the *Göttingische Anzeigen von gelehrten Sachen*. Johann Nikolaus Forkel, *Allgemeine Literatur der Musik* (Leipzig: Schwickert, 1792), 512.

[4] Thomas Young, "An Essay on Music," *British Magazine* (October 1800); reprinted in Thomas Young, *Miscellaneous Works of the Late Thomas Young* (London: J. Murray, 1855), 1:116. In his correspondence, moreover, T. Young noted that Walter had "treated the subject [of rhythm] in a very masterly manner, and his dissertation is well worthy [sic] the attention of critics as well as musicians." Thomas Young, "Letter to Andrew Dalzel of July 8, 1798," in *Memoir of Andrew Dalzel, Professor of Greek in the University of Edinburgh*, ed. Cosmo Innes (Edinburgh: 1861), 161.

[5] Verdicts veer from the rapturous (Omond enthused, "would that there were more such essays to cite"), to the dismissive (Fussell disparages Young as "a strikingly conservative metrist"). Thomas Stewart Omond, *English Metrists: Being a Sketch of English Prosodical Criticism from Elizabethan Times to the Present Day* (Oxford: Clarendon Press, 1903), 103–104; Paul Fussell, *Theory of Prosody in Eighteenth-Century England* (Hamden: Archon Books, 1966 [1954]), 50. See also George Saintsbury, *A History of English Prosody from the Twelfth Century to the Present Day* (London: MacMillan, 1908), 2:549; Fussell, *Theory of Prosody*, 41n, 46n, 50, 60, 85; Magdalena Sumera, "The Temporal Tradition in the Study of Verse Structure," *Linguistics* 8, no. 62 (1970): 46–48; Richard Bradford, *Silence and Sound: Theories of Poetics from the Eighteenth Century* (Rutherford: Fairleigh Dickinson Press, 1992), 60.

[6] See, e.g., John Mason, *Essays on Poetical and Prosaic Numbers and Elocution*, 2nd ed. (London: J. Buckland, 1761); Daniel Webb, *Observations on the Correspondence between Poetry and Music* (Dublin: James Williams, 1769); James Beattie, *Essays On Poetry and Music, as They Affect the Mind, on Laughter, and Judicious Composition, on the Utility of Classical Learning* (Edinburgh: W. Creech, 1776); John Walker, *The Melody of Speaking* (London: The Author, 1787); Anselm Bayley, *The Alliance of Music, Poetry, and Oratory* (London: J. Stockdale, 1789); William Mitford, *An Inquiry into the Principles of Harmony in Language*, 2nd revised ed. (London: T. Cadell, 1804). This list is adapted from footnote 41 in David Perkins's article "How the Romantics Recited Poetry," *Studies in English Literature, 1500–1900* 31, no. 4 (1991): 670.

[7] Kassler, *The Science of Music in Britain* 2:1096–1099. Brown also briefly discusses the role of the internal senses in the "Essay" in *Artful Virtue*, 156.

[8] The treatment of poetic meter takes up about one third of the total length of the "Essay." Saintsbury describes the work's originality as "purely musical." Saintsbury, *A History of English Prosody*, 549. There are precedents to this view: a short item on the "Essay" in *The Analytical Review* asserts, "We believe him to be guilty of few, if any mistakes; but neither does he appear to have made

158   HEARING WITH THE MIND

The "Essay" profoundly reflects its author's involvement with the social circles and philosophical discourses most characteristic of the Scottish
Enlightenment—particularly its ideology of a universal "science of man." This
is evident not only in Walter's interest in deducing universal principles of
rhythmic perception, but also in his employment of the methodology of conjectural history as a way of generalizing about human cognitive capacities across
time and place based on observations drawn from his own day and culture.

Walter was a figure of considerable prominence in late-eighteenth-century
Scottish musical life, "distinguished for his profound and scientific knowledge of harmony,"[9] and enjoying the reputation of being "the most splendid
private [that is, amateur] musician of his day."[10] Born in Haddington, near
Edinburgh, in 1745, he was the son of David Young (d. 1758), the rector of
the local grammar school, a distinguished local institution that catered to
the sons of "noble and honorable families," some of whom boarded with
the Youngs.[11] Sociologist-historian Charles Camic has proposed that the
beginnings of Scottish universalist approaches were cultivated by village
grammar schools, exactly like the one Walter's father directed, which in the
eighteenth century began to accept wealthier boarders while also teaching
local children.[12] It seems plausible to assume that Walter's worldview could

---

any discoveries," a verdict that makes more sense if one considers only the work's literary section.
The Analytical Review, or History of Literature, Domestic and Foreign 8 (September-December,
1790): 263.

[9] Philip A. Ramsay, *The Works of Robert Tannahill. With Life of the Author, and Memoir of Robert
A. Smith, the Musical Composer* (London and Edinburgh: A. Fullarton & Co., 1838), xlviii.
[10] Robert Walter Stewart, "Parish of Erskine," in *The New Statistical Account of Scotland: pt. 1–2
Renfrew, Argyle*, ed. John Gordon (Edinburgh and London: W. Blackwood and Sons, 1845), 8:514.
Walter taught the well-known Scottish composer Robert Archibald Smith harmony, and the latter's
collection, *Sacred Music . . . sung in St. George's Church, Edinburgh* (1825), contains three hymns
which he ascribes to Young. See Robert Archibald Smith, *Sacred Music . . . Sung in St. George's Church,
Edinburgh* (Edinburgh: R. Purdie, 1825), 55ff, 87ff, 98ff.
[11] David Rae, Lord Eskgrove (1729–1804), later Lord-Justice-Clerk of Scotland, was one such student boarder; see Alexander Stephens, *Public Characters, 1801–1802* (London: Richard Philipps,
1807), 570. Other distinguished alumni of the grammar school during David Young's tenure include the hydrographer Alexander Dalrymple F.R.S. and Robert Graham, Lord Fintry. See Robert
Chambers, *A Biographical Dictionary of Eminent Scotsmen*, vol. 2 (Glasgow, Edinburgh, and
London: Blackie & Son, 1853); Sophia Crawford Lomas, "The Manuscripts in the Possession
of Sir John James Graham of Fintry, K.C.M.G.," in *Report on Manuscripts in Various Collections* 5
(Hereford: His Majesty's Stationary Office, 1909), 222. This David Young should not be confused
with the Edinburgh-based writing master, who had worked as a music compiler and copyist for various Scottish dignitaries.
[12] In such settings, Charles Camic observes, "the heterogeneous group of pupils that populated
the town grammar schools repeatedly experienced treatment by uniform standards." Charles Camic,
"Experience and Ideas: Education for Universalism in Eighteenth-Century Scotland," *Comparative
Studies in Society and History* 25, no. 1 (1983): 61. One should note, of course, that schooling continued to be largely restricted to the middle and upper classes (with some exceptions), since poor

RHYTHM AS A UNIVERSAL "SCIENCE OF MAN"   159

have been influenced by the relatively new intermingling of classes he would have experienced at home and at school.

Walter's universalist outlook would have been further reinforced during his divinity studies in Edinburgh, where, around 1761, he joined one of the city's prestigious intellectual clubs, the Theological Society. This club, and other societies like it, were famously the hotbeds of the Scottish Enlightenment, in that they provided a regular forum where individuals— nearly always men—hailing from different social classes (though nearly always middle and higher) and of varying occupations could converse and debate in an egalitarian format.[13] The meetings of Scottish intellectual clubs generally featured a mix of formal lectures and open questions for debate proposed by members, invariably followed by dining and drinking. As historian David Daiches emphasizes, "However conscious of rank and social hierarchy in their normal behaviour, in the club life of eighteenth-century citizens there could be found varieties of social freedom not to be found in formal dinner parties."[14] Little is known today about the Theological Society other than that it met weekly at the college and drew its membership from the ranks of divinity students.[15] It seems to have functioned like comparable

families generally relied on the labor of their offspring from childhood on. See Robert A. Houston, *Scottish Literacy and the Scottish Identity: Illiteracy and Society in Scotland and Northern England, 1600–1800* (Cambridge: Cambridge University Press, 2002), 125–126.

[13] On the importance of social clubs to the Scottish Enlightenment see Wallace and Rendall, eds., *Association and Enlightenment*. The Pantheon Society admitted women as auditors from 1775, but they were prohibited from participating in the debates. See Rosalind Carr, *Gender and Enlightenment Culture in Eighteenth-Century Scotland* (Edinburgh: Edinburgh University Press, 2014), 85–86.

[14] David Daiches, *The Scottish Enlightenment: An Introduction* (Edinburgh: Saltire Society, 1986), 29. These exchanges, it should further be noted, took place at a time when stark social divisions affected not only the poor and working classes, but also the landowners and gentry; in Scotland in the eighteenth century, as Robert A. Houston reminds us, "only one adult male in a thousand could vote in parliamentary elections." Robert A. Houston, "Popular Politics in the Reign of George II: The Edinburgh Cordiners," *The Scottish Historical Review* 72, no. 194.2 (1993): 167. Gentry and landowners were prohibited from voting unless they possessed crown land valued at £400 Scots in 1743; see Roland G. Thorne, *The House of Commons, 1790–1820* (Woodbridge: Boydell & Brewer, 1986), 3:70–71.

[15] In the only surviving account of the club, the Scottish minister Thomas Somerville described it as "not only a school of mental improvement, but a nursery of brotherly love and kind affections." Thomas Somerville, *My Own Life and Times, 1741–1814* (Edinburgh: Edmonston & Douglas, 1861), 42. The Theological Society's rather austere name notwithstanding, the second goal seems to have overtaken the first, and the club dissolved five years after its founding, because, as Somerville recounted, the "tavern adjournments, which succeeded our weekly meetings in the college, were the cause of expense, and sometimes of excess and irregularity, unsuitable to our circumstances and professional views." Ibid., 44.

160 HEARING WITH THE MIND

Scottish institutions in that its chief purposes were at once intellectual inquiry and social exchange.[16]

In addition to the Theological Society, Walter joined the Glasgow Literary Society in 1779,[17] having been ordained minister of the village of Erskine, near Glasgow, seven years earlier. The Glasgow Literary Society—founded by Adam Smith, among others—met weekly to debate intellectual matters. It boasted some of the Scottish Enlightenment's leading lights among its members, including David Hume, Thomas Reid, John Robison, and John Anderson. Subjects of discussion ranged from queries concerning religion and literature to topics in chemistry, philosophy, medicine, politics, commerce, and education. The minutes record a question by Walter himself, namely "what is the difference betwixt the melody of Speech and music properly so called,"[18] as well as at least two discourses he delivered to the group, the first entitled "Observations on syllables and upon the rhythmical structure of language," and the second "On rhythm in continuation."[19] While Walter's contributions seem to be the only manifestly musical topics of inquiry taken up by the Glasgow Literary Society, related subjects pertaining to aesthetics and to psychology were frequently considered. Thus, for example, in the 1780s and 1790s, Adam Smith delivered his discourse "On the Imitative Arts" to the Society, and Archibald Arthur and James Moor delivered lectures on the fine arts. Moreover, Glasgow University professors Thomas Reid and George Jardine, as well as the Reverend Dr. William Taylor, explored themes such as free will, attention, and mental habits.

Over the next few decades, Walter attained the distinctions typically accorded to well-respected Scottish intellectuals.[20] In 1784 he was elected

---

[16] According to historian Peter Clark, late-eighteenth-century Edinburgh was home to over 200 social clubs of at least forty varieties; see Peter Clark, *British Clubs and Societies, 1580–1800. The Origins of an Associational World* (Oxford: Oxford University Press, 2000), 131.

[17] Roger Emerson's transcription of Glasgow Literary Society minutes indicates that Walter attended as a guest at least twice before his election, on April 17, 1778. Roger Emerson, *Neglected Scots: Eighteenth Century Glasgwegians and Women* (Edinburgh: Humming Earth, 2015), 108.

[18] Ibid. Walter did not deliver this question to the floor, however, and a different question by the philosopher John Millar was debated on that day.

[19] Ibid., 115, 124. Roger Emerson's transcription of the Glasgow College Literary Society's minutes include a number of additional questions and discourses that may have been delivered by Walter, but are attributed to John Young. These are: "On syllables and the rhythmical structure of language," (January 6, 1797), which is clearly a continuation of Walter's earlier discourse ("Observations on syllables and upon the rhythmical structure of language," delivered January 16, 1795), as well as questions such as "Is the present Notation of Music capable of Improvement," proposed on January 22, 1779, and "In what sense is it true that certain persons have no musical ear," proposed on April 16, 1779, which Emerson indicates as proposed by John, but which seem far closer to Walter's interests. Ibid., 118, 110, and 112, respectively.

[20] In 1793, Walter also contributed a description of the parish of Erskine to *The Statistical Account of Scotland Drawn Up from the Communications of the Ministers of the Different Parishes*, where he

to the Royal Society of Edinburgh, whose early membership represented "a cross-section of the Scottish political and cultural power-wielding elite," to quote Steven Shapin.[21] Walter's stellar religious and intellectual connections are further evident in that his sponsor for membership were the Reverend William Robertson, a distinguished historian and the Principal of Edinburgh University, the Very Reverend Alexander Carlyle, a leader in the Scottish Presbyterian Church, and Andrew Dalzell, professor of Greek at Edinburgh University. In 1796, he was awarded the honorary degree of Doctor of Divinity from Glasgow University.[22]

A Universal "Science of Man"

As a great many scholars have observed, the efflorescence of interest in the universal principles of mankind in eighteenth-century Scotland came about in large part as a reaction to the restrictions of the Scottish Reformation and especially the dogmas of Scottish Calvinism, which held that the Scottish church was the purest form of Christianity.[23] This strain of Scottish particularism continued well into the eighteenth century, only to be thoroughly rejected by the philosophers most closely associated with the Scottish Enlightenment, who abandoned the idea of "a world divided into non-Christians and Christians, unreformed and reformed Christians, non-Scottish and Scottish reformed Christians, and episcopalian and

---

reported on the parish's history as well as on its geographical, agricultural, economic, and demographic features. His piece is notable not only for including a brief mention of one of the last trials of witchcraft in Scotland in 1697, which saw criminal proceedings launched against seven members of the parish, but, more importantly, for featuring an account of a woman from the parish who initiated a profitable business of spinning fine yarn into string. Walter describes this initiative with considerable respect, an attitude which harmonizes with his support of his sister in her own entrepreneurial ventures (see Chapter 6). Walter Young, "Statistical Account No. V: Parish of Erskine," in *The Statistical Account of Scotland Drawn Up from the Communications of the Ministers of the Different Parishes*, vol. 9, ed. John Sinclar (Edinburgh: William Creech, 1793), 75.

[21] Steven Shapin, "Property, Patronage, and the Politics of Science: The Founding of the Royal Society of Edinburgh," *The British Journal for the History of Science* 7, no. 1 (1974): 37.

[22] Also in 1796, University of Glasgow professor John Anderson nominated Walter as professor of music in the university he intended to have founded with his bequest (see Chapter 1, note 77). This post does not seem to have been actually established with the founding of the Andersonian Institute later that year. John Anderson, "Will of John Anderson (so far as it relates to the institution) dated 7th May 1795," reprinted in *Local and Personal Acts of Parliament* (London: H.M. Stationery Office, 1877), 170. On the absence of a position in music see James Cleland, *Annals of Glasgow, Comprising an Account of the Public Buildings* (Glasgow: James Hederwick, 1816), 2:122.

[23] A review of the theological background relevant to this point is provided in Camic, "Experience and Ideas," 50–82.

162   HEARING WITH THE MIND

presbyterian Scottish reformed" in favor of a commitment to universal human brotherhood.[24] While the kinds of persons included in conceptions of universality varied between different thinkers, it is safe to say that the second half of the century saw the emergence of a new, "Enlightened" attitude in Scotland, one that, as Hume put it, assumed that "there is a great uniformity among the actions of men, in all nations and ages, and that human nature remains still the same, in its principles and operations."[25]

This regularity and consistency of human nature, in turn, was viewed as providing a sufficient justification for the emergence of conjectural history, the historical methodology most closely linked with Scottish Enlightenment thinkers between 1740 and 1800. Conjectural history, the Scottish strain of the *histoire raisonée* practiced by on the Continent by writers such as Jean-Jacques Rousseau and Johann Gottfried Herder,[26] was perhaps most canonically defined by the philosopher Dugald Stewart in 1794.[27] Stewart writes:

> In the want of direct evidence, we are under a necessity of supplying the place of fact by conjecture; and when we are unable to ascertain how men have actually conducted themselves upon particular occasions, of considering in what manner they are likely to have proceeded, from the principles of their nature, and the circumstances of their external situation.[28]

[24] Ibid., 54–55.

[25] David Hume, *An Enquiry Concerning Human Understanding* (1748), §8, 133. Hume's approach, which would prove so foundational to the Scottish Enlightenment—and influential in the formation of the modern disciplines of psychology, economics, and the social sciences—must, of course, be seen against the backdrop of its day, an era profoundly shaped by deep-seated notions of racial, national, religious, and class differences. Indeed, white supremacist ideas were advanced by Hume himself, who saw no conflict between universalism on the one hand and rigid racial hierarchies on the other (see, e.g., David Hume, "Of National Characters," in *Essays, Moral and Political*, 3rd ed., corrected (Edinburgh and London: A. Millar, 1748), 267–288. Richard Popkin has argued that the "the theories outlining what Hume called the science of man were transformed to meet eighteenth-century conditions," and in particular the desire of European "Enlighteners" to assure that Caucasian men remained at the top of the racial hierarchy by dehumanizing non-Whites and non-Christians. Richard H. Popkin, "The Philosophical Basis of Eighteenth-Century Racism," *Studies in Eighteenth-Century Culture* 3 (1974): 247.

[26] See, e.g., Rousseau's "Essai sur l'origine des langues" in *Collection complète des œuvres de J. J. Rousseau* (Geneva: Volland, 1780) and Herder's *Die Abhandlung über den Ursprung der Sprache* (Berlin: Voß, 1772).

[27] It might be helpful to provide a modern definition of the term as well: Roger Emerson views "conjectural history" as denoting "any rational or naturalistic account of the origins and development of institutions, beliefs or practices not based on documents or copies of documents or other artifacts contemporary (or thought to be contemporary) with the subjects studied." He concludes that "all histories of the origins and progress of language, civil society and government, the origin of ranks, sciences such as astronomy, fine arts like painting or of pagan religions are by this definition wholly or in part conjectural." Roger L. Emerson, "Conjectural History and Scottish Philosophers," *Historical Papers/Communications historiques* 19, no. 1 (1984): 65.

[28] Dugald Stewart, *Biographical Memoirs, of Adam Smith, LL. D., of William Robertson, D. D. and of Thomas Reid, D. D.* (Edinburgh: G. Ramsay, 1811), 48.

RHYTHM AS A UNIVERSAL "SCIENCE OF MAN"   163

The abundance of Scottish inquiries into the origins of institutions—moral, social, linguistic, and scientific—following the belief that their developmental trajectory could be ascribed to, and explained by, universal principles of human nature, rendered them a conspicuous feature of intellectual discourse in Edinburgh, Glasgow, and Aberdeen.[29]

Probably along the lines of the debates he would have frequently heard as a member of the intellectual societies in Glasgow and Edinburgh, Walter begins by situating his subject within a well-established philosophical *topos*: the role of the so-called internal senses, or "higher faculties of human nature," in determining the relationship between pleasure and sensory perception, specifically in vision and hearing.[30] He explains that his investigation will center on our internal sense for perceiving order and proportion, the "power" that enables us to recognize these external properties, which "are generally discovered, in a certain degree, in every thing which communicates immediate pleasure, either to our sight or hearing."[31] He observes that these capacities are unique to humans, and are differently distributed owing to ability, habit, and cultivation, which is to say, they are affected both by nature and by nurture.[32] Walter goes on to distinguish, with regard to order and proportion, the domains of pitch and time, regarding the latter as more important than the former, as we can obtain considerable enjoyment from the mere beating of a drum.[33] Moreover, variation in ability among individuals

---

[29] Conjectural history, as practiced by Scottish thinkers, often made use of a stadial framework, whereby humanity was regarded as preceding through various stages in its development. As Smith most canonically put it: "There are four distinct states which mankind pass thro: - 1st, the Age of Hunters; 2ndly, the Age of Shepherds; 3rdly, the Age of Agriculture; and 4thly, the Age of Commerce." Adam Smith, *Lectures on Jurisprudence (1763)*, ed. R. L. Meek, D. D. Raphael, and Peter Stein (Oxford: Clarendon Press, 1978), 14. Adam Ferguson, on the other hand, presented a three-stage model of progress from "savage" through "barbarous" to "polished." Adam Ferguson, *An Essay on the History of Civil Society* (Edinburgh: A. Kincaid and J. Bell, 1767).

[30] Young, "Essay," 55.

[31] Ibid., 56. By the eighteenth century, the traditional medieval understanding of internal senses had been replaced in Scottish philosophy by an approach, inspired by Locke, that regarded the internal senses as an integral part of aesthetic perception. Most famously, in his *Essay on the Conduct and Nature of the Passions and the Affections* (1728), the Scottish philosopher Francis Hutcheson regarded our internal senses as "pleasant perceptions arising from regular, harmonious, uniform objects; as also from grandeur and novelty." He elaborated upon the definition of these in *A System of Moral Philosophy* (1755) to include distinct inner senses of perceiving beauty, grandeur, harmony, novelty, order, and design. See Peter Kivy, *The Seventh Sense: Francis Hutcheson and Eighteenth-Century British Aesthetics* (Oxford: Oxford University Press, 2003), 34–36.

[32] Young, "Essay," 55.

[33] Here Walter may have been influenced by Descartes, who similarly maintained that "time in music has such power that it alone can be pleasurable by itself; such is the case with the military drum, where we have nothing [to perceive] but the beat." Descartes, *Compendium of Music*, 15. He could have also encountered a related claim in Alexander Gerard's *Essay on Taste* (1759). Gerard writes: "As the great force of proportion in time is evident from the universal attention that is paid

164 HEARING WITH THE MIND

is less drastic with regard to the sense of rhythm than to that of pitch, he asserts, and "though men possess, in different degrees, the power of feeling the proportional duration[s] of successive sounds, and of relishing an agreeable rhythm, there is perhaps no man altogether destitute of it."[34]

Walter next divides his "Essay" into two sections: the first examines our "rhythmical powers," i.e., the specific abilities that allow us to perceive "the proportional magnitudes of small intervals of time when these are marked out by motion, or by successive sounds."[35] The second explores musical practice, and specifically, the arrangement of sounds into what he calls phrases, as well as pleasurable divergences from regularity that are found in particularly ingenious works. In each of these sections, Walter treats both musical and poetic rhythm, though his main interest is clearly in the former, as can be seen in his bold importation of certain musical ideas onto poetic contexts.[36] The present chapter primarily treats his ideas about musical rhythm.

Recognizing, Dividing, and Compounding Equal Time Intervals

Walter commences his investigation by observing that our rhythmical powers depend on our ability to recognize some fixed unit or interval, of which the ensuing sounds are "either an aliquot part or a multiple."[37] Given that music, unlike poetry, deals with highly complex proportions, he argues that such a referential duration is necessary in order that "proportions may be felt, and that uniformity may be perceived amidst this variety."[38] The

---

to it in music of every kind; so the influence of variety of time appears particularly in the drum, the whole music of which is owing to it alone." Gerard, *Essay on Taste*, 62.

[34] Young, "Essay," 57.

[35] Ibid., 66.

[36] Most importantly, Walter distinguishes himself from any other eighteenth-century prosodist that I am aware of by proposing to do away with measures that begin with an unaccented foot in favor of measures that begin with accents. He asserts: "all iambic verses are to be considered as trochaics having a feeble syllable introductory to the measure, and ought always to be so scanned," and he even provides an idiosyncratic scansion of a famous line of iambic pentameter (five feet composed of weak-strong measures) by Alexander Pope that instead regards it as an unaccented syllable followed by four groups of trochees and ending with a strong syllable that reads thus: | To | wake the | soul by | tender | strokes of | art |. To my eyes, this scansion is highly reminiscent of the musical practice of beginning a phrase with an upbeat and subtracting the value of the upbeat at the end of a phrase. Ibid., 77–78. I thank Courtney Weiss Smith for bringing this point to my attention.

[37] Ibid., 59. An aliquot part is any integral factor of a number. (The number 1 is by definition such a factor of *any* integer and is therefore customarily not counted among its aliquot parts.)

[38] Ibid.

RHYTHM AS A UNIVERSAL "SCIENCE OF MAN"   165

ability to distinguish isochronous intervals of time is therefore essential to our experience of temporal order and proportion.

Walter next links our psychological capacity with embodied experience by surmising that this ability arises in two stages: almost everyone, he begins, can perceive the isochrony of intervals of time, within certain limits. He continues by suggesting that "we acquire this idea of equal intervals of time, from the motion of our own limbs, and of those of other animals, in walking or flying, which nature, for the purposes of ease and grace, has determined to be an uniform motion."[39] Walter here emphasizes how volitional movements undertaken by our bodies—rather than, say, the regular, involuntary pulse of the human heartbeat—provide a recurrent experience that shapes the kinds of knowledge available to our minds. His approach thus echoes that of his contemporary, the influential English prosodist Joshua Steele, who described this process at slightly greater length:

> Every species of rhythmical sound can be ascertained by the standard of our step. And though the various paces of quadruped furnish us with rhythmical movements of jig triples and double *cadences* such as the *ra ta pat* and *ra ta pa ta*, which are not naturally made by bipeds, yet our habit of riding makes us almost as familiar with the measures beaten by the paces of horses as if they were our own.[40]

Both Walter and Steele assume that our sense of rhythm arising from bidirectional interaction between our minds and our physical bodies, a view that is similar to embodied cognition approaches in our current moment, which likewise hold that our cognition is profoundly shaped by our physical, embodied interactions with the world.[41]

---

[39] Walter here differs from fifteenth-century music theorists such as Ramis de Pareja, Gaffurio, or Zarlino, who analogized between the tactus and the pulse; see Bonnie J. Blackburn, "Leonardo and Gaffurio on Harmony and the Pulse of Music," in *Essays on Music and Culture in Honor of Herbert Kellman,* ed. Barbara Haggh (Paris: Minerve, 2001), 128–149; Ruth I. DeFord, *Tactus, Mensuration, and Rhythm in Renaissance Music* (Cambridge: Cambridge University Press, 2015); see Dale Bonge, "Gaffurius on Pulse and Tempo: A Reinterpretation," *Musica disciplina* 36 (1982): 167–174; and Nancy C. Siraisi, "The Music of Pulse in the Writings of Italian Academic Physicians (Fourteenth and Fifteenth Centuries)," *Speculum* 50 (1975): 689–710.

[40] Joshua Steele, *Prosodia Rationalis: Or, an Essay Towards Establishing the Melody and Measure of Speech, to be Expressed and Perpetuated by Peculiar Symbols* (London: J. Nichols, 1779), 126. Original italics.

[41] A summary of current trends in embodied cognition can be found in Lawrence Shapiro, *Embodied Cognition* (New York and London: Taylor and Francis, 2011).

166 HEARING WITH THE MIND

Walter goes on to assert that we subsequently refine our notions of temporal equality through repeated exposure to timekeeping devices such as clocks and pendulums, noting that "by a habitual attention to these, men come by degrees to have a very accurate perception of small equal intervals of time."[42] On this point, too, he expresses similar sentiments to Steele, who writes:

> Though the INSTINCTIVE SENSE of *periodical pulsation* is certainly coeval with our animal frame, yet the invention of the *pendulum* has made the moderns more accurate and expert in divisions of time than those antients who had no such help.[43]

I take Walter and Steele here to refer to concepts relating to absolute time, such as a second, which would have indeed been foreign to cultures without technologically enabled conventions of precise temporal measurement.[44] By linking rhythm to properties of our shared physical and mental experience, both Walter and Steele attempt to provide a universal account of how humans come to perceive durations.

Unlike Steele, however, Walter emphasizes that the role of our own bodily experience in forming the very idea of a unit of duration also imposes constraints on the possible length of such a unit. He writes:

> Although we may conceive an interval to be divided into any number of equal parts, the number of parts into which we actually can divide it must depend upon the powers which we have of performing quick motions. These powers are very limited. The roll of a drum, the most rapid movement of a musician upon an instrument, does not divide a second of time into much more than sixteen or eighteen equal parts, hardly ever into twenty-four. Our power of dividing a small interval of time, equally and uniformly, and of perceiving such a division by the ear, is also confined to certain proportions.[45]

---

[42] Young, "Essay," 59.

[43] Steele, *Prosodia Rationalis,* 127. Original italics and capitalization.

[44] Ibid., 126–127. On the dual associations of isochrony with both mechanism and organicism in this period see Marcus Tomalin, "Pendulums and Prosody in the Long Eighteenth Century," *The Review of English Studies* 68, no. 286 (2017): 734–755.

[45] Young, "Essay," 59–60.

According to Walter, then, the interdependence between our bodily and mental affordances limits our ability to conceive small intervals of time: we can only perceive units of the dimensions that we—as humans—can actually perform. The durations Walter provides, which lie between 1/16 or 1/18 of a second, and not less than 1/24 of a second, are equivalent to range of between 62 and 41 milliseconds, an estimate confirmed by recent empirical work in music cognition.[46] His account thus foregrounds the role of our bodies in shaping both our perception of equal units of time and our ability to subdivide: all of these powers are formed by concrete physical interaction with the world.

Walter's treatment of our ability to mark time is reminiscent of similar ideas discussed by his fellow member of the Glasgow Literary Society, Thomas Reid. In his investigation of our ability to perceive short temporal intervals, Reid remarks:

> It may be observed, that one who has given attention to the motion of a second pendulum will be able to beat seconds for a minute with a very small error. When he continues this exercise long, as for five or ten minutes, he is apt to err, more even than in proportion to the time, for this reason, as I apprehend, that it is difficult to attend long to the moments as they pass, without wandering after some other object of thought. I have found, by some experiments, that a man may beat seconds for one minute, without erring above one second in the whole sixty; and I doubt not but, by long practice, he might do it still more accurately. From this, I think it follows, that the sixtieth part of a second of time is discernible by the human mind.[47]

Reid here describes an experiment known to contemporary psychologists as a "rhythmic continuation task," wherein a subject entrains to a regular beat and then continues to produce this beat by tapping. Reid remarks on the "drift," as it is termed today, whereby the tapping becomes increasingly irregular, and he ascribes this a gradual attenuation of the attention. However, Reid offers a much more generous assessment than Walter of our mental ability to recognize small units of time by suggesting that we can perceive durations of 1/60 of a second, a number which he ascribes to the fact that he has observed people executing the aforementioned "rhythmic continuation

---

[46] See London, "Cognitive Constraints," 529–550.
[47] Reid, *EIP*, 331.

168 HEARING WITH THE MIND

task" accurately for a whole minute. As the cumulated error does not arrive at a full second, Reid then retroactively assumes that we can perceive durations of approximately 16 milliseconds. This, too, is a duration borne out by experimental work today, but only in the domain of vision, where recent research has placed our perceptual boundary at 13 milliseconds.[48] Walter's more embodied approach thus seems more apt than Reid's in the context of discussing musical and poetic rhythm. Considered together, the works suggest that an interest in the human perceptual limits vis-à-vis timing was characteristic of at least some strands of Scottish Enlightenment thought.

Walter then turns to the division of durational units. He writes:

> Our power of dividing a small interval of time, equally and uniformly, and of perceiving such a division by the ear, is also confined to certain proportions. We can divide an interval into two or into four equal parts with almost the same ease. Having obtained either of these divisions, we can also consider each of the parts as an [*sic*] unit, and subdivide it into two or four, thus making a division into eight, or, if our powers of quick motion will admit, into sixteen. Beyond this we cannot carry the powers of two in the division of single intervals.

I take him here to assert that we can conceive directly of a half, a fourth, an eighth, and perhaps even a sixteenth of an interval, but the next power, 32, exceeds our ability to maintain a division based on the powers of two, due not only to the aforementioned constraint on our powers of motion, but also to the cognitive effort this entails. That is to say, while partitioning an interval into two or four requires nearly identical mental exertion, this effort increases as the numbers grow larger. More practically: the duration equivalent to a thirty-second note of the original interval in question would need to be conceived in relation to a shorter time span, to which it would stand in a simpler proportion. Triple divisions, Walter observes, are not as intuitively easy as duple, although they become so by practice.[49] These simple factors

---

[48] Mary C. Potter, Brad Wyble, Carl E. Hagmann, and Emily S. McCourt, "Detecting Meaning in RSVP at 13 ms per Picture," *Attention, Perception, & Psychophysics* 76, no. 2 (2014): 270–279.

[49] Walter lists the divisions produced by factorization into only 2, 3, and 4, namely: 8 is 2×4; 16 is 4×4; 6 is 2×3 or 3×2; 9 is always 3×3; and 12 is 3×4, 4×3, or 2×6, because conceiving it as 6×2 is more difficult. This is nearly identical to Holden's discussion of visual grouping in his *Essay towards a Rational System of Music* (1770), described in Chapter 2 of this book and in Holden, *Essay*, part 2, §11, 288–289.

are occasionally complemented by divisions into five, which require "a considerable effort of the attention."[50]

In addition to dividing beats into equal parts, Walter goes on to observe, we can also aggregate the beats themselves into spans of longer duration.[51] We do this by counting off equal groups of beats and granting more attention to the first beat of every group. Greater notice causes us to perceive that beat of a group as louder than its members, even when the group's remaining beats are all identical, as in the case of the ticking of a watch. Walter remarks that this experience—a phenomenon known today as subjective rhythmicization—can be had with groups of two, three, four, and other numbers. He writes: "In every case we uniformly imagine the first of each parcel to be more forcible than the others."[52]

The insight that we involuntarily group pulse streams into duple or triple groups by mentally accenting the initial pulse of each group was relatively new at the time: it was first discussed by John Holden in his *Essay* of 1770, as we saw in Chapter 2, and independently described in the second volume of Johann Philipp Kirnberger's *Die Kunst des reinen Satzes in der Musik* (1776). While Walter does not cite any secondary sources in his essay, there is no reason to believe that he knew of Kirnberger's work, as he does not refer to any of the German author's other ideas.[53] We can, however, safely assume that he was well aware of Holden's *Essay,* as he clearly relies on Holden's ideas and terminology to a remarkable degree. Moreover, Walter's sister, Anne, refers glowingly to Holden's treatise in her own *Introduction to Music* (1803), as discussed in Chapter 6.

Walter expands upon contemporaneous accounts of this phenomenon by observing that while the same effect should hypothetically be true of groups of five and seven, in these cases, we always feel that the "accented" beat comes too soon.[54] In order to maintain these combinations, he writes, we must alternately shift between counting units of two and units of three (or units of three and units of four), an act which "requires such a constant and even painful effort of the attention, as is inconsistent with that ease and

---

[50] Young, "Essay," 60.

[51] Walter uses the terms "groups," "combinations," or "parcels" synonymously to describe these units.

[52] Young, "Essay," 62.

[53] See Holden, *Essay,* part 1, §95, 83–84 and part 2, §16, 293–294 and Kirnberger, *Die Kunst des reinen Satzes in der Musik,* II, 114–115.

[54] Young, "Essay," 62.

170 HEARING WITH THE MIND

simplicity of conception and operation, which is essential to every thing that is agreeable."[55]

The wording of this passage is reminiscent of the account of temporal proportions presented in the Scottish music theorist Alexander Malcolm's *Treatise of Musick, Speculative, Practical & Historical* (1721). Malcolm states:

> That the Proportion of the *Time* of Notes may afford us Pleasure, they must be such as are not difficultly perceived: For this Reason the only *Ratios* fit for *Musick*, besides that of Equality, are the double and triple, or the *Ratios* of 2 to 1 and 3 to 1; of greater Differences we could not judge, without a painful Attention; and as for any other *Ratios* than the multiple Kind (*i.e.* which are as 1 to some other Number) they are still more perplext.[56]

While Walter does not cite Malcolm, his reference to the "painful" effort of the attention required to ascertain proportions beyond the duple and triple, and his linking of the "ease and simplicity of conception" to the aesthetic experience of pleasure makes it seem very likely that he encountered his countryman's musical writings.

Walter, however, goes on to assert that the same holds true for attending to individual beats as single events and not as part of a group; this is possible but requires "a considerable exertion of the mind."[57] Grouping into twos, threes, or fours is thus a cognitive strategy that reduces the demands on our attention. Walter's repeated references to the faculty of the attention indicate some awareness of a key subject in Scottish psychological philosophy at the time, possibly acquired via Holden's *Essay*, or perhaps in conversation at the Glasgow Literary Society with eminent thinkers who explored the topic such as Reid, Jardine, and Smith.

---

[55] Ibid., 63.

[56] Malcolm, *A Treatise of Musick*, 390. Original italics. The wording of Malcolm, in turn, is clearly influenced by the account of rhythm found in Descartes's *Compendium musicae* (1651), which states, "Time in sound must consist of equal parts, for these are perceived most easily ... or it must consist of parts which are in a proportion of 1:2 or 1:3; this progression cannot be extended, for only these relations can be easily distinguished by the ear ... If time values were of greater inequality, the ear would not be able to recognize their differences without great effort, as experience shows." Descartes, *Compendium* trans. Walter Robert, 13. Malcolm quotes lengthy passages from Descartes's treatise, including the philosopher's preliminary axioms of perception, see Malcolm, *A Treatise of Musick*, 196–199. Moreover, a comparison of Malcolm's text with the 1653 English translation of Descartes makes it clear that he is translating directly from the Latin, as he indeed claims.

[57] Young, "Essay," 67.

Deducing Bars, Phrases, and Rhythmical Measures

Beyond the powers of dividing and aggregating beats and their sub-divisions, Walter notes that we also have the ability to consider—and to *feel*—a group of beats as "something separate and distinct from what went before and what is to come after."[58] These groups are musical bars, and they are composed of aggregates of two or three beats where the first beat of each bar is accented, or four beats where the first and third beat are accented, with the second accent weaker than first.[59] He introduces the notion of musical accent in a matter-of-fact manner, simply commenting in a footnote that "I have here used the term accent in its musical acceptation, to denote that *imaginary* degree of force or emphasis which a sound acquires *from the circumstance of its being the first of a parcel* in a rhythmical succession."[60] A similar claim can also be found in *An Inquiry into the Fine Arts* (1784), an idiosyncratic treatise composed by Walter's colleague, the Scottish minister Thomas Robertson whom we met in Chapter 4, who writes, "The beginning, and the divisions in the middle [of the measure] occupy the attention. Weight and stress are mentally laid upon them. The imagination comes in, and fancies she hears those instants sounding louder than the rest."[61] Both Walter and Robertson here echo Holden, who wrote that "it is not so much the superior loudness of the sound, as the superior regard which a hearer is led to bestow upon it, that distinguishes one part of the measure from another."[62] An awareness of the phenomenon of subjective rhythmicization seems to have directly led these Scottish thinkers to the insight that the accentual pattern within a bar is psychological rather than acoustic.[63]

---

[58] Ibid., 63.

[59] Walter writes: "In parcels of four, the third, being the first of a pair, is also accented, but not so strongly as the first." Ibid., 63. A few pages later, he observes that we can also recognize the different parts of a measure when these include rests. He writes: "After a distinct impression has been obtained of the units of which a rhythmical succession is composed, and of the parcels according to which it is constructed, we do not lose that impression, although the succession should stop, or no sound be heard, during the time of one or more of the units. These vacant or silent times, if they are not too long continued, we reckon with nearly the same ease and certainty, as if they had all been expressed by sounds; and we clearly perceive the particular part of the measure at which the succession of sounds recommences." Ibid., 70.

[60] Ibid. My italics. Later in the "Essay," Walter offers a different definition of poetic accent, *viz.*, "When I apply the term accent to syllables, I use it in its grammatical acceptation, to denote that superior force of articulation, and that inflection of the voice, with which we always mark in our pronunciation some particular syllable or syllables of every word." Ibid., 76.

[61] Robertson, *Inquiry*, 54.

[62] Holden, *Essay*, part 1, §95, 83.

[63] Note that all three differ from Descartes, who regarded such accents as an actual increase in volume, "accomplished in vocal music by stronger breathing and on instruments by stronger pressure, so that at the beginning of each measure the sound is produced more distinctly." Descartes,

172   HEARING WITH THE MIND

Walter observes that the procedure of dividing and compounding small units of time that are related to each other in simple proportions seems highly complex, and yet this activity takes place apparently effortlessly whenever we hear a piece of music. He notes that this experience is only one of many complex processes that our minds execute effortlessly without the intervention of our consciousness, and—in what may be an oblique reference to Holden's theory of pitch perception, which likewise involves the division and compounding of (miniscule) proportional durations—surmises that the perception of pitch relationships is more complicated still.[64]

When listening to a piece of music, Walter maintains, we first identify a certain temporal unit, and then, using cues such as an accentual pattern or other circumstances in the musical surface, we very swiftly ascertain how many of these units are contained in a group. This determination allows us to follow the performer and "feel the proportional duration of every note,"[65] and results in a duration that constitutes a standard—the rhythmical measure—to which all divisions and aggregates are referred. That is to say, the measure is something we intuit, and similar in this way to Holden's module. It is scaled to our own capabilities, to that which is most congenial to our mental and corporal affordances. Walter thus uses the term "measure" in a way that does not quite map onto its contemporary usage as a synonym for "bars." (While Walter does not use the term "meter," I use this word in its modern sense when his reference to "measure" seems directly to imply our modern concept. In other, more ambiguous cases, I use the term "rhythmical measure" in order to retain some distance from our contemporary usage of the word.)

The meter, Walter observes, is most easily perceived when bars combine into twos and fours, and when these combinations, which he terms musical strains or phrases, form larger sections of the piece, as is the case in minuets, marches, or gavottes. He observes that we derive pleasure from assigning these units to appropriate groups: for a rhythm to be "perfectly agreeable and satisfactory, [it] must be constructed according to some measure; the whole succession must be made up of parcels of some determined number of units, and must be so contrived as that the hearer may be instantly led to

---

Compendium, 14. While Malcolm does not discuss accentuations in his treatise, he does discuss Étienne Loulié's recently invented *chronomètre*, presenting perhaps an early and necessary stage of the trajectory of thinking that would eventually allow Holden and his followers to understand subjective rhythmicization. See Malcolm, *Treatise*, 407–408.

[64] Young, "Essay," 64. Holden's theory is discussed in Chapter 2 of this book.
[65] Ibid., 64.

adopt that number and retain it."[66] Walter thus recommends that composers "teach" listeners to attend to the meter by featuring the return of similar sounds, combinations, and divisions. Such reinforcements, he suggests, help the listener retain the meter throughout smaller or complicated divisions or syncopations. He cites the first movement of Johann Stamitz's Symphony no. 6 as an example: the piece is in quadruple meter (4/4) and it begins with a two-bar unit in which the first bar consists of hammerstrokes on the first three beats and the second bar divides each beat into halves (i.e., eighth notes); this two-bar unit is then repeated exactly in mm. 3–4, as shown in **Figure 5.1**:

Figure 5.1 Stamitz, Symphony no. 6, mm. 1–5, violins. Throughout his "Essay," Walter merely cites works by name and bar number; I have included his musical examples in order to illustrate his ideas.

Having ascertained the grouping of a series of beats within bars, Walter observes, we will also want to combine the bars into larger phrases consisting of two, three, or four such units. He analogizes between the experience of grouping beats into bars and that of grouping bars into phrases, regarding them both as mental projections. He writes,

> When we hear a minuet, or any piece of music, which is constructed according to regular rhythm, we have the impression of a pause at the end of every bar; we have the same impression more strongly at the end of every phrase; and yet we are certain, that at many of those passages, no real pause is made.[67]

This idea is similar to the modern concept of hypermeter in the claim that we feel a stronger sense of a break between individual phrases than in between those bars that make up the phrase. However, Walter does not pursue this

---

[66] Ibid., 84–85.
[67] Ibid., 80.

174 HEARING WITH THE MIND

line of inquiry further, nor does he discuss the relative weight or importance of these imaginary pauses.

Walter next remarks that the grouping of bars into phrases is accomplished "chiefly by pauses, and by the return of similar sounds, or of similar combinations and divisions."[68] Note that his reasoning here is largely melodic: the grouping of bars into phrases is determined by rests and by melodic parallelism.[69] These cues significantly augment our listening pleasure by providing us with "certain stages or resting places, and [we] are enabled to count off the parcels with more steadiness and with a smaller effort of the attention."[70] Pieces that consist of what he terms "a simple and agreeable number" of these combinations—that is, an even number divided consistently into units corresponding to powers of two—provide the impression of a complete whole, one that affords "a lively perception of that proportion and arrangement of parts, which is essential to every thing that can be accounted beautiful or pleasing."[71]

In cases where the meter is not immediately apparent, Walter continues, regular bodily motions greatly assist us in feeling and perceiving the beat. He links the beating of time to the origin of dancing, suggesting that "we are naturally disposed . . . to accompany [rhythmical measures] with corresponding motions of the body."[72] Beating time thus supports both the performer, charged with executing a piece in regular time, and the listener, tasked with perceiving the accented notes and deducing the meter.

---

[68] Ibid., 86.

[69] According to Caleb Mutch, theories of form almost exclusively address and reproduce examples of melodies, rather than bass lines or harmonies, until well into the twentieth century. Private conversation, October 19, 2021. For a discussion of a surprisingly late example of this tendency in the work of the Austrian musicologist Wilhelm Fischer (1886–1962) see Caleb Mutch, "The Formal Function of Fortspinnung," *Theory & Practice* 43 (2018): 1–32.

[70] Young, "Essay," 86.

[71] Ibid. Here, too, Walter may implicitly refer to Descartes, who writes, "Two types of mensuration are used in music; namely, division into three units or into two. This division is indicated by the bar or measure, as it is called, to aid our imagination, so that we can more easily apprehend all the parts of a composition and enjoy the proportions which must prevail therein. This proportion is often stressed so strongly among the components of a composition that it aids our understanding to such an extent that while hearing the end of one time unit, we still remember what occurred at the beginning and during the remainder of the composition. This happens when the entire melody consists of 8 or 16 or 32 or 64 units, etc., i.e., all divisions result from a 1:2 proportion. For then we hear the first two units as one, then we add a third unit to the first two, so that the proportion is 1:3; on hearing unit 4, we connect it with the third, so that we apprehend them together; then we connect the first two with the last two, so that we grasp those four as a unit; and so our imagination proceeds to the end, when the whole melody is finally understood as the sum of many equal parts." Descartes, *Compendium*, 14.

[72] Young, "Essay," 65. Here again Walter may echo Descartes, who observed that "we accompany each beat of the music by a corresponding motion of our body; we are quite naturally impelled to do this by the music." Descartes, *Compendium,* 14.

Walter goes on to remark that unlike the ancients, whose dramatic music required the band to "beat time with great force and noise," the moderns have set such actions almost entirely aside by "construct[ing] their music by equal and regular measures."[73] There is no doubt, he claims, that our enjoyment is greater when we "perceive the measure from the real accent and expression of the notes than when we must be assisted by the continual repetition of unmusical sounds."[74] Here, too, Walter emphasizes that we deduce the meter from accent and expression, without explicitly addressing harmonic movement.

The Perception of Time and Rhythmic Proportion

Walter then turns to consider why our perception of temporal units differs at distinct durations. He observes that musical experience suggests that there are certain notes that one easily considers as single durations that are perceived neither as compounds of shorter units nor as being themselves divisions of a longer unit. So, he argues, our minds must have a specialized idea of a unity or unit of time within certain temporal limits (this is what we would today term a tactus). Returning to an embodied explanation, Walter surmises that we consider units of time to be potentially single, undivided durations when they occupy between two seconds and half a second, a range he ascribes to "those natural uniform motions from which our original impressions of rhythmical movement are derived, and particularly the motion of our own limbs in walking or running."[75]

This undivided (though divisible) unit of time, Walter claims, is essential to our process of perception, which makes simultaneous use of the operations of division and composition. Appealing to experience, he offers two arguments in support of this idea: when we first learn to sustain longer note values, we typically count off single beats marked by "uniform motions of the hand." With experience we learn to hide the traces of this process (such as swells or accents), yet we never stop counting while sustaining the note.[76]

[73] Young, "Essay," 65.
[74] Ibid., 66.
[75] Ibid., 67. Of course, Walter is not the first to link the tempo of a musical tactus to our physical movements; see DeFord, *Tactus, Mensuration and Rhythm*, 209.
[76] Such counting, Walter remarks, can also take the form of attending to an accompaniment line which articulates these smaller note values. Although he does not discuss inexplicit counting in this particular passage, he repeatedly references the experience of unconsciously "feeling" proportional durations in the treatise. See, e.g., Young, "Essay," 63, 64, and 68.

176   HEARING WITH THE MIND

A related process occurs at the opposite end of the spectrum when dealing with fast notes. In music featuring more virtuoso figurations, we frequently find the same duration divided once into four and immediately after into three (or vice versa).[77] We must take the longer duration as the unit in such cases, as otherwise we would have to instantly recalculate our unit in the relatively complex proportion of 4:3. Selecting a durational unit that stands approximately in the middle of our temporal range thus allows us to have a stable reference point between the slowest and fastest notes (say, between a whole note and a thirty-second note), that typically occur in music.

Walter's insight that simultaneous acts of division and combination are essential to our mental processing is very likely another debt to Holden, who discusses a number of cognitive constructs (including the bar) that enable the division and combination of durations into fractions and multiples, as discussed in Chapter 2. Like Holden, moreover, Walter sets a limit on the kinds of proportions our minds can conceive of directly, writing:

> We frequently meet with semibreves and semiquavers in the same piece of music; notes which are to one another in the proportion of one to sixteen. This proportion is too great to be conceived and felt by a single operation of the mind. When, however, the crochet is accounted the unit, we are enabled to estimate, and accurately to express those distant times, without going beyond the simple and familiar proportion of four to one, on either hand.[78]

Maintaining a tactus level thus constitutes a cognitive strategy for comprehending large proportions, as our minds cannot directly conceptualize complex proportions from a top-down or bottom-up perspective, which is to say, because the relationship between 1:16 cannot be grasped directly, it must be mediated by an intermediate level corresponding to 1:4. Indeed, Walter argues, "These opposite operations of division and combination facilitate and simplify the process of rhythmical perception, and, at the

---

[77] Walter does not provide an example here, but we can consider, for example, the first movement of Haydn's violin concerto in G major.

[78] Young, "Essay," 69. Descartes similarly writes: "these numbers [four to one or eight to one] are not prime to each other and, therefore, do not create new proportions. They simply multiply the two-to-one ratio; the proof for this is that they can be used only in pairs; for I cannot write such notes individually if the second is a fourth [the duration] of the first one." Descartes, *Compendium*, 13. Compare with Holden, *Essay*, part 2, §35, 119, who argues that we cannot conceive of the proportion of 9:8 directly (see also Figure 2.7).

Part 2 of Young's "Essay": Rhythm and Pleasure

Walter begins the second part of his "Essay" by briefly recapitulating the rhythmical powers he described earlier, culminating in our ability to fix our attention upon a given duration (approximately the speed of walking), of which we then regard all other durations as either combinations and or subdivisions. The act of following a piece of music in this way, he then proposes, gives rise to pleasure:

> As the exercise of every power, which we possess, conveys a certain degree of pleasure, we obtain a gratification when we hear a succession of sounds justly proportioned in duration to one another, and are able, at the same time, to go along with, or to feel the several proportions which they bear.[80]

Indeed, if the unit itself were unstable, or its divisions unfamiliar, "we should instantly feel the difference . . . our pleasure would be sensibly diminished or altogether destroyed."[81] A definition of what he calls "regular and perfect rhythm"—a combination of what we call the "metrical hierarchy" and "symmetrical phrase rhythm"—then summarizes these ideas:

> A regular and perfect rhythm . . . is a succession of measured sounds, all of which are either equal to, or are certain multiples or certain parts of some determined portion of time, which may be called a unit, and are so arranged and disposed so that the hearer is easily led to count off those units by equal parcels of some simple number, and also to combine two, three, or four of those parcels together, the whole succession containing a small determined number of those larger aggregates.[82]

Walter's claim here indicates that he regards the metrical hierarchy and phrase rhythm as parts of the same phenomenon: a perceptual continuum

---

[79] Young, "Essay," 69.
[80] Ibid., 84.
[81] Ibid.
[82] Ibid.

178   HEARING WITH THE MIND

wherein beats relate to their subdivisions as whole phrases relate to measure groups and measures. This view differs from the one commonly held today, which distinguishes between phrase rhythm and hypermeter as indeed between rhythm and meter, and it assumes that events within the measure may be governed by different rules than those beyond it.[83]

Nonetheless, Walter observes that we also derive pleasure from "productions of human genius" that diverge from regular and perfect rhythm.[84] To account for this apparent violation of the principle he has just stated, he offers a conjectural history in which the development of musical and poetic conventions mirrors the development of culture as well as mankind.[85] That is to say, the methodology of conjectural history enables Walter to apply the allegedly universal mental preferences that he himself has elicited from the repertoire of his day to try to understand the development of musical practice across human history. He begins by noting that in our infancy we can only obtain pleasure from "that which is most simple and most easily conceived."[86] Thus we derive our initial idea of equal motions of time from motions such as walking, and likely obtain the idea of pairing such motions from similar circumstances whereby "the first of each alternate pair is made particularly to attract the attention."[87] This notion of joining intervals into groups will gradually lead us to join two or four units, and then to consider these groups as units and join them yet again at a higher hierarchical level. These actions also enable us to conceive of dividing a single interval into half, and those halves into halves, respectively. Hierarchies of binary divisions characteristic of the march or gavotte, he argues, are therefore those rhythms which immediately—and initially—pleased the human mind.

---

[83] For a view that rejects the opposition of rhythm and meter see Christopher Hasty, *Meter as Rhythm* (Oxford: Oxford University Press, 1997).

[84] Young, "Essay," 87.

[85] By offering a conjectural history of rhythm, even if in a highly abbreviated form, Walter's "Essay" discloses itself as a product of the Scottish Enlightenment, which saw a proliferation of similarly conjectural accounts about human culture and behavior, including David Hume's *The Natural History of Religion* (London: A. Millar, 1757), Adam Smith's *The Theory of Moral Sentiments. To Which is Added a Dissertation on the Origin of Languages*, 3rd ed. (London: A. Millar, 1767), and James Burnett and Lord Monboddo's *Of the Origin and Progress of Language*, 6 vols. (Edinburgh: J. Balfour, 1773–1792). Note that Walter's preface to MacDonald's *A Collection of Highland Vocal Airs* (1784) also includes a conjectural account of the emergence of chromatic passing tones and the minor mode out of the major mode. See MacDonald, *A Collection of Highland Vocal Airs* (1784), 4–5. Walter here adopts nearly verbatim much of Holden's description of the qualities of the scale degrees but does not cite him.

[86] Ibid.

[87] Ibid.

In support of this claim Walter refers to Charles Burney's account of the child prodigy William Crotch (1775–1847), whose keyboard improvisations at the age of two Walter describes as typically marches, "proceeding chiefly by the dactylus and spondaeus."[88] Taking a conjectural history approach, he concludes that such simple rhythms were thus also likely characteristic of the earliest verses of the ancients.[89] Walter's appeal to the highly gifted Crotch should be seen within the context of the Enlightenment fascination with wild or feral children, who were regarded as containing within themselves the seed of ancient languages.[90] In Walter's own milieu, the eccentric Scottish judge and philosopher, James Burnett, Lord Monboddo, referred to feral children—Peter the Wild Boy (ca. 1713–1785) and Marie-Angélique Memmie Le Blanc (ca. 1712–1765)—at considerable length in many of his works, including in the first volume of his *Of the Origin and Progress of Language* (1773) and the third and fourth volumes of his *Antient Metaphysics* (1784, 1795).[91]

A modern reader might assume that a celebrated child prodigy who was presented to the King at the age of three would have been the very opposite of a feral child. Yet it appears in fact that Crotch's lowly background as the son of a carpenter and his resulting lack of training set him apart from other musically talented children at the time, and that he was, in significant ways, understood as somehow close to nature. Thus, for instance, Burney describes his improvisations, or voluntaries, as "little less wild than the native notes of a lark or a black bird,"[92] and declares that "little Crotch, *left to nature*, has not only been without instructions but good models of imitation; while Mozart

---

[88] Ibid., 88. A dactyl is a long-short-short rhythm, such as quarter followed by two eighth notes, whereas a spondee is a long-long rhythm, such as two quarter notes. See also Charles Burney, "Account of an Infant Musician," *Philosophical Transactions of the Royal Society of London* 69 (1779): 183–206.

[89] Young, "Essay," 88.

[90] This view preceded the eighteenth century: already in antiquity, according to Herodotus in part 2 book 2 section 2 of *The Histories*, Pharoah Psamtik (probably Psammetichus I) had two children raised in complete silence by a shepherd in order to ascertain what language they would first speak; their spontaneous vocalizations were determined to sound most like the Phrygian word for bread. Herodotus, *The Histories*, translated by Robin Waterfield (New York: Oxford University Press, 2008), 95–96.

[91] Monboddo, who met Peter and interviewed Memmie Le Blanc during their adulthoods, was instrumental in bringing feral children to the forefront of developmental discourse in the Anglophone sphere. He also wrote a preface to his secretary's translation of Marie-Catherine Homassel-Hecquet's biographical account of Memmie Le Blanc's life, published in 1768 as *An Account of a Savage Girl*. For more on Peter and Le Blanc see Julia V. Douthwaite, *The Wild Girl, Natural Man, and the Monster: Dangerous Experiments in the Age of Enlightenment* (Chicago: University of Chicago Press, 2002). A longer history of the fascination with feral children can be found in the second chapter of Karl Steel's *How Not to Make a Human: Pets, Feral Children, Worms, Sky Burial, Oysters* (Minneapolis: University of Minnesota Press, 2019), 41–74.

[92] Burney, "Account of an Infant Musician," 201.

180 HEARING WITH THE MIND

and Samuel Westley [*sic*], on the contrary, may be said to have been nursed in good music."[93] The parallel between feral children and the boy musician is made explicit in his final peroration, where Burney reports:

> It is the wish of some, that the uncommon faculties with which this child is endowed might be suffered to expand by their own efforts, neither refrained by rules, nor guided by examples; that, at length, *the world might be furnished with a species of natural music, superior to all the surprizing productions of art* to which pedantry, affection, or a powerful hand, have given birth. But, alas! such a wish must have been formed without reflexion. . . . He might as well, *if secluded from all intercourse with men, be expected to invent a better language than the present English*, the work of millions, during many centuries, as a new music more grateful to the ears of a civilized people than that with which all Europe is now delighted.[94]

It appears from this passage that certain people were tempted to maintain Crotch's untutored status as a kind of experiment for learning about the emergence of music "in the wild," as it were. Fortunately for Crotch and for his contemporaries, the boy was very well educated and would go on to become a professor of music at Oxford University.

There are, moreover, some important differences between Walter's and Burney's interpretations of Crotch's musical practices. Burney reports that

> no adherence to any particular measure is discoverable in his voluntaries; nor have I ever observed in any of them that he tried to play in triple time. If he discovers a partiality for any particular measure, it is for dactyls of one long and two short notes, which constitute that species of common time in which many street-tunes are composed, particularly the first part of the Belleisle March, which, perhaps, may first have suggested this measure to him, and impressed it in his memory.[95]

Whereas Walter interprets Crotch's penchant for dactyls as reflecting a less advanced developmental state (and adds, it should be noted, a proclivity for spondees that is not present in the original account), Burney ascribes this partiality to the boy's exposure to street music, and specifically a popular march he

---

[93] Ibid., 203. My italics.
[94] Ibid., 204–205. My italics. This is, of course, a notably anti-Rousseauian point of view.
[95] Ibid., 196.

RHYTHM AS A UNIVERSAL "SCIENCE OF MAN"   181

had often heard.[96] That is, following the tenets of Scottish Realist philosophy, Walter has no difficulty in assuming that certain cognitive preferences are inborn, whereas Burney regards them as acquired via a more Lockean process of association. It is notable, however, that Walter seems to have been influenced by Burney's assertion that Crotch made no use of triple meter in assigning such rhythms to a later developmental stage, as will shortly be demonstrated.

Shifting back from his example of young Crotch to the species-general domain, Walter observes that with the maturation of our powers of perception, it is no longer sufficient for a piece to be simple and agreeable—our minds can now appreciate, and indeed demand greater complexity.[97] The same holds true at the level of musical culture, where originally, additional complexity was likely accomplished through occasionally adding another pair (or two pairs) of bars to arrangements of two, four, or eight measures. As such triple combinations gradually became familiar, they would have given rise to the notion of triple meter. Such a grouping of three beats "gave more exercise to the rhythmical powers; at the same time, it did not fatigue them."[98] Triple meter is less uniform and solemn than duple meter, and hence the minuet is "at this day accounted the most elegant and pleasing movement in music."[99] These groups of threes, in turn, would suggest the option of gradually diminishing the unit, and thus pairs of three-beat bars would, over time, develop into compound triple meters, in which units of three beats would be grouped into pairs or fours, as is the case with 6/8 and 12/8 meter, respectively.[100] Over time, the same process would give rise to triple groups of triple meter, as in the case of 9/8 in music. This genealogy is sketched out in **Figure 5.2.**

[96] The third measure of the Belleisle March contains triplets, rendering Burney's conclusion surprising: it seems odd to conclude that the child was only influenced by the first two measures, but perhaps his performances also seemed to draw melodically on the opening of the tune.

[97] Walter was not the first to theorize the relationship between psychological development and the requirement of commensurate musical complexity. One early account can be found in David Hartley's aforementioned *Observations on Man* (1749), where the author, a physician and philosopher, suggests that a newborn infant finds all intervals unpleasant. Octaves and fifths, intervals with the simplest ratios, appeal first to her developing ear, then fourths, thirds, sixths, etc. However, this early familiarity ultimately breeds ennui, as her ear searches for more complicated relationships within sound, and therefore she tires of the simplest intervals first, as she continually learns to enjoy more complex intervals, chords, and harmonies. The ever-increasing desire for novelty thus enables a trained listener to learn to find beauty even in dissonant passages: "those concords in which the ratios are simplest would become pleasant first, and the others would continue to excite pain, or to border upon it. It is agreeable to this, that discords become at last pleasant to the ears of those that are much conversant in music, and that the too frequent recurrence of concords cloys." Hartley, *Observations on Man,* 226–227.

[98] Young, "Essay," 88.

[99] Ibid., 89.

[100] In a footnote, Walter alternately suggests that the rhythmical measure of the gigue may have arisen "from the sound of a horse's feet, when running at full speed." Ibid.

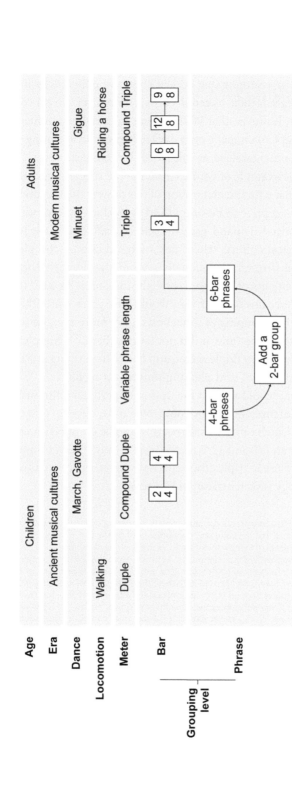

Figure 5.2 Schematic of Walter's Account of the Development of "Rhythmical Measures."

RHYTHM AS A UNIVERSAL "SCIENCE OF MAN"   183

Walter's genealogy moves between different hierarchical metric levels: the embodied experience of walking gives rise to the idea of bars of two, which in turn give rise to bars of four. The same process then recurs, leading to phrases of two, four, and eight. It is only with an additional expansion of these phrases by means of surplus groups, however, that the notion of triple divisions comes into play, and this idea then propagates downwards toward the beat level, informing first a basic triple division of the bar and then, recapitulating the initial steps of duple divisions for the case of compound triple (compound division of the triple meter into two and four, ultimately giving rise to division into three). This is an extraordinary—and extraordinarily flat—theory that collapses the distinction between meter at the beat level and grouping at the bar level, so that instead of progressing from what we would today regard as simple (groups of beats) to more complex (groups of groups of beats), it categorically distinguishes between simple (duple) and triple (complex), regardless of where they occur in the metric hierarchy.

Pleasurable Deviations

A regular (but not necessarily perfect) rhythm, therefore, is one in which the bars—whether duple or triple—are equal in length, as are the phrases to which they accrue, provided the lengths of these are what Walter calls "simple and obvious," presumably, that is, an even and relatively small number of bars.[101] The most basic departure from this scheme involves extending the piece beyond the aforementioned limits by adding supplementary bars. Walter speculates that these alterations are necessary because longer pieces require more diversity in their metrical organization: "In proportion as rhythmical measures become more an object of attention and are more frequently presented to the ear, the necessity of variety becomes the greater."[102] He continues, "We are often pleased with a bold deviation from what is strictly regular. The very surprise which it causes is agreeable."[103] This desire for novelty, he concludes, is the reason why composers occasionally decide to avoid regular rhythmical structures.

Walter now surveys a number of common deviations from regular rhythmic organization. His first and most paradigmatic example concerns the case of the minuet, form he regards as exemplifying the normative principles

---

[101] Ibid. Walter does not specify what this length might be.
[102] Ibid., 90.
[103] This claim is reminiscent of David Huron's exploration of the experience of pleasure as arising from musical surprise; see Huron, *Sweet Anticipation.*

184   HEARING WITH THE MIND

of the music of his day. In a standard minuet, he observes, two-bar groups combine into four-bar phrases marked by a cadence, with two of these phrases in turn making up an eight-bar strain, and the typical minuet featuring two such strains. Walter notes, moreover, that this structure is often lightly altered: the number of four-bar phrases in a strain can be increased from two to three, four, or six, or a four-bar phrase may be expanded into a six-bar phrase. Both of these changes occur frequently without the minuet losing its distinctive character.[104] In contrast, Walter emphasizes, if a piece includes phrases composed of three or five bars, it "ceases to be a minuet."[105] That is to say, he regards the minuet's two-bar groupings—and not combinations at the phrase and strain level—as essential to its function as a dance. However, Walter continues, in the case of instrumental music, "modern musicians, especially those of the German school," often include second minuets, or trios, that make use of groups of three bars.[106] However, as the three-bar groupings, like the other combinations in the piece, typically occur in pairs, the overall number of bars in a strain almost always remains even.[107]

Next, Walter turns to longer pieces, in which such metrical deviations, when used in moderation, are often unapparent to the listener. Nonetheless, he speculates, these irregularities serve a number of important functions in the piece, including rousing "the attention of the hearer, which is apt to flag in a long piece, when the rhythm uniformly proceeds by equal combinations."[108] Another advantage to varying the length of the groups, he continues, is that doing so enables the composer to highlight formal features by giving "a more emphatic introduction, or a more striking and distinct close, to some remarkable strain."[109] Walter here suggests that changes in phrase structure, even if we are not actively conscious of where and how they appear, can have productive psychological effects in terms of retaining the listener's interest and clarifying the rhetorical function of a section of the work. The insight that the listener's attention needs to be explicitly engaged through unpredictable musical elements corresponds to similar trends at the time, perhaps most famously embodied in Haydn's *Surprise* Symphony (1791), composed and premiered in London.[110]

---

[104] The minuet dance is composed of a sequence of four steps transpiring over six beats, hence the requirement for two-bar hypermeter. See Eric McKee, *Decorum of the Minuet, Delirium of the Waltz: A Study of Dance-Music* (Bloomington & Indianapolis: Indiana University Press, 2012), 22.

[105] Young, "Essay," 91.

[106] Ibid.

[107] As examples of this, Walter cites two minuet trios: the Trio from Haydn's String Quartet op. 2, no. 2, and the Trio from Haydn's String Quartet op. 1, no. 1.

[108] Ibid.

[109] Ibid., 91–92.

[110] On this point see Mathew, "Interesting Haydn: On Attention's Materials," 655–701.

RHYTHM AS A UNIVERSAL "SCIENCE OF MAN" 185

Diversifying the organization of bars has additional advantages. In longer pieces, the initial duple organization is often altered, so that a bar that previously appeared to start a two-bar group is displaced. Walter terms these cases inversions, and he observes that they can happen in five ways.[111] The first is when a phrase that ends on the beginning of an odd bar immediately elides either with a repetition of itself or with a new phrase. He maintains that such arrangements help emphasize the new theme, as "a bold and animated strain . . . is thought to produce a greater effect, when it is introduced in this sudden and abrupt manner."[112] He illustrates this by referring to bar 31 of the fourth movement of Stamitz's Symphony no. 11, indicated with an asterisk in **Figure 5.3**. Here, rather than extending the half-cadential V-chord throughout an additional bar in order to complete the paired bar structure and complete the transition, a new theme begins instead. While m. 32 in fact acts as an upbeat to m. 33, and thus falls within the two-bar hypermetrical span initiated by m. 31, Walter appears to make no distinction between phrase and hypermeter, observing only that a new "strain" starts at m. 32. That is to say, he does not remark on that measure's metrical character as an upbeat, suggesting that he did not perceive m. 32 as simultaneously grouping with m. 31 (metrically) *and* with mm. 33–39 (as a phrase). This theme repeats twice, after which the pair structure aligns both metrically and in terms of phrasing; however, the repetition means that the second appearance of the subordinate theme is two bars shorter, with an elision at m. 47 (shown with two asterisks in Figure 5.3).

Walter's second strategy is the repetition of a bar in the middle of a phrase, which results in a temporary inversion of the two-bar pairing. This, he suggests, can be used to "give the appearance of greater bustle and confusion to music that is impetuous and rapid."[113] As an illustration, he cites m. 15 and m. 56 in the last movement of Haydn's String Quartet op. 1, no. 1.[114] The former is shown in **Figure 5.4**.

---

[111] Walter's strategies for breaking up the regular duple pairing of measures has some similarities with the phrase alteration strategies proposed by Joseph Riepel and Heinrich Christoph Koch, in that all three theorists call upon compositional techniques such as repetition, insertion, and elision to account for unusual phrase structure; however, Walter's more modest goal in the "Essay" is to explicitly account for cases that deviate from a duple phrase structure, rather than to discuss varying phrase structure more broadly. On Riepel, see John Walter Hill, *Joseph Riepel's Theory of Metric and Tonal Order, Phrase and Form: A Translation of his Anfangsgründe zur musicalischen Setzkunst, Chapters 1 and 2 (1752/54, 1755) with Commentary* (Hillsdale, NY: Pendragon Press, 2015), 370ff; on Riepel and Koch, see Elaine R. Sisman, "Small and Expanded Forms: Koch's Model and Haydn's Music," *The Musical Quarterly* 68, no. 4 (1982): 444–475.

[112] Young, "Essay," 92.
[113] Ibid.
[114] Walter calls this String Quartet op. 1, no. 3; however, this numbering is specific to J. Hummel's 1765 edition, reissued in London by Bremner between 1765 and 1768.

Figure 5.3 Irregular bar structure in Stamitz's Symphony no. 11, fourth movement, mm. 1–47, first violin. Walter's interpretation is indicated with solid brackets; mine is shown in dashed brackets. The m. 32 upbeat is shown at the first asterisk; the m. 47 elision by two asterisks. (Note that a similar metrical structure occurs already at m. 24, which serves as an upbeat to m. 25, although this part of the work does not feature the changes of key, texture, and dynamic level that likely attracted Walter's attention to mm. 31–32.)

RHYTHM AS A UNIVERSAL "SCIENCE OF MAN" 187

Figure 5.4 Mm. 1–17 of the last movement of Haydn's String Quartet op. 1, no. 1. The asterisk marks the "added" bar (m. 15).

Figure 5.4 Continued

Walter then examines a third way of superseding strict two-bar groups: the addition of a bar that features the key note of a piece or a phrase, thus serving as an introduction to the theme that immediately follows.[115] "This first bar being as it were set aside, or considered as standing by itself, the music afterwards proceeds by regular pairs, commencing at the second bar."[116] He provides two examples of this type of metrical deviation, both taken from Haydn's *Divertimento* in E-flat Major, Hob. II:6, at m. 23 of the first movement and at the very beginning of the last movement.[117] These are shown in **Figures 5.5** and **5.6**, respectively, where, to my ears, Walter's reading (shown

---

[115] This is similar to a technique illustrated in both Riepel and Koch but not named, which Hill terms a "Leader." See John Walter Hill, *Joseph Riepel's Theory of Metric and Tonal Order, Phrase and Form: A Translation of his Anfangsgründe zur musicalischen Setzkunst, Chapters 1 and 2 (1752/54, 1755) with Commentary* (Hillsdale, NY: Pendragon Press, 2015), 385–386.

[116] Young, "Essay," 93.

[117] Walter calls this String Quartet op. 1, no. 1.

RHYTHM AS A UNIVERSAL "SCIENCE OF MAN" 189

Figure 5.5 Mm. 1–35 of Haydn's *Divertimento* in E-flat Major, Hob. II:6. The solid brackets indicate Walter's interpretation; dashed brackets show my interpretation. Note his "extra" bar at m. 23, indicated with an asterisk.

Figure 5.5 Continued

Figure 5.6 Violins 1 and 2, mm. 1–9 of the last movement of Haydn's *Divertimento* in E-flat Major, Hob. II:6. The solid brackets indicate Walter's interpretation; dashed brackets show my interpretation. Note his "extra" bar at the very beginning, indicated with an asterisk.

in solid brackets) is plausible if one examines the first violin part alone. Considered in the context of the overall harmony and the other string parts, both of his interpretations seem highly idiosyncratic, and I have included my own interpretations in dashed brackets. In the case of Figure 5.5, the eight measures (mm. 15–22) that precede the imitative section clearly articulate a strong-weak two-bar pairing that prepares us to hear m. 23 as the first beat of a two-bar group. Next, the cello and viola lines seem to articulate an implicit movement from V to I every two bars in mm. 23–32. Finally, the cadential pattern that starts in m. 31, and in particular the deceptive cadence on the downbeat of m. 33, strongly suggests that the "extra" bar is not in m. 23, but rather in m. 35. I likewise disagree with his interpretation of the passage shown in Figure 5.6. Nowadays, we would consider these bars as composing a standard eight-bar sentence, with a hypermetrical downbeat at m. 1 and again at m. 9.

Walter's fourth method concerns the pairing of phrases comprising an odd number of bars, so that a bar that at first appears to start a two-bar group is immediately reinterpreted as ending the preceding phrase it, because that preceding phrase's first bar immediately follows. He provides two such examples, shown in **Figures 5.7** and **5.8** respectively: in Haydn's

Figure 5.7 Haydn's String Quartet op. 2, no. 1 begins with two three-bar phrases.

RHYTHM AS A UNIVERSAL "SCIENCE OF MAN" 193

Figure 5.8 Five-bar phrases in the last movement of Erskine's first overture op. 1, no. 1, mm. 66–75. The final sixteenth note (B) in m. 5 of the second violin is most likely a typographical error in the printed edition.

String Quartet op. 2, no. 1, the first movement features two three-bar phrases, while the second part of the last movement of the first overture by Thomas Erskine begins with two five-bar phrases. Walter remarks that while the frequent pairing of odd-bar phrases ensures that the broader even-numbered structure is generally preserved, pieces sometimes nevertheless end up with an odd number of bars. He praises such "occasional interruptions of uniform movement . . . [for giving] an additional relish to the regularity that is observed in other parts of the composition,"[118] and he compares their effect to "discords in the harmonical structure of music."[119]

Walter's final strategy pertains specifically to the case of fugue, a genre that he regards as stuffy and outmoded, with "combinations . . . sometimes so various and obscure, that the hearer can scarcely retain the impression of them."[120] Here, staggered entrances mean that the subject frequently enters

---
[118] Ibid.
[119] Ibid.
[120] Ibid., 93.

194   HEARING WITH THE MIND

in unequal and unpredictable intervals, often in the middle of the bar. In a brief conjectural aside, Walter surmises that

> the ancient music . . . had its rise at the time when the chief professors and improvers of the musical art were churchmen, and when, of course, that kind of music was chiefly cultivated, which was thought to be best calculated to compose and elevate the mind, and to inspire devotion. A simple, regular and distinct rhythm was probably thought by them to give the music a light and airy cast, inconsistent with the effect which they wished to produce.[121]

Walter here reads the aesthetic values of his day, which prized simplicity and regularity, back into the church music of the preceding century in order to conclude that the clergymen of that time intentionally made their music complex in order to elevate their audiences and render them receptive to a religious meditation. Skeptical that an untrained listener can enjoy such intricacy, he notes that "of all music it gives least pleasure to one who has not been accustomed to it."[122]

Concluding his summary of atypical phrasing, Walter observes that while the grouping of bars allows for a considerable degree of variability, this flexibility does not pertain to the length of the bars themselves or the duration of their groups. While changes in the division of groups can be made, this should only occur after a given uniform division has been established for a substantial length of time and brought to a natural stop. "Such changes, when skilfully managed, enliven the music, surprise the hearer, and excite his attention."[123] But when such devices are used too often, they leave the hearer "in a state of continual suspense and uncertainty," and detract from their satisfaction.[124] Walter speculates that "we may bear to be, in some degree, offended a certain number of times, when such offence has the effect to stimulate and surprise, and when it is quickly compensated by some striking beauty," but warns that this approach fails if too often attempted, as the piece will sound ridiculous.[125]

---

[121] Ibid., 94.
[122] Ibid.
[123] Ibid., 95.
[124] Ibid.
[125] Ibid.

At the end of the treatise, the universal principles that the "Essay" has labored to establish are applied to a case that is dear to Walter's heart, namely an attempt to explain reports of the music of those paragons of the Enlightenment, the ancient Greeks. Here Walter once again appeals to conjectural history, using the allegedly universal mental preferences that he himself has elicited from the repertoire of his day, to try to understand and evaluate reports of musical practice in antiquity.

The feature of ancient Greek music that Walter finds truly bewildering is the fact that accounts of that music consistently report on the regular employment of different rhythmical measures within a single phrase. He writes:

> The verses to which they adapted music, were composed of unequal feet, such as trochees and spondees, which they respectively considered as measures of three and of four equal times, and these occurring sometimes alternately; and we are told that the music rigidly observed the measure of the verse.[126]

The existence of non-isochronous meters, of course, strongly contradicts Walter's firm conviction, described in his initial account of the spontaneous mental grouping of watch ticks, that our minds cannot regard bars composed of groupings of twos with threes, or threes with fours, which is to say, groups of fives and sevens, as the basis for the larger hierarchical aggregates he believes are essential components of the process of perceiving music. Pursuing such groups, he observes, will generate in the listener a strong desire to complete the collection to six or eight, and this "will not be resisted without some great and constant effort of the attention, and even some degree of force and constraint."[127]

The case of Greek musical practice is particularly perturbing, since not only do reports from antiquity describe groups of seven, but these often occur irregularly, interspersed with other groupings. Walter notes that while performing groups of seven might be possible at a very slow tempo (where each beat could be conceived as a separate event), doing so at a faster rate would be impossible for either the performer or listener to distinctly perceive without assistance.[128]

---

[126] Ibid., 96.
[127] Ibid., 97.
[128] Ibid., 93.

196 HEARING WITH THE MIND

Walter offers several conjectures with regard to the actual performance of ancient Greek choral music. To begin with, he admits that it is possible that their choruses, led by a coryphaeus marking the *arsis* and *thesis* of a measure, "did truly and accurately express contiguous unequal parcels of rhythmical times."[129] Nonetheless, he soon calls this into question, asserting that for the Greeks to have done this would mean that they had flouted "a strong propensity of nature, in order to attain an object, the agreeable or happy effect of which we cannot now so much as conceive."[130] Walter here seems to be grudgingly admitting the possibility that ancient Greek music simply affected its hearers in a way we cannot fathom today.

It becomes clear, however, that this is not a position with which he is comfortable, for he next goes on to ask whether we can be *certain* that the individual units within these unequal groups were in fact performed uniformly, that is, as beats equal in duration. We can assume they did, he says at first, in view of "the strong natural propensity which all men feel to express such small times equally and uniformly, when it is not their professed intention to do otherwise."[131] Professing himself skeptical that this arrangement was adhered to on all occasions, Walter hypothesizes that the ancient Greeks' intentions may have been strongly counteracted by their "natural propensity to assemble [small] times into equal parcels," and that "it is not unnatural to suppose that sometimes that other propensity might preponderate, and that some inequality might be admitted amongst the singular times which were marked by the crepitacula [a rattle used to beat time], in order to bring the feet or parcels more nearly to equality."[132] After all, Walter speculates, it is possible that *they themselves* were not aware of the extent to which their theory and practice diverged, and that the coryphaeus, or chorus leader, being "a musician of distinguished talents," might unconsciously have favored more equal measures. Walter concludes by dismissing the possibility that the length of a unit within a single bar could be varied. This, he argues, "has never been attempted in written music, and can hardly be done without almost entirely destroying every impression of rhythm or measured sounds."[133]

---

[129] Ibid., 97.
[130] Ibid., 98.
[131] Ibid.
[132] Ibid., 98.
[133] Ibid., 99.

Concluding his discussion of the music of antiquity, Walter admits that he has always wondered at the apparent fact that "the delicate ears of the ingenious and enlightened Greeks should not only bear, but even be delighted, with what a modern cannot hear without pain and disgust" and reiterates his hypothesis that "their feet or bars were sometimes unequal, more in theory than in practice."[134] That is to say, confronted with historical counterevidence to the principles he is developing, which he regards as natural and hence universal, Walter either questions the reliability of his sources by hypothesizing that the Greeks tacitly regularized those meters in performance, or simply asserts that, if they did so, it must have been unpleasing, even to the Greeks themselves.

It is worth noting that elsewhere Walter takes a similar approach to Scottish Highland music, a tradition in which he was an acknowledged expert.[135] In the 1780s, he assisted Patrick MacDonald in editing the *Collection of Highland Vocal Airs* (1784), a set of 219 Scottish folksong transcriptions collected by MacDonald's late brother Joseph (1739–1763). The MacDonalds had been educated at David Young's school in Haddington and would have met Walter as children.[136] The preface of Patrick MacDonald's *Collection*, written by Walter, includes the following explanation of an editorial decision to alter some of the song collector's meticulous transcriptions by presenting the melodies, and specifically the "slow plaintive tunes," with regular barring.[137] Walter writes:

> These are sung by the natives, in a wild, artless and irregular manner. Chiefly occupied with the sentiment and expression of the music, they

---

[134] Ibid.

[135] In 1787, the laird John Ramsay of Ochtertyre supplied the great Scottish poet Robert Burns with a letter of introduction to Walter, mentioning that Burns would be visiting his vicinity and was eager to learn more about a traditional Scottish musical genre, the sung Highland *Luinags*. See "Letter No. XXXII. From Mr. J. Ramsay, to the Reverend W. Young, at Erskine, Ochtertyre, October 2, 1737" in James Currie, ed., *The Works of Robert Burns: With an Account of His Life, and Criticism on His Writing; to which are Prefixed Some Observations on the Character and Condition of the Scottish Peasantry* (Liverpool: J. M. M'Creery, 1800), 2:115–119.

[136] Walter Young, "Preface," in MacDonald, *A Collection of Highland Vocal Airs*, 1.

[137] Musical expertise for MacDonald's project was provided by both Walter and by Ramsay, but the extent of their contributions is not entirely known. While the collection's preface and appended "Dissertation on the Influence of Poetry and Music on the Highlanders," are unsigned, Ramsay reported that Walter had written the former and he himself the latter, a conclusion supported by the style and content of the writing in question. John Ramsay, *Scotland and Scotsmen of the Eighteenth Century from the mss. of John Ramsay, esq. of Ochtertyre*, ed. Alexander Allardyce (Edinburgh: W. Blackwood and Sons, 1888), 1:411. Note that the entry on Walter in the *Fasti Ecclesiæ Scoticanæ* (1894) erroneously lists him as the author of the Dissertation and omits any mention of the preface.

198 HEARING WITH THE MIND

dwell upon the long and pathetic notes, while they hurry over the inferior and connecting notes, in such a manner as to render it exceedingly difficulty for a hearer to trace the measure of them. They, themselves, while singing them, seem to have little or no impression of measure. It would appear, that in his notation of these airs, in place of reducing them to regular time, Joseph had attempted, as nearly as he could, to copy and express the wild irregular manner, in which they are sung, and, without regarding the equality of the bars, had written the notes, according to the proportions of time, that came nearest to those, which were used in singing. It was judged improper, to lay them before the public, in that form.... All music, however, is now written in just measure. This is necessary, in order to point out the accented and emphatical notes, without attending to which, it is impossible to enter into the meaning of the piece.... To render these airs therefore more regular, especially in their measure, is, in fact, bringing them nearer to their original form.[138]

In Walter's view, regularizing Joseph's careful record of the originally performed rhythms of the airs was, in fact, to restore them to something approaching their "original form," a sort of Platonic ideal unblemished by the rustic performance practices of the Highlanders on their pipes.[139] Walter's views as expressed in the quoted passage indicate that his procrustean attempt to fit the rhythms of ancient Greek music to his timeless rhythmic principals was part and parcel of his broader universalizing worldview.

*

Hew Scott, *Fasti Ecclesiæ Scoticanæ* 2, no. 1 (Edinburgh: William Paterson, 1894), 246–247. With regard to Walter's authorship see Karen McAuley, *Our Ancient National Airs: Scottish Song Collecting from the Enlightenment to the Romantic Era* (New York: Routledge, 2013), 23–28.

[138] Young, "Preface" 2–4.

[139] There is some additional nuance in Walter's account: he regards these slower tunes as the oldest in the collection and hence originally composed for harp, rather than bagpipe. As a result of the adoption of the bagpipe by the Scots, "an instrument more congenial to the martial spirit of the country ... many of the pieces, that had been originally composed, and had been chiefly performed or accompanied by the harpers, are irrecoverably lost: and those, which have been preserved by tradition, may naturally be supposed to have been gradually degenerating." Ibid., 4. Similar ideas about the degeneration of traditional Highland music from the harp to the bagpipe can be found in Thomas Robertson's *Inquiry*, discussed briefly in Chapter 4 ("The History of Highland Music, when we advance to the present times, is the History of a declining people. At first, no doubt, as every where else, the music might be very rude, but so far back as History conducts us, the Fine music has come foremost, and the rude last in order; the Harp going before the Bagpipe.") Robertson, *Inquiry*, 409.

Walter's "Essay" articulates a significant number of insightful claims that are now familiar from the modern-day field of music cognition. These include the acknowledgment that humans share the same set of basic capacities pertaining to the perception of rhythm, and the attribution of the individual differences to both nature and to nurture. Such capabilities have their origins in concrete bodily affordances but are also affected by (or dependent upon) higher cognitive faculties such as attention. Additional innovations include his account of subjective rhythmicization, his intuition that our process of perception makes simultaneous use of the operations of division and composition, his portrayal of the tactus level as akin to a "referent time period" (to borrow Mari Reiss Jones's term), as well as his description of the motor and temporal constraints on entrainment.[140] His findings are supplemented by his analysis of the pleasure afforded by musical surprise arising from unpredictable phrase groupings, and his attempt to offer something like an evolutionary account of the emergence of different meters. The quantity and quality of Walter's insights—particularly when viewed in the context of a fifty-five-page essay on musical and poetic meter—is truly remarkable.

To a reader schooled in Western music theory and its history, much of what Walter says about meter and phrase rhythm will seem both reminiscent of the rhythmic theories of his German contemporaries Kirnberger, Joseph Riepel, and Heinrich Christoph Koch, and intuitively convincing, since repertoire from Walter's historical moment would subsequently become the touchstone against which later rhythmic developments would be evaluated. A contemporary reader of the "Essay" may well be reminded of Rick Cohn's claim that the way music theorists generally think about, and teach, rhythm has remained essentially unchanged from the 1770s.[141]

Alongside its many astute observations regarding the psychology of rhythm perception, Walter's pioneering attempt to locate the origins of aesthetic preferences in specific aspects of the musical repertoire of his day, aspects which he then ascribes to properties of our shared physical and mental experience, can be regarded as an early example of a fallacy that would continue to plague psychology in general, and music theory and cognition in particular, until the present. For well over a century, as a number

---

[140] See Carolyn Drake, Mari Riess Jones, and Clarisse Baruch, "The Development of Rhythmic Attending in Auditory Sequences: Attunement, Referent Period, Focal Attending," *Cognition* 77, no. 3 (2000): 251–288.

[141] Richard L. Cohn, "Why We Don't Teach Meter, and Why We Should," *Journal of Music Theory Pedagogy* 29 (2015): 5–23.

200 HEARING WITH THE MIND

of scholars have recently demonstrated, experimental psychologists inferred universal cognitive principles from the empirical results of a culturally biased sample of the human race, with disconcerting results.[142] Investigations of rhythm perception in both music theory and music cognition have, in particular, suffered from this prejudice with regard to non-isochronous rhythms, which, until recently, were largely ignored outside of ethnomusicological studies.[143]

In the same way, Walter's fervent commitment to the establishment of universal features of perception rendered him unaware of his own biases and assumptions, as is particularly evident in his conjectural history of ancient Greek musical rhythm. The "Essay" thus represents a pioneering attempt to relate universal principles of cognition to distinct features of cultural productions—one that foreshadows debates that still dominate our contemporary cultural and intellectual landscape.

[142] A highly influential formulation of this phenomenon can be found in Joseph Henrich, Steven J. Heine, and Ara Norenzayan, "The Weirdest People in the World?," *Behavioral and Brain Sciences* 33, nos. 2–3 (2010): 61–83.

[143] Rhythm's obviously intimate relationship with embodiment may have long allowed only certain kinds of bodies, and only certain kinds of rhythms, to be regarded as representative of the human race. See Kofi Agawu, "The Invention of African Rhythm," *Journal of the American Musicological Society* 48, no. 3 (1995): 380–395.

6

What's in a Game?

Rediscovering the Music Theory of Anne Young

In 1801, Anne Young of Edinburgh was granted a royal patent for a set of *Musical Games* accompanied by "an Oblong Square Box, [with] two faces or Tables, and various dice, pins, counters ... to render familiar and impress upon the memory, the fundamental principles of the science of music."[1]

The game itself, **shown in Figure 6.1,** is a luxury item, with a box made of brass, satinwood, and mahogany, and nearly two hundred pieces—including pins, dice, plates, and counters—in ebony and ivory.[2] The materiality of the game itself unsurprisingly reflects the trade patterns of the British empire: most of the ivory pieces originate from West African elephants, whose tusks would have been shipped to Scotland on slave ships.[3] In addition to thirty-two ivory dice with different music-notational symbols, some of which are specific to each player and some of which are shared, each player has three ivory plates engraved with a clef, twenty-four large and fourteen small pins, fourteen pins with accidentals, and twelve large and twelve small counters. The game board features the circle of fifths around the circumference of each of its halves, with the sharps clockwise to the left, and the flats counterclockwise to the right. The left half of the board features two different representations of a five-octave gamut, one on a grand staff, and the other on keyboards, while the right half displays a grand staff with various slots for clefs, notes, and accidentals. There

---

[1] Anne Young, "Specification of the Patent granted to ANN YOUNG, of St. James's Square, in the City of Edinburgh, for an Apparatus consisting of an oblong square Box . . .," *The Repertory of Arts, Manufactures, and Agriculture*, vol. 16 (London, G. and T. Wilkie, 1802), 9.

[2] In his review, John Wall Callcott huffed that "the idea might have been implemented and perhaps executed for less than *seven guineas* (the price of the box)." Callcott, "Review of Instructions," 45; Original italics.

[3] DNA analysis of material taken from one of the remaining sets of Young's *Musical Games,* conducted by Elena Essel and Benjamin Vernot, suggests that most of the ivory used in their manufacture can be traced to West African elephants; however, a small piece—a relatively short pin featuring an accidental on its head—appears to have originated from a bovine, either a tooth or a horn. Private communication, April 30, 2024. In the eighteenth century, mahogany was only harvested in the Americas and exported abroad. Unfortunately, the origins of the ebony and satinwood used in the *Musical Games* could not be determined via DNA analysis.

Hearing with the Mind. Carmel Raz, Oxford University Press. © Oxford University Press 2025.
DOI: 10.1093/9780197786208.003.0007

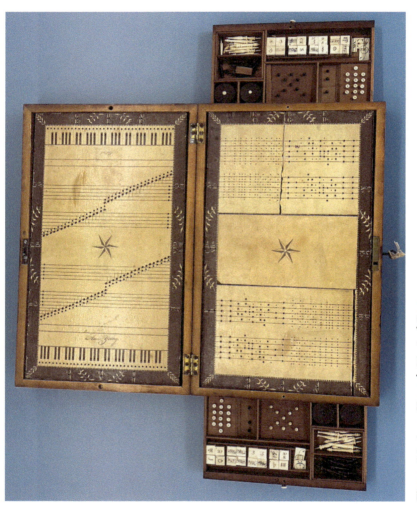

Figure 6.1 Anne Young's *Musical Games.*

are seven primary games, each of which employs distinct combinations of pieces and regions of the board, and there are several optional variations, so that at least twenty-one different games can be played on the set.

Anne's patent for the *Musical Games* specifies that they are designed for two players and suitable for children from the age of eight. The first British patent ever granted for a board game, it was also the only patent awarded that year to a female inventor.[4] In 1803, two years after the launch of the *Musical Games*, Anne published a companion treatise entitled *An Introduction to Music*, which explained music theory fundamentals via detailed instructions for playing the games. A final invention, a set of *Musical Cards* designed to teach rhythm, appeared a year later.

Little documentary evidence has thus far been uncovered about Anne's life: in 1801 she is mentioned briefly in newspaper accounts of her patent and in advertisements for the *Musical Games* themselves. In 1802, she married the distinguished Scottish cellist and flutist, John Gunn (1766/7–1824), himself an esteemed instrumental pedagogue and a fervent advocate of her work.[5] Two years later she published the *Introduction* under her married name, Anne Gunn. The *Musical Games* were noticed in a number of publications and reviewed by the eminent music theorist John Wall Callcott in *The British Critic and Quarterly Theological Review* in 1803, as was her treatise two years later.[6] The last mention of her in print seems to have occurred in 1812, when her husband praised her musical cards in his *Essay … Towards a More Easy and Scientific Method of Commencing & Pursuing the Study of the Piano Forte*.[7]

Nineteenth-century biographical lexica likewise offer little information about Anne. Perhaps the most detailed account of her life appears in John Graham Dalyell's *Musical Memoirs of Scotland* (1849), in an entry pertaining to her husband, which notes that he married "Miss Young, the inventor of the musical game—a method of acquiring knowledge of the theory of music"

---

[4] More precisely, David Ghere and Fred Amram note that "of the 4,090 patents issued by British monarchs between 1617 and 1816, Ann [*sic*] Young's patent was one of only forty that included the name of a woman. Of these forty, 33 patents reflect a female inventor." David Ghere and Fred Amram, "Inventing Music Education Games," *British Journal of Music Education* 24 no. 1 (2007): 55.

[5] Edinburgh marriage bans, July 20, 1802. In *The National Records of Scotland, Old Parish Registers: Marriages* 685/2 180, St Cuthbert's, 51.

[6] See Anon., "Review of New Musical Publications," 428; Johann Christian Hüttner, ed., *Englische Miscellen* 6 (1802): 82; Callcott, "Review of *Instructions*," 40–45; [John Wall Callcott], "Review of *Introduction to Music*," *The British Critic* 25 (January, 1805): 64–72; 163–171.

[7] John Gunn, *An Essay … Towards a More Easy and Scientific Method of Commencing & Pursuing the Study of the Piano Forte* (London: Preston, 1812), 10.

204   HEARING WITH THE MIND

and remarks that "supervening indisposition did not admit of their living together."[8] François-Joseph Fétis's *Biographie universelle des musiciens*, vol. 4 (1837) provides different entries for "Gunn (Jean)" and "Gunn (Madame Anne)," but his 1862 revision of the *Biographie universelle* carelessly implies that Anne had died between 1820 and 1827.[9] This error was taken over by *Grove's Dictionary of Music and Musicians* (1879), which provides no entry for Anne in her own right but briefly describes her under her husband's entry as "an eminent pianist," the inventor of the musical games, and "the authoress of a work entitled *'An Introduction to Music . . .* illustrated by musical games and apparatus and fully and familiarly explained' (Edinburgh about 1815)."[10] The entry goes on to note that "A second edition [of the Introduction] appeared in 1820, and a third (posthumous) in 1827."[11]

The scanty biographical information available about Anne has proven challenging for scholars of her work.[12] However, the cellist and scholar George Kennaway recently made the important discovery of a cache of letters by the aforementioned Gunn to his friend and patroness, Lady Margaret Clephane of Torliosk, documents which provide important insight into the last decade of Anne's life.[13] In what follows, I examine new evidence from Kennaway's finding as well as from other archival sources in an attempt to enhance our knowledge of Anne's biography, while also closely reading her games and her writings in order to better understand the music-theoretical, pedagogical, and social contexts from which they emerged.

*

---

[8] John Graham Dalyell, *Musical Memoirs of Scotland with Historical Annotations and Numerous Illustrative Plates* (Edinburgh: T.G. Stevenson, 1849), 235–236.

[9] In the 1837 edition of the *Biographie universelle*, Fétis notes that he did not know when the first edition of Young's *Introduction* was issued, and that the second was posthumous. This information is obscured in the 1862 edition of the *Biographie universelle*, where Fétis interpolates new details about the third (1827) edition of the *Introduction* without noting that the second edition had already been posthumous. See François-Joseph Fétis, "Gunn (Madame Anne)," in *Biographie universelle des musiciens* 4 (Brussels: Meline, Cans et Compagne, 1837), 470–471; François-Joseph Fétis, "Gunn (Madame Anne)," in *Biographie universelle des musiciens*, 2nd ed. (Paris: Didot, 1862), 4:164. On his ownership of the second edition see *Catalogue de la bibliotheque de F. S. Fetis: acquise par l'Etat Belge* (Brussels: C. Muquardt, 1877), 865.

[10] "Gunn, John" in *A Dictionary of Music and Musicians*, ed. George Grove (Boston: Ditson & Co., 1879), 1:641.

[11] Ibid.

[12] In *The Science of Music in Britain,* Kassler attempts to piece together a biography as well, but this contains some major errors, as does my own earlier publication on Anne. Carmel Raz, "Anne Young's Musical Games (1801): Music Theory, Gender, and Game Design," *SMT-V: Videocast Journal of the Society for Music Theory* 4, no. 2 (2018), goo.gl/ZXR6Cv.

[13] These letters appear in an appendix entitled "Clephane Letters" in George Kennaway, *John Gunn: Musician Scholar in Enlightenment Britain* (Woodbridge: The Boydell Press, 2021).

WHAT'S IN A GAME? 205

Anne Young (1756–1813) was born in Haddington, a village near Edinburgh, to a musical and scholarly family. Anne was particularly close with her older brother, Walter (1745–1814), the subject of Chapter 5 of this book. At least two other siblings survived to adulthood, Mary (1753–1827) and Isabella (1740–1820).[14]

The grammar school at Haddington was renowned for the quality of its musical instruction.[15] It was also distinguished by the practice, relatively unusual at that time in Scotland, of putting on full-length plays with all the parts performed by the schoolboys.[16] At Haddington, this tradition had been established by David Young's predecessor, the well-known pedagogue John Leslie (rector 1720–1730), and it was enthusiastically supported by the city council as well by the students' parents.[17] The practice of staging student plays was continued under David Young; in 1731, for instance, the boys put on Nicholas Rowe's play, *The Tragedy of Jane Shore* (1714).[18]

Anne's reputation as a music teacher was doubtless bolstered by her family connections to an elite school and her immersion in an environment that valued creative pedagogy. She appears in the Edinburgh city directories of 1794 and 1797 as Miss Young, "a teacher of music [at] 1, St. James Court," and the municipal records confirm a person of this name was living at that address in a flat owned by the Reverend Walter Young in 1799.[19] Later that year, however, she moved to the more upscale location of 7 St. James Square, a fashionable area popular with music teachers, and henceforth no

---

[14] See Young, "Essay," 55–110. As we will see in note 38, Mary appears repeatedly in Gunn's correspondence.

[15] Keith Sanger, "A Letter from the Rev. Patrick MacDonald to Mrs. Maclean Clephane, 1808," *Scottish Gaelic Studies* 26 (2010): 24.

[16] Attitudes, including ambivalence and even outright opposition, to student plays at different moments in early eighteenth-century Scotland are discussed in J. McKenzie, "School and University Drama in Scotland, 1650–1760," *The Scottish Historical Review* 34, no. 118 (1955): 103–121.

[17] On the support of the town council and a 1730 letter from the father of two students to Leslie, see ibid., 106. See also see James Miller, *The Lamp of Lothian, Or, The History of Haddington* (Haddington: James Allan, 1844), 451–452. In 1729, for example, the boys mounted a double bill comprising *Julius Caeser* and the celebrated Scottish poet Allan Ramsay's pastoral, *The Gentle Shepherd* (1725). The latter was subsequently developed by Ramsay into the first Scottish ballad-opera, and while there seems to be no definite evidence linking the Haddington schoolboys to the musical version of the piece, they may have been indirectly involved in its gestation. For further details on these productions, see Peter Holman, "A Little Light on Lorzeno Bocchi: An Italian in Edinburgh and Dublin," in *Music in the British Provinces, 1690–1914*, ed. Rachel Cowgill and Peter Holman (Milton Park, Abdington: Routledge, 2007), 77.

[18] James C. Dibdin, *The Annals of the Edinburgh Stage with an Account of the Rise and Progress of Dramatic Writing in Scotland* (Edinburgh: R. Cameron, 1888), 42.

[19] Edinburgh Dean of Guild Court Records, Box 1799/8.

206 HEARING WITH THE MIND

longer appears in the city directory as a music teacher, but simply as "Miss Young." We might speculate that this move indicated an improvement in her circumstances, and that perhaps she no longer felt the need to publicly advertise her services.[20]

Anne appears to have catered to the daughters of aristocratic families. In 1787–1788 she appears as the "music mistress" in the accounts of the Palace of Dalkeith, where she would have taught three of the young daughters of Elizabeth Montagu, Duchess of Buccleuch, a prominent patron of music. We learn from the palace accounts that they would send a chaise for her and pay for two five-hour visits a week, indicating that she spent substantial time with the young ladies. These accounts also specify that Anne supplied sheet music for her charges.[21] We know of one additional student, likely a woman of her own social class, who took out an advertisement in the *Caledonian Mercury* of 1816 stating that "Miss Ocheltrie, Pupil of Mrs. Gunn, begs leave most respectfully to intimate her Friends and the Public, that she [Miss Ocheltrie] continues to teach the piano forte, singing, and musical games, on the most approved principles."[22]

Anne's enterprising attitude toward her profession is evident in an early publication, entitled *Elements of Music and of Fingering the Harpsichord,* which first appeared ca. 1790 and was reprinted in numerous editions.[23] Comprising four pages of music-theoretical and fingering rudiments followed by 116 short pieces, this collection—one of only a handful of published musical works compiled by a woman in the eighteenth century—features minuets by Haydn and other contemporary composers, as well as many Scottish and Welsh dances. Notably, it also incorporates three pieces by female composers: two are by the Scottish aristocrat Lady Semphill, and the third is a jig composed by a seven-year-old girl (see **Figures 6.2–6.4**).

The inclusion of these pieces suggests that Anne was interested in promoting the talents of women to an extent that would have been unusual

---

[20] According to John Cranmer, "Shakespeare Square and St. James's Square were particularly well populated by music teachers." John Cranmer, "Concert Life and Music Trade in Edinburgh ca. 1780–1830" (PhD diss., University of Edinburgh, 1991), 149. We can be sure that the James Square address was inhabited by our Miss Young, as she lists this address on her patent application of 1801.

[21] I am grateful to Jeanice Brooks for this observation. The palace accounts list a set of concertos by the London-based German composer Johann Samuel Schroeter (ca. 1752–1788) by name, as well as an additional music book.

[22] Advertisement [Miss Ocheltrie], *The Caledonian Mercury* 14694, February 5, 1816, 1. As both Cranmer and Kennaway note, Miss Ocheltrie had taken out a similarly-worded advertisement in the *Edinburgh Evening Courant* for May 25, 1811; this notice stated that "Miss Ocheltrie (pupil of the late Mrs Gunn) respectfully intimates to her Friends and the Public, that she has moved from 59 South Bridge to 2 Lothian Street, where she continues to teach the Piano Forte and Musical Games on the most approved principles." For a discussion of the term "late" in 1811 see Kennaway, *John Gunn*, 12.

[23] On the various editions of Young's first publication see Kassler, *The Science of Music*, 1:431.

Figure 6.2 *Captain Bosvill's March* by Lady Semphill (Young, *Elements of Music*, 59).

in her day. Later in her life, two marches of her own composition were published by the London-based firm of Thomas Preston, although no copies appear to have survived.[24]

---

[24] Thomas Preston advertised "A March, inscribed to the Memory of the late brave, noble, and ever to be lamented general Sir Ralph Abercromby, Bart.; composed by Anne Gunn, for the Piano-Forte or Harp. 1s. 6d." Preston, "Correct List of New Publications," *The Literary Magazine or Monthly Epitome of British Literature* (June 1805): 335. General Ralph Abercromby, a Scottish solider celebrated for extinguishing the 1796 rebellion in Jamaica and capturing Trinidad and Tobago, died in the Battle of Alexandria in 1801. Kennaway discusses this advertisement briefly and notes that no copy has survived. Kennaway, *John Gunn,* 17. A few months later, Preston also advertised "A March

Figure 6.3 *Miss Maxwell's Delight* by Lady Semphill (Young, *Elements of Music*, 59).

In August 1802, Anne Young married the Scottish music teacher John Gunn. Gunn—who had been born in the Scottish Highlands around 1766 but had left the country as a child—had only recently moved from London to Edinburgh (probably in early 1802).[25] At the time of her

for the Piano Forte, composed by Anne Gunn. 1s. 6d." *The Literary magazine; or, Monthly Epitome of British Literature* (October 1805): 555. It is possible that both advertisements refer to the same piece.

[25] On Gunn's early years see Kennaway, *John Gunn*, 1–39.

Figure 6.4 *A Jigg Composed by a Young Lady of 7 Years of Age* (Young, *Elements of Music*, 60).

marriage, Anne would have been forty-six years old—eleven years older than Gunn—and certainly far beyond the age at which she could have possibly expected a suitor. Their relatively brief courtship would not have been unusual at the time, and upon marriage she would have ceased to work as a music teacher.

However, Gunn appears to have enthusiastically and consistently supported his wife's theoretical pursuits. Even prior to their marriage, his enthusiasm for Young's *Musical Games* was publicized by the Edinburgh-based

210 HEARING WITH THE MIND

firm Muir Wood & Co., which manufactured and sold them. Thus, for example, we read in an advertisement from January 2, 1802, that:

> The patentee [Miss Young] was lately favoured with an extract from a letter, written by Mr John Gunn, professor of Music in London, to Bethel Cox, Esq. which they beg leave to offer to the public, having the writer's permission. "I was, in the beginning of this week, most wonderfully delighted and astonished (forgive me the enthusiasm of the expression), to find all that you and myself had so frequently talked of, the possibility of conveying every musical information by moveable symbols, under the form of an amusement, at last carried into effect, infinitely beyond my utmost expectation; ... by means of (what is really wonderful) a few highly amusing, and in themselves interesting Games."[26]

In his treatise on cello playing published later that same year, Gunn praised the *Games* further and included the following disclaimer:

> Having so often mentioned this musical Curiosity, with apparent enthusiasm, I think the high degree of responsibility I stand in with the public ... obliges me to declare, that without having ever previously heard the name of the ingenious and scientific Inventor, I was struck with the merits of these Games."[27]

This partiality was noticed by at least one critic: in his 1803 review, Callcott could not resist observing that, "Since the publication of these games, the inventress has bestowed her hand upon Mr. Gunn, whose flattering encomiums on the scheme were excited by a stronger passion than science alone can impart. These we shall perhaps hereafter notice, with all due allowance for the lover's partiality."[28] Yet Gunn was undaunted: in his 1812 *Essay... Towards a More Easy and Scientific Method of Commencing & Pursuing the*

---

[26] Muir, Wood, & Co., Advertisement, *The Caledonian Mercury*, January 2, 1802, 3. Richard Bethel Cox, esq., was a wealthy and well-connected magistrate; he may have been Gunn's student or patron.

[27] John Gunn, *An Essay, Theoretical and Practical, with Copious and Easy Examples on the Application of the Principles of Harmony, Thorough Bass, and Modulation, to the Violoncello* (London: Preston, 1802), 49. Gunn also praises the games at length on page 5 of the treatise.

[28] Callcott, "Review of Instructions," 45.

Study of the Piano Forte, he averred that "no contrivance has ever appeared, so simple, so elegant, and beautiful, as the last production of the inventor of the Musical Games; namely, Mrs. Gunn's Musical Cards."[29]

We might wonder why an apparently successful teacher and pedagogue might want to leave her hard-won career behind for marriage to a much younger man she hardly knew. While marriage was generally considered an improvement in social standing over spinsterhood, women at the time did decide against the institution (Jane Austen's rejection of Harris Bigg-Wither's proposal in 1802 is one such well-known example).[30] Of course, Anne's union with Gunn may well have been a love match: the two certainly shared common interests in music theory and pedagogy, at least. His letters shed no light on this point, although he kept in close touch with her brother and sister, and he mentioned "the esteem and regard I have for that family" in his correspondence.[31]

Unfortunately, the two musicians did not have long to enjoy each other's company. In or around 1805, Anne seems to have fallen prey to an illness of some kind, possibly depression.[32] In a letter to Lady Clephane dated November 1, 1811, Gunn reveals that he and his brother-in-law Walter had placed Anne in an asylum a few weeks earlier:

> Twice since you left Edinburgh did I proceed far on my way to Erskine, but there I found letters from Dr. [Walter] Young assigning sufficient reasons for a longer prorogation of the time of removal of poor Mrs. Gunn. I was on that business and absent from this place for more than the whole month of October having just returned here in time to receive your letter, after having settled our distressed friend in the asylum agreed upon: the change has as yet produced no alteration whatever not even any regret at leaving her brother & sisters, except during the Journey. The kind & humane Mistress of the house, says she never met with a Case so distressing—but she has been little more than a week as yet in that change of situation.[33]

---

[29] John Gunn, *An Essay ... Towards a More Easy and Scientific Method of Commencing & Pursuing the Study of the Piano Forte* (London: Preston, 1812), 10.

[30] George Holbert Tucker, *Jane Austen the Woman: Some Biographical Insights* (London: Palgrave Macmillan, 1995), 36.

[31] Gunn, *Clephane Letters*, November 1, 1811, in Kennaway, *John Gunn,* 174.

[32] Recollecting a number of distressing events in 1805, Gunn mentions "the melancholy change in my wife." Gunn, *Clephane Letters*, August 9, 1813, in Kennaway, *John Gunn,* 184. It is unclear whether this comment refers to Anne's state of melancholy or her unfortunate condition more generally.

[33] Gunn, *Clephane Letters*, November 1, 1811, in Kennaway, *John Gunn,* 171.

212   HEARING WITH THE MIND

At some point in 1813, we learn from Gunn's letter, she died, aged fifty-six or fifty-seven; her death went unnoticed by the press, and the location of her grave remains unknown.[34]

Gamifying Education

Anne and her sisters would have expected to spend their adult lives as wives or as teachers. Teaching was, at the time, regarded as the only honorable occupation for unmarried women of the middle (or higher) classes. As Mary A. Maurice observed in *Mothers and Governesses* (1847), a good education provided young women with the ability to earn their own keep, were some dire change of circumstance to demand it. Otherwise, it was possible that as a consequence of family misfortunes, a young lady might be

> forced to become a governess. She then discovers her incompetence for the undertaking; her parents, who gave her an expensive education, are miserably disappointed; she has been too genteely brought up for any other employment, and what is to be done?[35]

We can glean some insight into Anne's experience of teaching music from the remarkably candid preface of her *Introduction*. She begins by observing that teaching young people "has always been found to be a work of much difficulty and delicacy" and recommends that "the drudgery of the elementary parts should be got over in early life; as grown persons are seldom disposed to bestow the necessary attention and application."[36] Indeed, she continues, "Frequent repetition of the same instructions, though often necessary and unavoidable, are irksome and discouraging to both teacher and pupil. The former cannot always refrain from betraying some undue degrees of fretfulness and impatience; nor can the latter avoid feeling mortification and disgust."[37] The games thus worked to save both teacher and student from these difficulties. It is surprising to see Anne's sentiments expressed with such frankness in her treatise. As she rather poignantly confesses, it was the design

---

[34] On the little we know regarding the Anne's death, see ibid., 14.

[35] Mary A. Maurice, *Mothers and Governesses* (London: John W. Parker, 1847), 27.

[36] Anne Gunn (née Young), *An Introduction to Music . . . as Illustrated by the Musical Games and Apparatus* (Edinburgh: C. Stewart & Co, 1803), vi.

[37] Ibid., vi–vii.

of the *Musical Games* that had "for some years . . . employed her most anxious thoughts, and been the most interesting occurrence of her life."[38]

While Anne seems to have been able to support herself as a music teacher, her older sister Mary was employed as a governess to the children of various aristocratic families; an experience that was, as we learn from letters written by Gunn (her brother-in-law), rife with frustrations.[39] The life of a governess was precarious and lonely. As the governess Ellen Weeton poignantly wrote in her journal of 1814–1825, "I sit alone in the evening, in the schoolroom . . . there is nobody in the house with whom I can be on equal terms, and I know nobody out of it, so I must make myself contented."[40] As if that were not enough, the governess had to contend with the resentment to which the mere fact of her presence in the household was likely to give rise. As Maurice observes,

> the husband dislikes the restraint, which will be imposed on the hours of domestic comfort, by the presence of a stranger; the children dislike the idea of one, who will enforce stricter discipline, than has hitherto been exercised over them; the servants dislike the idea of a new resident, who will add to their trouble; and the nurse not wishing for a diminution of her authority, inspires the younger ones with the idea, that their new governess will be cross, or at least will not indulge them as she has done. The mother sees the necessity of the step she is about to take, but in addition to the anxieties of the rest, she feels that she will be in a degree separated from her daughters.[41]

---

[38] Ibid., viii.

[39] Archival evidence suggests that Mary was employed as governess to the Maclean Clephane family around the year 1808. See Sanger, "A Letter from the Rev. Patrick Macdonald to Mrs Maclean Clephane," 32. She stayed in touch with the family, and they seem to have eagerly awaited news of her from Gunn, who reported on her travails with her new employers. In one such letter, he writes, "although Miss Young's situation is in great measure such as you apprehend it to be, from the injudicious direction of studies & plans prescribed by one Lady that is not mistress of the subject, and by the officious interference of another Yet she is more and more resigned to it . . . after my month's absence in the West I found Miss Young more reconciled to the harshness of these proceedings, & her opinions of her pupil, more favorable—her progress in Music, even exceeded my expectation— but it is little prized by the female part of the family, & the least possible time allowed for it." Gunn, *Clephane Letters,* November 1, 1811, in Kennaway, *John Gunn,* 172. A few months later, he reports that "I have written to & heard from poor Miss Young, matters are no better in that house," but a subsequent missive reveals that "Matters are going on rather more pleasantly in that Quarter, & I suppose she will have told you of the credit got from the Child's progress in Music etc." Gunn, *Clephane Letters,* February 4, 1812 in Kennaway, *John Gunn,* 178. Punctuation modernized.

[40] Ellen Weeton, *Journal of a Governess 1811–1825,* ed. Edward Hall (London: Oxford University Press, 1939), 2:68–69.

[41] Maurice, *Mothers and Governesses,* 5

214   HEARING WITH THE MIND

Jane Austen summed the matter up in *Emma* (1815) by characterizing the position of governess as entailing an abdication from "all the pleasures of life, of rational intercourse, equal society, peace and hope," entailing a life of "penance and mortification for ever."[42]

The governess's inherently isolated and socially precarious position—trapped alone in the fraught liminal space between family and servants, and often concerned as well about employment security—made it advisable for her to make the experience of study as enjoyable as possible both for her charges and for herself. As Anne herself makes clear, it was chiefly this that inspired the creation of her *Musical Games*.[43] Moreover, some governesses or schoolmistresses were able to leave their positions and maintain themselves financially by writing and publishing popular manuals of instruction—although that of course meant that, as historian Kathryn Hughes notes, they were "dependent on selling their books if they were to avoid returning to [the schoolroom]."[44] For women with the right combination of literary ability, self-confidence, and motivation, this was a possible, if certainly a daunting, course. It may be, therefore, that Anne saw in the *Musical Games* the prospect of eventual emancipation from the labor of teaching.

Anne's *Musical Games* thus reflect the increasing interest in pedagogy, and, in particular, in the creation and publication of marketable pedagogical devices aimed at affluent families, in the first few decades of the nineteenth century.[45] This trend is evident in the appearance of a spate of didactic toys and inventions, a number of which were designed by women. Thus,

---

[42] Jane Austen, *Emma: An Authoritative Text, Backgrounds, Reviews and Criticism*, ed. Stephen M. Parrish (New York: W. Norton, 2000), 106. A subplot of the novel describes the horror that various characters in the book feel about Jane Fairfax's impending employment as a governess.

[43] See Gunn (née Young), *Introduction*, vi–vii.

[44] Kathryn Hughes, *The Victorian Governess* (London: Bloomsbury, 2003), xii. Hughes observes that "Three of the foremost writers on middle-class female education during the first half of the Victorian period—Anna Jameson, Mary Maurice, and Elizabeth M. Sewell—had all spent part of their working lives as schoolteachers or governesses." Ibid., 207.

[45] As Jill Shefrin reminds us, in *Mansfield Park* Jane Austen uses Fanny Price's ignorance of educational games to depict the socioeconomic difference between her and her wealthy cousins: "her cousins . . . thought her prodigiously stupid, and for the first two or three weeks were continually bringing some fresh report of it into the drawing-room. 'Dear mama, only think, my cousin cannot put the map of Europe together?'" as quoted in Jill Shefrin, "'Make it a Pleasure and Not a Task': Educational Games for Children in Georgian England," *Princeton University Library Chronicle* 60, no. 2 (1999): 251. This passage refers to the ubiquitous geographical puzzle games pioneered by the Abbé Gaultier, to which I will shortly turn.

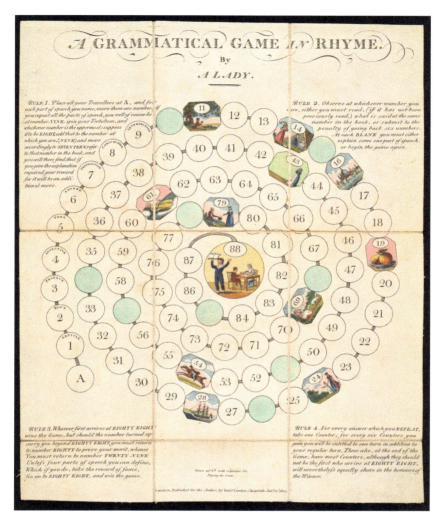

Figure 6.5 Rowse, *A Grammatical Game in Rhyme, by a Lady* (1802).

for example, in 1802, Elizabeth Rowse invented a board game entitled *A Grammatical Game in Rhyme, by a Lady*, followed two years later by a board game called *Mythological Amusement*, designed to help children learn the characters of the Greek myths (Figures 6.5 and 6.6). 1804 also saw the publication of an astronomy-themed board game called *Science in Sport*, adapted from the well-known game of Goose by Margaret Bryan (ca. 1760–1816), a writer and science educator who directed a girls' school in

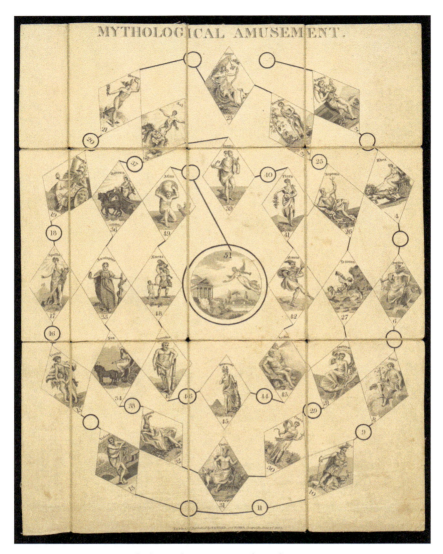

Figure 6.6 Rowse, *Mythological Amusement* (1804).

Blackheath (**Figure 6.7**). **Figure 6.8** shows *The Study of the Heavens at midnight during the winter solstice, arranged as a game of astronomy, for the use of young students in that science,* designed by the novelist and educator Alicia Catherine Mant (1788–1869) in 1814, to whom we will also return.

And even in cases where pedagogical games were not manufactured or sold, they were sometimes used to demonstrate resourcefulness: in *Letter*

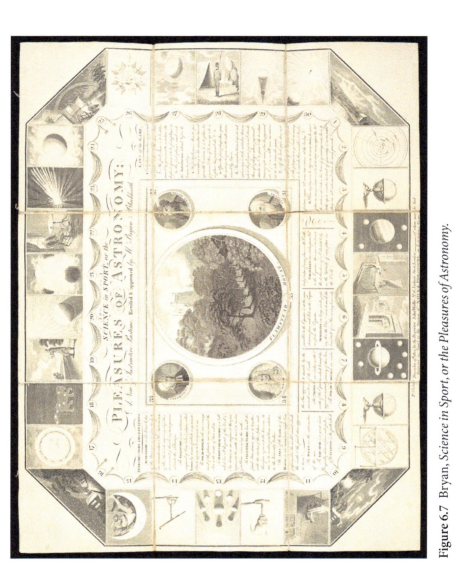

Figure 6.7 Bryan, *Science in Sport, or the Pleasures of Astronomy.*

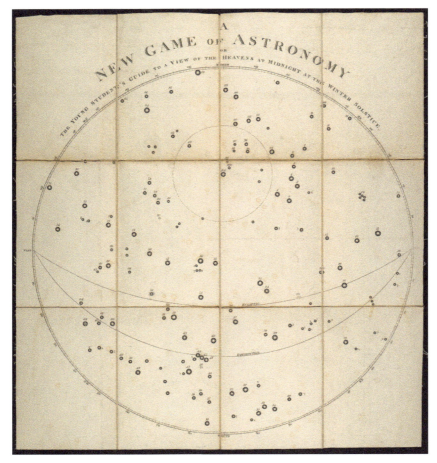

Figure 6.8 Mant, *The Study of the Heavens at Midnight.*

of a Village Governess; descriptive of rural scenery and manners (1814), a collection of anecdotes, the Scottish pedagogue Elizabeth Bond includes an account of an impromptu game concerning the rudiments of music, in which

> the children are seated round a table; each girl has six counters; and a sheet of music lies before them. My first question is—'How many lines; then how many spaces, &c. &c. &c., till I go through the whole musical catechism. Every false answer pays a forfeit; and, of course, the girl who retains the last counter wins the pool, which consists of fruit, gingerbread, or any tempting

morsel that may be had. Thus, for once, the spirit of gambling is converted to a useful end, and raises a degree of emulation that soon gains the wished-for point.[46]

The sizable increase of female teachers went hand-in-hand with the emergence of a new pedagogical philosophy dedicated to enhancing learning by play. The origins of this shift were generally ascribed to the educational writings of John Locke, as Callcott observed in his 1803 review of Anne's *Instructions for Playing the Musical Games:*

> It has been the fashion for several years past, to communicate science to the infant mind through the medium of play: and, according to Locke's advice, "children are to be cozened into knowledge." To this system the Abbé Gaultier has largely contributed; and his works have a degree of analytical merit, which is not to be found in many of his imitators. The present attempt is a favourable exception to the last remark.[47]

Locke, in *Some Thoughts Concerning Education* (1693), had considered how elements of play could encourage children to regard the act of learning as entertaining and enjoyable. By using specially designed pedagogical toys, Locke maintained, children could be "taught to read, without perceiving it to be anything but a sport."[48] As a possible example of such a device, he proposes the following:

> What if an ivory-ball were made like that of the royal-oak lottery, with thirty-two sides, or one rather of twenty-four or twenty-five sides; and upon several of those sides pasted on an A, upon several others B, on others C, and on others D? I would have you begin with but these four letters, or perhaps only two at first; and when he is perfect in them, then add another; and so on, till each side having one letter, there be on it the whole alphabet. This I would have others play with before him, it being as good a sort of play to lay a stake who shall first throw an A or B, as who upon dice shall throw six or seven.[49]

---

[46] Elizabeth Bond (of Fortrose), *Letter of a Village Governess; Descriptive of Rural Scenery and Manners; with Anecdotes of Highland Children* (London: The Author, 1814), 1:170.

[47] Callcott, "Review of *Instructions*," 41.

[48] John Locke, *Some Thoughts Concerning Education (1693), Critical Edition*, ed. John W. Yolton and Jean S. Yolton (Oxford: Clarendon Press, 2000), §149, 209.

[49] Ibid., §150, 209–10.

220   HEARING WITH THE MIND

Using play in a pedagogical context could become an effective method of encouraging the acquisition of—and reinforcing the retention of—higher forms of knowledge. Furthermore, it would allow children to channel the intense energies that they brought to memorizing the rules of recreational games to the acquisition of useful knowledge. Locke remarked on having seen "little girls exercise whole hours together, and take abundance of pains to be expert at dibstones [jacks]," and noted that "it wanted only some good contrivance to make them employ all that industry about something that might be more useful to them."[50]

For older children, the heightened emotions linked to the vicissitudes of competition were an integral part of pedagogical play. The London-based French educational reformer Abbé Aloïsius Gaultier (1745?–1818), also mentioned by Callcott in his review, was an important pioneer of more advanced didactic games that took advantage of the strong emotions induced by competition. Gaultier achieved particular renown with his geographical games, which made use of maps, as well as his games for learning grammar, history, and manners.[51] As a 1787 commission of *L'Académie Royale des Inscriptions* reported, Gaultier's inventions were particularly successful in encouraging knowledge, as elements of competition and reward mobilized children's "interest, pride, emulation, glory, [and] shame."[52]

At this point it will be helpful to briefly depict a few of the actual games involved in Anne's creation in more detail. These generally require the player to set up various musical configurations on the game board as a response to throws of specially designed musical dice. For example, the first game, "Key Signatures," has players throwing a die with the letters of the twenty-four major and minor sharp and flat keys (one player takes the sharps, and her dice are the major keys with one to six sharps, indicated in capital letters, the major keys with seven to twelve sharps, indicated in capital letters, and two more dice with the respective relative minors; a corresponding set of four dice with flats is available to the other player).[53] Each player throws one die and sets up their keys on the staves in response to the die. Throws resulting in relative, parallel, or alphabetically adjacent keys to the outcome on the other player's die are awarded with double

---

[50] Ibid., §152, 210.

[51] On Locke and Gaultier, see Shefrin, " 'Make it a Pleasure and Not a Task,' " 251–275.

[52] "L'interêt, l'amour-propre bien ordonné, l'émulation, la gloire, la honte, sont autant de mobiles qu'elle met en action." *Rapport de l'académie des inscriptions* (27 August, 1787), as cited in Abbé Gaultier, *Leçons de grammaire* (Paris: 1787), x.

[53] Anne's inclusion of keys with up to twelve sharps and flats is discussed later in this chapter.

points, as are other various relationships between the two players' accidentals. The first player to achieve twelve points wins.

In "Intervals," the third game, the object is to race up or down the keyboard. Each player first rolls a set of two dice with key signatures and notes. They must correctly describe the key signatures and the intervals between the notes and identify whether they are consonances or dissonances. They must then transpose these intervals onto the note upon which their counter currently stands and move accordingly. Consonant intervals afford movement in the desired direction; dissonant intervals set the player back in the opposite direction. Rolling parallel or relative keys results in bonuses of an extra octave and a fifth, but only if correctly identified.

The fourth game, entitled "Rule of the Octave," teaches the by-then standard thoroughbass harmonization of the ascending and descending scale. Players throw a set of four dice marked with various thoroughbass figures and set up a scale degree on the board's staff only if the thoroughbass figure corresponding to that scale degree is thrown. The correct figure will usually emerge gradually over several throws rather than in a single throw. If this fails to occur after three throws, the appearance of any of the figures involved in the harmonization of that scale degree means that the player can go ahead and set up her chord on the board. The last game of the set, "Modulations," is another racing game. The object of the game is to complete a full revolution around the circle of fifths by modulating. Each player throws a die with various accidentals and numbers, along with a die showing movement of the bass, and sets up their modulation in response.

In spite of the abundance of new educational games at this time, the design and execution of the *Musical Games* far exceed those of comparable products. This is evident if we compare Anne's invention to other musical games of the time, which tended to repurpose extant games to teach music theory. These included John Wallis's *Musical Dominos* from 1793, E. E. Thomas's *New Musical Games* [1821?], a set of cards that used musical symbols for the purpose of playing the popular card games Pope Joan, Casino, and Commerce,[54] or L. M. Drummond's *Musical Game for Children* of 1805, which likewise borrowed a staple template of educational games— the traditional boardgame of Goose.[55]

---

[54] For more information on these games see David Parlett, *Penguin History of Card Games* (London: Penguin UK, 2008).

[55] See Adrian Seville, *The Cultural Legacy of the Royal Game of the Goose: 400 Years of Printed Board Games* (Amsterdam: Amsterdam University Press), 2019.

222   HEARING WITH THE MIND

We can even compare the *Musical Games* to another production by the author herself, her *Musical Cards* of 1804.[56] The cards, a pack of forty-eight, are divided into four color-coded suits named after the seasons: Spring is green; Summer, red; Autumn, lilac; and Winter, black. They feature various figures such as a knight in armor, a lady, a boy, a girl, a shepherd, and a shepherdess, etc., which correspond to different rhythmic values (notes and rests indicated separately), with two cards taking on the dual role of demisemiquaver (sixteenth note) or the addition of a dot to the temporal values of the other cards. The booklet describes seven possible games. In the first, one party plays the highest value note or rest in her possession, and the other must attempt either to exceed this value or match it with two or more cards in her possession. The second game is a round game, and the third is based on the rules of whist. Other games involve placing the cards with various note values into bars of varying time signatures according to various rules.

Unlike the *Musical Cards* or the other inventions described above, Anne's *Musical Games* were not a musical adaptation of a pre-existing game. Some of the individual games employ the accumulation of points, while others exploit the competitive device of racing to explore musical spaces ranging from the circle of fifths to the keyboard and the rule of the octave. Moreover, the board itself offers different ways to visually represent musical concepts— for example, the circle of fifths can be pursued on a schematic diagram, on a blank musical stave, on a representation of a six-octave gamut, or on a keyboard—and certain games can be played on more than one part of the board. This affordance helps reinforce the various interconnections between the different domains, and the central metaphors, of music theory.

Anne's *Introduction*

Anne opens the preface to her treatise with the following observation:

> To communicate knowledge to young persons, to engage and fix their attention, to render the subjects of instruction simple, and commensurate to the powers of their understanding, and to impress them upon the mind, so as they may be readily and distinctly recollected, and thus constitute a real

---

[56] The *Musical Cards* can be viewed at https://catalog.princeton.edu/catalog/9941125343506421.

WHAT'S IN A GAME? 223

and permanent acquisition, has always been found to be a work of much difficulty and delicacy.[57]

This carefully structured statement breaks down the process of knowledge acquisition into distinct stages: in order to communicate knowledge, the teacher must engage and fix the student's attention with simple and commensurate materials in order to make a clear and durable impression upon the pupil's mind. No mere rhetorical flourish, this psychologistic interest in the materials and mental capacities necessary for learning music is a consistent feature of the book.

Anne regarded the games as offering a distinctive complement to more traditional methods of acquiring music-theoretical knowledge. A student embarking upon the course of study outlined by her treatise, she explained,

> ought to practise the different exercises upon the pianoforte, before he proceed to work them upon the game tables, or what may be called the dumb clavier. By the former, his hand will acquire a readiness in executing the several combinations and successions, and his ear will be habituated to the sound of them; by the latter, the knowledge of them will be imprinted in his mind.[58]

Here we have a division of music-theoretical skill into three domains: the embodied, the aural, and the mental. While playing an instrument could reinforce the first two, playing the games, Anne argues, could provide essential support for the latter.

Later in the preface, she remarks on the importance of disentangling the ear from the mind and recommends that students practice their harmony exercises not only upon the piano but also silently, by placing pins upon the piano keys illustrated on the game tables, in order to reinforce embodied skill with intellectual knowledge. Possession of the former on its own, she cautions, is insufficient:

> A young person who has a good ear, and is either of an impatient or of an indolent disposition, if he makes a mistake in striking a chord upon an instrument, is immediately apprised by his ear, that something is wrong. In

---

[57] Gunn (née Young), *Introduction*, i.
[58] Ibid., xii.

224 HEARING WITH THE MIND

place, however, of endeavouring to correct his error by rule or principle, he searches about the instrument, tries one key after another, until he obtain a combination, that gives the sound which he expected. Although, however, he may succeed in this, he has gained no advantage, and is not secure against falling into the same mistake, at the next occurrence. But upon the game tables, having no assistance or direction from his ear, he can only perform correctly by the efforts of his mind; and if he commit errors, he must hazard the loss of his party.[59]

Here we find Anne recapitulating the traditional distinction between *sensus* and *ratio*, with the ear enabling the musician to bypass explicit, rational knowledge in favor of corporeal happenstance. In unexpected echoes of older speculative ideas, the games thus constitute a domain where music-theoretical knowledge has silent consequences.

Although nominally designed for children, the games require a musically literate adult to referee between any two novices playing the game, as Anne does not provide comprehensive answer keys in her treatise. At least some of the more advanced games are explicitly intended for students to play with their instructors. The very last section of the book, for example, sketches out ways in which the seventh game might be extended to practice distant modulations. She elaborates that

when this game is played by a master and an advanced pupil, the different keys which are presented by the dies to be modulated into may afford various subjects of instruction and practice, as each throw may be considered as proposing an harmonical problem, in the solution of which, knowledge and ingenuity may be displayed.[60]

Anne's psychologistic approach appears elsewhere in the preface as well. A comprehensive understanding of the principles of music theory is essential, she maintains, in order for a "person to perform a musical composition, with that readiness, and accuracy, with that just expression, and with that perception of the meaning and connection of its several parts, which alone can produce the intended effect, or communicate real satisfaction to an intelligent hearer."[61] Here, Anne outlines an explicit progression of knowledge

---

[59] Ibid., xxii.
[60] Ibid., 256.
[61] Ibid., iv.

from technical competence to technical accuracy, to expressivity, and finally to the ability to comprehend and perceive musical meaning via an understanding of the kind of musical structure that only theory can illuminate, arguing that only possession of such understanding can enable a performer to convey full satisfaction to informed listeners. Rather than residing in sung words or recognizable affect, here meaning can also be found in intrinsic structural relationships of the kind revealed by music theory.

Anne goes on to caution that "destitute of this knowledge, a performer is often in the situation of a person who rehearses, or reads, from the mere knowledge of characters, what has been composed or written in an unknown language."[62] This analogy is reminiscent of a passage in Holden's *Essay*. Holden, the only author she mentions by name in her treatise, compares an unskillful listener, who, having missed the beginning of a piece, is unable to discover its key, to a schoolboy who has missed the first part of a mathematical demonstration and "can neither discover the connection, nor the design of the reasoning, for want of knowing what is the principal proposition intended to be demonstrated."[63] Seen in this context, her passing remarks may indicate an awareness of an aesthetic discourse that seems largely unique to Scotland at the time, one that placed an emphasis on the necessity of understanding music-theoretical structures in order to fully comprehend and enjoy music.

Finally, traces of Anne's psychologistic approach also crop up in the treatise itself. Thus, for example, she provides an explanation for the standard, five-line staff in a footnote, surmising that "the number five was probably pitched upon, not only because it is easily surveyed by the eye, but also as without leger lines, it comprehends an octave and 4th, the usual compass of the human voice."[64] Perhaps she had discovered that this question frequently came up in her experience instructing children and included these two explanations as a result. At the same time, her invocation of what is "easily surveilled by the eye" suggests an interest in perception, one that is reminiscent of Holden's theorizing, and in particular his notion of visual grouping, discussed in Chapter 2 of this book.[65]

---

[62] Gunn (née Young), *Introduction*, iv.

[63] Holden, *Essay*, part 1, §30, 26–27.

[64] Gunn (née Young), *Introduction*, 17.

[65] As readers will recall, in Chapter 2 we encountered Holden's claim that "where equal and equidistant objects affect our senses, that there is a certain propensity in our mind to be subdividing the larger numbers into smaller equal parcels; or as it may be justly called, compounding the larger numbers of several small factors, and conceiving the whole by means of its parts." Holden, *Essay*, part 2, §11, 288–89.

226 HEARING WITH THE MIND

A similar explanatory approach can also be discerned in her account of rhythm, a topic not addressed by the *Musical Games* but included in the *Introduction* so that it "may, in some degree, answer the purpose of an elementary treatise."[66] Describing rhythmic notation, Anne remarks that "as none but the more simple proportions of duration can be distinctly perceived, or accurately expressed in musical sounds; the notes of time, which are employed in music, proceed from the longest to the shortest, by a regular succession of degrees, each of which is, in duration, exactly the half of the one immediately preceding."[67] As in the preceding instance, this is a cognitive explanation for a fact that is almost always presented without any attempt at elucidation, and may reflect a response to the incessant questioning of children as well as Anne's own interest in psychology.

The Theory

The *Introduction* intersperses ten chapters of theory fundamentals with instructions for playing eight different kinds of musical games. Chapters 1–5 appear first, then a description of the game table and pieces, followed by an introductory game for practicing key signatures, succeeded by games dealing with (1) key signatures, (2) clefs, and (3) musical intervals. These games are followed by Chapters 6–7 and then game (4), cadences; this is succeeded by Chapter 8 and Game 5, which treats the rule of the octave in major and minor; Chapter 9 and Game 6a treats the resolution of discords, while Chapter 10 and Game 6b treats chords by license. Game 7 then deals with modulation, which is explained by text preceding, and intermixed with, the game instructions.[68]

In the treatise, Anne makes no reference to any other musician or music theorist—with the single aforementioned exception of Holden, whose *Essay* she describes in her preface to the 1803 edition as the model for her presentation of the theory.[69] The body of the treatise mentions neither scientific ideas such as the overtone series nor theoretical concepts such as the fundamental bass, and it does not provide a theoretical explanation of the conceptual

---

[66] Gunn (née Young), *Introduction*, xiv.

[67] Ibid., 23–24.

[68] The second (1820) and third (1826) editions of the book feature another five games which treat more elementary subjects: note names, the number of semitones and tones in each interval, building the major scale, interval inversions, and ledger lines.

[69] Gunn (née Young), *Introduction*, xviii.

WHAT'S IN A GAME? 227

materials it does teach. However, a careful examination of the authorial decisions that Anne takes in order to present this simplified account reveals her keen awareness of, and engagement with, contemporaneous British and French music theory.

To modern eyes, one of the most unusual features of Anne's *Introduction* is her prominent inclusion of theoretical keys in the design of both the games and the game board. The latter features the circle of fifths around the outer rim of the game boxes, a representation that corresponds to **Figure 6.9** in her *Introduction*. Unlike any previous representation of the circle of fifths of which I am aware, however, including the depiction offered in her own earlier *Elements of Music*, Anne extends the circle through twelve accidentals in each direction, resulting in a space that includes theoretical keys such as E-sharp major with eleven sharps (seven sharps and four double sharps) and B-double-flat minor with twelve flats (seven flats and five double flats).[70] That is, the enharmonic substitution only occurs at the origin point (C Major = B-sharp Major = D-double-flat Major), and not approximately halfway through the circle (at C-sharp Major = D-flat Major; A-sharp minor = B-flat minor) as precedent and practice would suggest.

It is possible that Anne designed her circles so to better correspond with the games to be played on the game board, and specifically Game 1, where she assigns one player to the sharps and the other to the flats, inviting each to move their pieces around the outer rim of the game board. A closer look at some of Anne's contemporaries, however, reveals that such an expansive approach to distant keys was widespread among British theorists in her day and later. Thus, for example, Charles Burney provides a sketch that includes all twelve sharp-side keys in his entry on music in Rees's *Cyclopedia*, written around 1803 but published posthumously in 1819 (**Figure 6.10**).[71]

We find a similar extension of the circle of fifths in William Crotch's *Elements of Musical Composition* (1812), which includes major and minor keys ranging from E-double-flat to C double-sharp (**Figure 6.11**).

---

[70] In the introductory portion of her *Elements of Music,* Anne offers the standard circle of fifths, with an enharmonic substitution at F-sharp Major/G-flat Major. Young, *Elements of Music and of Fingering the Harpsichord* (Edinburgh: ca. 1790), 5.

[71] Charles Burney, "Music. Thoroughbass or Accompaniment—Plate II," in *The Cyclopædia; or, Universal Dictionary of Arts, Sciences, and Literature*, ed. Abraham Rees (London: Longman, Hurst, Rees, Orme, & Brown, 1819), 24:unpaginated.

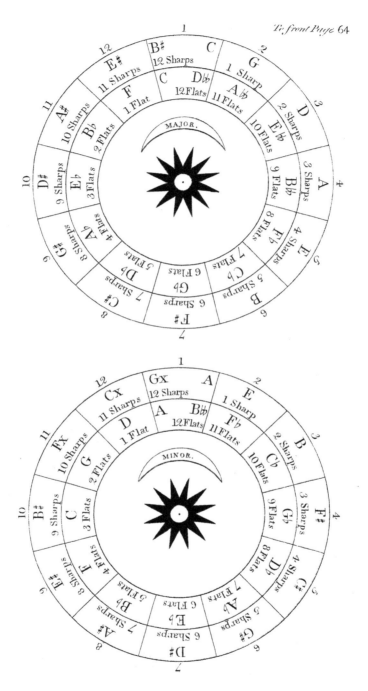

Figure 6.9 Anne's Chromatic Circle of Fifths (Gunn (née Young), *Introduction*, facing page 64).

Figure 6.10 Burney's Circle of Fifths: the extreme sharp keys appear on the upper staff; an enharmonic shift is indicated in the seventh measure in the lower staff.

Figure 6.11 Crotch's examples 9 and 11, showing the accidentals involved in twenty-five major and minor keys.

Other striking representations of theoretical keys include *a Musical Balloon Containing 52 Different Keys* (ca. 1810), by one "Schultze," shown in **Figure 6.12**, and, in **Figure 6.13**, John Smith's *Musical Sacred Globe, Containing 50 Different Keys, & How to Transpose Them* (ca. 1814).

We might ascribe this apparent exuberance with regard to theoretical keys to the resistance to equal temperament in Britain at the time.[72] Anne, like many of her contemporaries, regarded just intonation as the theoretical ideal.[73]

---

[72] Robert Palmieri observes that "[u]nlike France and, above all, the areas of German influence, Great Britain put up a strong resistance to equal temperament (which was viewed as a novelty from the Continent, and from Germany in particular)." Robert M. Palmieri, *Piano: An Encyclopedia*, 2nd ed. (New York: Routledge, 2003), 391. As late as 1806, the Earl of Stanhope related the following: "I had the curiosity to converse with sixteen or eighteen of the most eminent musicians in England upon this subject. Half of them did then approve of what is called THE EQUAL TEMPERAMENT. ... The other half, on the contrary, reprobated that mode of tuning, as never satisfying the ear perfectly in any one key whatsoever." Charles Stanhope, "The Science of Tuning Instruments with Fixed Tones," in *The Repertory of Patent Inventions, and Other Discoveries and Improvements* 9, no. 2 (1806): Appendix, 3. Original capitalization. Similarly, in 1812 John Southern described Robert Smith's meantone tuning (1749) as one that "continues in common use." John Southern, "On the Vibrations of Musical Strings; with a Mode of Ascertaining the Sound Producible by any Given Number of Vibrations," *The Philosophical Magazine* 40, no. 175 (1812): 335.

[73] The direct consequence of this attitude meant that in practice, many musicians preferred meantone or circulating temperaments to equal temperament. Equal temperament, according to mathematician and acoustician Alexander Ellis, would only be adopted by British piano and organ manufacturers in the 1840s and 1850s, respectively. Alexander Ellis, "On the History of Musical Pitch," *Journal of the Royal Society of Arts* 28, no. 1,424 (1880): 295.

Figure 6.12 Schultze, *Musical Balloon Containing 52 Different Keys.*

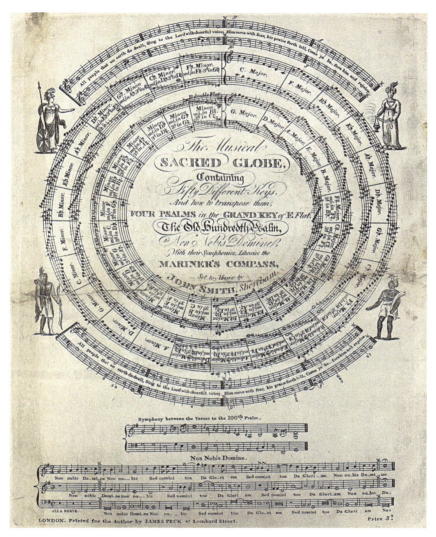

Figure 6.13 Smith, *Musical Sacred Globe, Containing 50 Different Keys, & How to Transpose Them* (ca. 1814).

In the treatise, she repeatedly emphasizes that the equal-tempered approach she adopts is a compromise.[74] This suggests that she, like most musicians of her day, would have been operating with at least two profoundly different tuning systems in mind: the ideal of just intonation on the one hand, and on

---

[74] Anne notes that "all semitones, although not strictly equal, are accounted as equal in the practice of music," 42.

232 HEARING WITH THE MIND

the other, equal temperament, a concession to fixed-pitch instruments (and pedagogy). Because, in just intonation, notes that are enharmonically equivalent actually have differing pitches, Anne may have regarded it as pedagogically important to include such exotic keys as C-double-sharp major and minor as an indication of the far-flung distances that various modulations might traverse, and perhaps as a way of honoring the fact that the series of fifth-related notes and their associated keys form an infinite spiral. Her passing allusions to potential features of just intonation imply that she felt obligated at least to hint at her true theoretical principles.

Despite her awareness of the complexity of the music-theoretical landscape, Anne takes care to emphasize that keyboard instruments do not afford a sounding difference corresponding to just intonation's distinction between major and minor semitones. She writes: "Although it is certain, in theory, that C C-sharp is a smaller interval than C-sharp D; yet, in the composition and practice of music for keyed instruments, all the semitones of the clavier are used, as if they were of equal extent."[75] Anne here threads a rather unusual needle: she at once presents elementary music-theoretical principles pragmatically focused around fixed-pitch keyboard instruments, while implicitly indicating an allegiance to an always unattainable just intonation.

There are, moreover, practical explanations for Anne's decision to incorporate theoretical keys into her pedagogy. For one thing, her example demonstrates the consequences of stacking twelve ascending fifths from C to B-sharp (or descending fifths from C to D-double-flat), showing her pupils exactly how one would arrive at the "enharmonic" equivalent of the departure note after twelve modulations. Another, far more prosaic reason, has to do with the design of the game table itself, which depicts these expanded circles of fifths around the border of each half (see **Figure 6.14**). It would be easier and more consistent to include twelve "positions" on each side, each with one type of accidental (that is, one side with sharps and one side with flats), than to use enharmonically equivalent accidentals on both sides. Anne's design thus allows one player to take the sharps and the other the flats, and for the key positions to be equally spaced from each other around the edges of the game tables.[76]

---

[75] Gunn (née Young), *Introduction*, 48.

[76] For a discussion of how the symmetrical constraints of antagonistic gameplay appear to have influenced later music theory, see Callum Blackmore, "Berton's Ludic Pedagogy and the Subdominant Otherwise: Tension and Compromise in the Early Paris Conservatoire Curriculum," *Current Musicology* 104 (Spring 2019), https://doi.org/10.7916/cm.v0i104.5395.

WHAT'S IN A GAME?   233

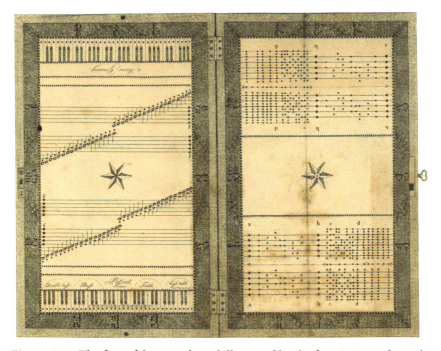

Figure 6.14 The face of the game board illustrated by the frontispiece of Anne's 1803 treatise. Note the extended circles of fifths along each margin.

Another point of interest concerns Anne's treatment of the perennial question of the consonance or dissonance of the perfect fourth. Anne regards the perfect fourth as a concord, even though it "sometimes calls to be resolved, which many theorists consider as the characteristic of a discord. The same thing, however, may be occasionally affirmed of the perfect 8ve, and perfect 5th."[77] The perfect fifth, of course, does indeed present itself as an interval requiring resolution in any minor seventh chord, though of course, in such cases it is really the co-presence of the chord's root that renders the upper note of that fifth dissonant. Anne's claim that the octave, too, on occasion, requires resolution is harder to substantiate. Perhaps she had in mind a case such as the opening of Mozart's Sonata no. 12 in F Major, K. 332 (**Figure 6.15**). At the downbeat of m. 4, the top-voice F forms an octave with the bass, but because the bass here is a pedal point dissonant at that moment with the entire diminished seven chord above it, the top-voice F's consonant

---

[77] Gunn (née Young), *Introduction*, 41.

234   HEARING WITH THE MIND

Figure 6.15   Mm. 1–4 of Mozart's Sonata no. 12 in F Major, K. 332.

relation with the bass F is negated by its dissonant character as an accented passing tone. Thus, paradoxically, Anne here adheres to the traditional view that the perfect fourth is a concord but does so because she questions the almost equally traditional association between discord and the requirement of resolution.

Perhaps the most significant issue raised by Anne in her *Introduction*, however, concerns the subdominant chord. Having introduced the concepts of the triad, the fundamental, chord inversion, and different types of cadences, she remarks that

> the first of the scale, or key note, is by some musicians called the Tonic, and its chord, the chord of the tonic. The 5th of the scale is called the dominant, and the 4th the subdominant, and their chords are respectively named accordingly. The 3rd and 6th of the scale have been called the Mediants. The major 7th is named the Sensible note.[78]

Here we should observe that only the terms "tonic," "dominant," and "subdominant" are explicitly said to be names of chords, and that scale degree II is not mentioned here at all due to the fact that it has no name beyond its numerical designation. It is with regard to the seventh chord built on scale degree II, however, that things become interesting. Like a number of other eighteenth-century theorists, Anne sees what we call the II^7 chord as the third inversion of the very chord that *we* would usually consider *its* first inversion, the chord that Rameau called "the chord of the large sixth" and which Anne calls "the discordant harmony of the subdominant."[79] This is the chord we call II6/5. In her own words:

---

[78] Ibid., 136–137.
[79] Ibid., 144.

The only discording interval which appears in this chord is that of major 2nd, subsisting betwixt the 5th of the chord and the added 6th, or C and D, the key and 2nd of the scale. This discord of 2nd, is in the 3rd position of the chord, inverted into a minor 7th. If now, D, the lowest note of that position, be considered as the fundamental of the chord, the intervals to it will be f minor 3rd; a, perfect 5th; and c, a minor 7th. This chord therefore may be considered as having two fundamentals; as two of its notes have a 3rd, and perfect 5th among its intervals. These are the 4th of the scale, in which the discord is the added 6th; and the 2nd of the scale, in which the discord is an added 7th. To this last, the term subdominant chord, has by some authors been appropriated; and it has been called the minor chord of the 7th; as the dominant chord, above treated of, has been named the major chord of the 7th.[80]

Here, Anne notes that the same term—subdominant—can be applied to chords built on II as well as on IV, suggesting that she views the term not as relating to two versions of the same *chord,* but rather as a kind of a sublation of the two chords into a category that contains them both. Implicit in this passage is a new understanding of *classes* of chords as versions of the same function. It is unclear to whom Anne refers to by "some authors," as I can think of no precedent to this claim. This phrase occurs only in her treatment of the subdominant category and may thus indicate her own theoretical contribution.

Anne's treatment of II7 here implicitly invokes an established theoretical tradition of considering the minor seventh chord as possessing two fundamentals—namely, the notes we call its root and its third—in order to justify regarding it as a kind of a subdominant. This tradition, discussed at greater length in Chapter 2 of this book, stems from the writings of the Swiss music theorist Jean-Adam Serre in 1753 and should not be confused with Rameau's *double emploi.* With the term *double emploi,* Rameau set out to explain how the requirement for fundamental bass motion by harmonic interval, that is fifths and thirds and their inversions, can be fulfilled when it comes to scale degree VI in an ascending and descending major scale. Serre's double fundamentals are assigned to individual notes within a chord, and are part of his larger, speculative system of chord generation.[81] The double

---

[80] Ibid., 144–145.
[81] On this point see Pau, "The Harmonic Theories of Jean-Adam Serre," 1–4.

236   HEARING WITH THE MIND

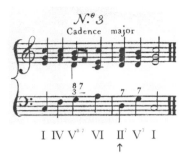

Figure 6.16 Gunn (née Young), *Introduction to Music* Plate XII, no. 3. The scale degree annotations are my own; the chord in question is likewise indicated by the addition of an arrow.

fundamentals that we encounter here are far less ambitious in their scope, and reminiscent of the way in which double fundamentals are introduced in the writings of Holden, whose treatment of chords Anne cites in the preface to her treatise as a model for her own approach.[82] However, Anne here differs from Holden, who regarded the double fundamental as expressing a psychological state of divided attention, an ambiguity resolved by determining the given chord's governing fundamental according to doubling, chord voicing, and compositional agency.

Similarly, in her explication of the progression I-IV-V-VI followed by what *we* would call II7 leading to a "regular" cadence, V7 to I (shown in **Figure 6.16**), Anne writes, "For the sake of variety, the subdominant chord is here taken in what we have called its third inversion, or what by many is considered as its erect form."[83] Here, II7 is the subdominant chord because it is the third inversion of the IV with added 6th, the proper subdominant. Yet many, she says, regard it as a distinct chord. This claim reinforced in the next paragraph, when Anne observes that, in the II7 chord in minor, that chord's bass note, "the second degree of the scale, has by some musicians been accounted the fundamental of the subdominant chord."[84]

It is in the light of this shift in the extension of the term "subdominant" that, I think, we should understand Anne when she writes, "perhaps there are few real chords in music, which are not either perfect chords, or to be referred to the class of dominant or subdominant chords, although they may

---

[82] See Serre, *Essais sur les principes de l'harmonie*, 47; Holden, *Essay*, part 1, §160, 144–158.
[83] Gunn (née Young), *Introduction*, 164.
[84] Ibid., 145.

be disguised by suspending or altering one or more of the notes."[85] The context here makes it clear that, by perfect chords, Anne specifically has in mind the tonic. In effect, this is a new approach to the very notion of subdominant, in which the term no longer refers to a single scale degree, but rather to a class of chords with a predominant quality. Her talk of "the class of dominant or subdominant chords," however, seems to be an early, if not the earliest, intimation of the much later theory of the three harmonic functions of tonic, dominant, and subdominant.

Anne's innovation with regard to considering functional types as categories was noticed by Callcott in his review of her treatise. Observing that the section in question "contains some very important doctrines, which we have never seen before so well expressed,"[86] he continued,

> we are not only heartily inclined to subscribe to this theory, but are convinced by long study and experience, that all chords are reducible to the tonic, dominant, or subdominant, with some slight exceptions in the case of sequences and licences [*sic*]: and, even in these, one of those three abovementioned is generally implied, if not decidedly audible.[87]

Callcott's immediate acceptance of Anne's theoretical idea that all chords potentially belong to one of three classes suggests that the first inklings of the notion of functional chord categories was both intuitive and not necessarily obvious to musicians at the time.

Although, as mentioned earlier, the only theorist Anne explicitly invokes in the *Introduction* is Holden, Callcott lists a number of other theorists whose influence he recognizes in her work. She is "orthodox in her opinions of theory, and agrees with Dr. Pepusch concerning the minor semitone," he approvingly observes, responding to her comment on the different theoretical sizes of the semitone.[88] He also identifies Anne's presentation of cadences—which includes motion up by fifth (an irregular cadence), motion down by fifth (a regular cadence), the false cadence (moving from V to VI) and the "gradation of the fourth" (IV to V)—as the "perfect and imperfect cadences of the moderns, . . . the false cadence of Rameau, and the mixed

---

[85] Ibid., 160.
[86] John Wall Callcott, "Review of A. Gunn's *Introduction to Music*," *The British Critic* 25 (January 1805): 70.
[87] Ibid., 71.
[88] Ibid., 67.

238   HEARING WITH THE MIND

cadence of Tartini,"[89] suggesting that she was aware of French and Italian music theory. Moreover, in both his 1803 review of the *Musical Games* and in his 1805 review of the *Introduction,* Callcott links Anne's treatment of the "doctrine of gradations"—the notion that a 6/5 chord on scale degree IV can simply be followed directly by V, rather than requiring the interpolation of fundamental-bass progression by fifth—to the writings of Jean-Baptiste Mercadier de Belesta (1750–1816), whose treatise, *Nouveau système de musique théorique et pratique* appeared in 1776.[90]

Given the tantalizing suggestions of speculative thought in Anne's *Introduction,* we can only regret that she restricted herself to presenting a practical, pared-down version of music theory in what is essentially akin to a thorough bass manual for playing her games. Yet for the reasons to be discussed shortly, she may have felt ambivalent about presenting too many original ideas in her treatise itself, as such an act would have disclosed speculative, rather than pedagogical, aspirations. Such a concern could explain another unusual feature of the treatise: its heavy reliance on grammatical constructions in the passive voice, many of which appear in conjunction with anonymizing terms such as "some," "some authors," or "some musicians." For example, the term "some authors" appears only twice in her treatise, each time in connection with her intervention with regard to the subdominant/scale degree II. This rhetorical maneuver allows Anne to avoid cumbersome references to other theorists in the interests of simplicity but also has the effect of obscuring her own contributions.

Simplifying Science

The fashion for gamifying education, exemplified by Anne's *Musical Games,* had reached such heights in the early nineteenth century that Sir Walter Scott included a remark in *Waverley* (1814) to the effect that his was "an age in which children are taught the driest doctrines by the insinuating method of instructive games."[91] Alongside, and related to, the growing interest in

---

[89] Ibid., 68.

[90] Ibid., 70, John Wall Callcott, "Review of A. Gunn's *Introduction to Music,*" *The British Critic* 25 (February 1805): 164–166; John Wall Callcott, "Review of *Instructions for Playing the Musical Games,*" *The British Critic* 21 (January 1803): 43.

[91] Walter Scott, *Waverley; Or, 'Tis Sixty Years Since* (Edinburgh: James Ballantyne & Co., 1814), 1:34. The full context of his observation is of some interest: "An age in which children are taught the driest doctrines by the insinuating method of instructive games, has little reason to dread

WHAT'S IN A GAME?   239

gamifying education, the cusp of the nineteenth century saw female science writers—many of whom also designed pedagogical games—emerge on the intellectual scene in Britain as the authors of popular expositions aimed mostly at children and at lay audiences.[92] The distillation and simplification of scientific ideas was regarded as the prime goal of such popularizations. For example, in a review of the aforementioned Margaret Bryan's *Lectures on Natural Philosophy* (1806), a book which treated optics, hydrostatics, magnetism, electricity, mechanics, and astronomy, the critic approvingly remarked that "the depths of these sciences are not sounded by the aid of analytic formulæ, yet those of their truths which are most useful, most simple, and most easy of apprehension, are culled for the acceptance and information of the reader."[93]

Anne's treatise participates in a similar kind of "feminine" scientific discourse of simplification and popularization. This characteristic of her work was noted by at least one critic—the violinist and composer Samuel Wesley, who mentions a number of recently published music-theoretical writings in the opening passages of his review of Callcott's *Musical Grammar*, and concludes with Anne's contribution:

> Within the last ten years several excellent treatises have appeared. Mr Kollman has displayed very sound and extensive knowledge in his "Essays on Musical Harmony and Composition;" Mr. Shield has given the result of his long and intimate acquaintance with the science in his "Introduction to Harmony;" Mr. King has published a useful compendium of practical

the consequences of study being rendered too serious or severe. The history of England is now reduced to a game at cards, the problems of mathematics to puzzles and riddles, and the doctrines of arithmetic may, we are assured, be sufficiently acquired by spending a few hours a-week-at a new and complicated edition of the Royal Game of the Goose. There wants but one step further, and the Creed and Ten Commandments may be taught in the same manner, without the necessity of the grave face, deliberate tone of recital, and devout attention hitherto exacted from the well-governed childhood of this realm. It may in the mean time be subject of serious consideration, whether those who are accustomed only to acquire instruction through the medium of amusement, may not be brought to reject that which approaches under the aspect of study; whether those who learn history by the cards, may not be led to prefer the means to the end; and whether, were we to teach religion in the way of sport, our pupils might not thereby be gradually induced to make sport of their religion." Ibid.

[92] See Ruth Watts, "Scientific Women Their Contribution to Culture in England in the Late Eighteenth and Nineteenth Centuries," in *Women, Education, and Agency, 1600–2000*, ed. Jean Spence, Sarah Aiston, and Maureen M. Meikle (New York and Milton Park, Abingdon: Routledge, 2009), 49–65.

[93] Anon., "Review of Mrs. Bryan's *Lectures on Natural Philosophy*," *The Monthly Review, or, Literary Journal* 51 (December 1806): 380.

240 HEARING WITH THE MIND

information, and Mrs. Gunn has successfully endeavoured to simplify the most important principles of music in her "Introduction to Music."[94]

The passage above demonstrates that Anne's work was seen as having a different purpose from the more "scientific" writings of Kollman, Shield, and even King, whose slim volume had no pretensions of scientific novelty.[95] Of considerable interest, therefore, are the rhetorical strategies Anne deploys in her preface to position herself and her work, tactics she shares with contemporary female science writers.

To the modern reader, perhaps the most jarring feature of Anne's preface is the prominent inclusion of "an apology, for obtruding a work, which, so far as it goes, may be considered as a Treatise on Music, and which, from the number of excellent books upon that subject, already published, may to some appear as superfluous."[96] It was only on the recommendation of certain renowned musicians, she goes on, that she had even embarked upon such a project, and this only because her games, "which may undoubtedly be comprehended and played by children, were often considered by grown persons as intricate and puzzling."[97] By assigning the idea of writing a treatise to an anonymous group of well-known musicians, while also emphasizing the fact that the games themselves were extremely simple, Anne seems to be attempting to forestall any recriminations to which the presence of her book might give rise.

A similarly self-deprecating gesture appears in prefaces by other female science writers. Thus, for example, Jane Marcet (1769–1858) opens *Conversations on Chemistry* (1805) with an apology and disavowal of her own expertise:

> In venturing to offer to the public, and more particularly to the female sex, an Introduction to Chemistry, the author, herself a woman, conceives that some explanation may be required; and she feels it the more necessary to

---

[94] This evaluation appears in Wesley's review of John Wall Callcott's *Musical Grammar* (1807) [Samuel Wesley], "Fine Arts. Art. I. *A Musical Grammar in Four Parts*," *Annual Review* 5 (1807): 701. Wesley's authorship is confirmed by Kollmann in "A Chronological List of Periodical Musical Works," reproduced in Kassler, *Kollmann's Quarterly Musical Register*, 5.

[95] King's *General Treatise on Music, particularly on Harmony or Thorough Bass* (1800), a mere seventy-seven pages, was deemed an "elementary work" by the reviewer of *The Monthly Review, or, Literary Journal*, who pointed out its numerous flaws. Anon., "Review of *A General Treatise on Music*," *The Monthly Review, or, Literary Journal* 32 (August 1800): 382.

[96] Gunn (née Young), *Introduction*, xiii.

[97] Ibid., xi.

apologize for the present undertaking, as her knowledge of the subject is but recent, and as she can have no real claims to the title of chemist.[98]

Margaret Bryan, on the other hand, begins the preface of her first book, *A Compendious System of Astronomy . . . in which the Principles of that Science are clearly elucidated, so as to be intelligible to those who have not studied the Mathematics* (1799), by asserting that her work, aimed solely at educating the young, is an unoriginal compilation of the insights of [men of] genius, and apologizing if she has failed to convey their ideas clearly:

> I am well aware, that, to write on subjects which have been so extensively considered, and fully delineated, by the ablest Mathematicians and by Philosophers of the most penetrating genius, will not procure me any honour on the score of originality; yet, I trust, I shall not incur censure by publishing a Compendium, which, according to my ideas, will render subjects, generally thought obscure, clear to the understanding of young people. If I have failed in the attempt through the imbecility of my judgement, I hope the motive may be my apology.[99]

Alicia Catharine Mant, also mentioned above, likewise launches her preface to *The Study of the Heavens at Midnight During the Winter Solstice, Arranged as Game of Astronomy for the Use of Young Students in that Science* (1814) by admitting that "in the present favourable age for the instruction of youth, considering the variety of methods pursued to blend that instruction with amusement, it may possibly be thought superfluous to offer any new plan to the public, and certainly it would be presumptuous to do so, with any pretensions to superiority."[100]

The defensive apology is of course a trope that dates back to antiquity, and mildly apologetic gestures *can* on occasion be found in other prefaces around 1800.[101] However, the abject professions of unworthiness that

---

[98] Jane Marcet, *Conversations on Chemistry* (London: Longman & Co., 1805), iii.

[99] Margaret Bryan, *A Compendious System of Astronomy . . . in which the Principles of that Science are clearly elucidated, so as to be intelligible to those who have not studied the Mathematics* (London: J. Wallis, 1799), vii.

[100] Alicia Catherine Mant, *The Study of the Heavens at Midnight During the Winter Solstice, Arranged as Game of Astronomy for the Use of Young Students in that Science* (London: J. Harris, 1814), 2.

[101] On the earlier history of the "affected modesty" trope, see Ernst Robert Curtius, *Europäische Literatur und lateinisches Mittelalter* (Bern und München: Francke: 1948), 2:91–93. I thank Nathan Martin for this reference.

242   HEARING WITH THE MIND

Anne and other women writers make in justifying the very existence of their books—of writing while female—reflects the precarious position of a women science writer at the time.[102] As a comparison, we can consider the apology that Holden includes in the preface to his *Essay*: "I shall therefore make no other apology for the faults of the following Piece than this, which seems to be the only proper one an author can make, I have done my best."[103] Callcott's preface to *A Musical Grammar* (1806) concludes with its author's similarly understated apology, namely, the hope that "the various professional occupations in which he has been incessantly engaged, will be an excuse for any small inaccuracies which may strike those who are conversant with the subject."[104] The tone that Holden and Callcott take here could not be more different than that of Anne or Bryan.

Anne also takes some pains to advertise her association with the education of upper-class girls. Both the *Musical Games* and the *Introduction* are dedicated to the young Princess Charlotte "by permission." The same strategy of obtaining the princess's royal patronage was also followed by Bryan in her *Lectures on Natural Philosophy* and by the author Elizabeth Semple's *Summer Rambles, Or, Conversations, Instructive and Entertaining: For the Use of Children* (1801). The "permission" referenced in their dedications suggests that all three authors received an acknowledgement and possibly some financial reward for their dedication; naturally, given that Princess Charlotte was born in 1796, this would have been handled through various intermediaries.

Although Anne's *Musical Games* were aimed at musical amateurs as well as children, the emphasis her treatise places on the youthfulness of its intended audience is also a feature shared with other educational works authored by women at the time. In the preface, Anne emphasizes her experience "in conducting the musical education of young ladies"[105] and notes that she had tested early versions of the games on her own students. The *Musical Games* had, she continues, "since that time, been constantly and successfully employed by her, in the course of her own teaching, as an instrument or vehicle of musical instruction."[106] Similarly, the subtitle of Bryan's first book, which includes the line "so as to be intelligible to those who have not

---

[102] See also Watts, "Scientific Women," Ann B. Shteir, "Elegant Recreations? Configuring Science Writing for Women," in *Victorian Science in Context*, ed. Bernard Lightman (Chicago: University of Chicago Press, 1997), 236–255.

[103] Holden, *Essay*, vi.

[104] John Wall Callcott, *A Musical Grammar: In Four Parts* (London: McMillan, 1809), viii.

[105] Gunn (née Young), *Introduction*, v.

[106] Ibid., ix.

studied the Mathematics,"[107] makes it very clear that her writing is aimed at the female sex, and its preface stresses the author's love and "almost parental tenderness" for the pupils at her girl's school, along with other sentimental effusions.[108] The emphasis that Anne and Bryan place on the gender and juvenility of their audience seems designed to deflect critics such as the aforementioned reviewer of Bryan's second book, who thought it necessary to note that the work was not intended for "those rough males who migrate from the north to learn Mathematics and Wranglership in the University of Cambridge," but rather for "the softer and fair sex" for whom "it will serve profitably to relieve the duties of the needle, the lessons of the harpsichord, and the conjuagations [*sic*] of Chambaud [a writer on French grammar]."[109]

Prefaces composed by female science authors also employ another strategy that seems intended to assuage potential concerns about incursion into the (male) sphere of intellectual inquiry, namely, the performance of explicit self-situating within domestic and intellectual patriarchal structures. Using the third person, Anne relates that in her study of music theory, "she received occasional assistance from one, upon whom it had devolved, in early life, to act towards her the part of a protector and father, as well as of a brother."[110] Anne here gives credit to her brother Walter, a reference which might have been included at least in part in order to dispel fears about the propriety of an unmarried woman overly devoted to music and games, as well as to locate her in relation to a respectable man. An even more unabashedly domestic gesture features in Bryan's preface to *A Compendious System of Astronomy,* which includes a frontispiece entitled "Mrs. Bryan and Children," featuring the author with her two daughters and various astronomical devices (as shown in **Figure 6.17**). The unusual

---

[107] *A Compendious System of Astronomy* was praised in *The Lady's Monthly Museum* as "a concise, perspicuous, and masterly compendium of astronomical knowledge ... [which] we may confidently recommend to those who have the care of the rising generation;" Anon., "Mrs. Margaret Bryan," *The Lady's Monthly Museum* 7 (August 1801): 74; a reviewer in *The British Critic* likewise pronounced "from all that we see in this production, we are inclined to felicitate those parents who have placed daughters under the care of an instructress so judicious and intelligent." Anon., "Review of *A Compendious System of Astronomy*," *The British Critic and Quarterly Theological Review* 11 (May 1798): 539.

[108] Bryan, *A Compendious System of Astronomy,* iv. These declarations include the following: "Could it be possible to receive the repeated proofs of your affection which I have received, and not love you? No! can I ever forget the gratulations you have annually offered me, when each countenance, glowing with affection or dissolving in tenderness, as the grateful ideas flowed in your breasts, expressed more than your words? Believe me, at those moments I have felt the most sensible sympathy thrill at my heart; such emanations as I then received glowed too warmly in my bosom ever to be extinguished." Ibid., iv–v.

[109] Anon., "Review of *Lectures on Natural Philosophy*," 382.

[110] Gunn (née Young), *Introduction*, v.

Figure 6.17 Frontispiece, Bryan, *A Compendious System of Astronomy* (1799).

WHAT'S IN A GAME? 245

inclusion of this image seems likewise intended to reassure readers of Bryan's maternal respectability.[111]

In addition to depicting herself as a mother, Bryan also placed her work under the aegis of male intellectual patrons, namely "my very learned friend and patron [the English mathematician Charles Hutton], whose approbation of my ideas could alone induce me to give them currency," and the "very excellent divine Dr. Paley," whose work was the model for "the whole of the anatomical parts of these Lectures."[112] Marcet, who published her *Conversations on Chemistry* anonymously, was married to a physician and fellow of the Royal Society who would later devote his entire attention to chemistry, and her preface mentions the happy consequence of having "frequent opportunities . . . of conversing with a friend on the subject of chemistry, and of repeating a variety of experiments" prior to "attending the excellent lectures delivered at the Royal Institution, by the present professor of chemistry [Humphry Davy]."[113]

Anne also goes to great lengths to locate her theoretical writing explicitly in the lineage of Holden. In contrast to her brother, whose treatise on rhythm bears the unmistakable imprint of Holden's thought (although without citation), Anne explicitly emphasizes that her treatment of rhythm, chords, and cadences, is "nearly that of the late ingenious Mr. Holden, [whose] . . . method . . . seems to give the most distinct and connected views of the subject."[114] She also provides further information to readers about where this "excellent" work could be purchased. The propitiatory gesture of locating her work in Holden's tradition seems intended to support her disavowal of any theoretical originality.

By preemptively apologizing, renouncing any explicit claim to having original music-theoretical ideas, emphasizing the gender and youthfulness of their audience, and locating themselves and their work within patriarchal structures, female science popularizers hoped to carve out an uncontested place in the intellectual sphere where they could thrive. Their anxieties were not unfounded. After all, as the aforementioned reviewer of Bryan's *Lectures*

---

[111] This image combines two distinct genres: that of the scholar in (his) study surrounded by accoutrements of profession, with a portrait of a mother and her daughters. Perhaps unwittingly, the image embodies the tension between these two aspects of Bryan's life. Seven years later, with the publication of her *Lectures on Natural Philosophy*, Bryan included a simple portrait of herself without scientific instruments or children as a frontispiece, suggesting that she no longer felt the same need to market herself as an expert while simultaneously defending herself as a mother.

[112] Margaret Bryan, *Lectures on Natural Philosophy: The Result of Many Years' Practical Experience of the Facts Elucidated* (London: Thomas Davison, 1806), preface unpaginated.

[113] Marcet, *Conversations*, iii.

[114] Gunn (née Young), *Introduction*, xviii.

246   HEARING WITH THE MIND

on Natural Philosophy cautioned in his otherwise sympathetic account of her book:

> Our jealousies are not hitherto excited; yet we shall guard our peculiar provinces with care and watchful suspicion. Their borders may be visited for curiosity and amusement, but against a formal inroad and invasion of female Philosophers we shall take arms. In our code, we have written that Politics, Greek, and Analytics, are generally forbidden to the ladies: too much study will spoil their engaging faces and their fascinating manners.[115]

As this admonishment illustrates, the stakes were high for women who wanted to write about areas of inquiry traditionally regarded as distinctly masculine pursuits.[116] A woman writing about the "science of music" would have come under less scrutiny at the time due to the general view of music as a typical feminine accomplishment (though we should keep in mind that there are vanishingly few—if any—other comparable works authored by women in the early nineteenth century). Indeed, Anne takes a number of steps to position her book in this way, ranging from her dedication to Princess Charlotte, which wishes her "improvement in every virtue, and in every useful and ornamental accomplishment,"[117] to the ending of her preface, which expresses the hope that her book will "facilitate the acquisition of a liberal art and an elegant accomplishment."[118] In her account of her

[115] Anon., "Review of *Lectures on Natural Philosophy*," 382.

[116] Bryan expressly hopes that her work will be read without bias: "I expect some countenance from those whose extensive learning and liberality lead them to judge impartially; the false and vulgar prejudices of many, who suppose these subjects too sublime for female introspection, (ascribing to mental powers the feebleness which characterizes the constitution,) invalidate the idea by affording all laudable exertions their avowed patronage,—acknowledging truth, although enfeebled by female attire." Bryan, *A Compendious System of Astronomy*, x. Marcet also notes, "In writing these pages, the author was more than once checked in her progress by the apprehension, that such an attempt might be considered by some, either as unsuited to the ordinary pursuits of her sex, or ill justified by her own recent and imperfect knowledge of the subject." Marcet, *Conversations*, iv. The three reviews of Marcet's work which I found make a point of mentioning her gender, but all praise the work without reservations (one reviewer even describes overcoming his initially negative, biased response: "We acknowledge that we took up this production with some degree of prejudice . . . we candidly confess that, notwithstanding our respect for the talents of our country-women, we apprehend that chemistry is a science by no means adapted to their acquirement. . . . As we advanced in the perusal of the volumes, however, we found our feelings of disapprobation gradually diminished; and before we arrived at the conclusion, we were convinced that they were, in a great measure, what we have already styled them, prejudices.") "Review of *Conversations on Chemistry*," *The Monthly Review, or, Literary Journal* 50 (July 1806): 330. See also "Review of *Conversations on Chemistry*," *The Annual Review and History of Literature* 4 (1806): 883; "Review of *Conversations on Chemistry*," *The British Critic* 28 (December 1806): 635–641.

[117] Gunn (née Young), *Introduction*, unpaginated dedication page.

[118] Ibid., xxiii.

brother's assistance, moreover, she describes him as one "who has been in use to unbend his mind, from the duties and occupations of a serious profession, by indulging a constitutional propensity to the study and practice of music."[119] Music, as Anne strategically emphasizes, is the preserve of masculine relaxation and feminine accomplishment.

*

Anne Young's music-theoretical works represent a remarkable and original achievement in the domains of educational games, music theory pedagogy, and speculative music theory. While ludic explorations of the seemingly infinite combinatorics of musical syntax certainly predate the *Games*—we might think here of Ghiselin Danckert's chessboard canon *Ave Maris Stella* (1535) or of the many *Musikalisches Würfelspiele* composed by the likes of Johann Philipp Kirnberger, C. P. E. Bach, and Wolfgang Amadeus Mozart in the late eighteenth century—Anne's invention also differs from these predecessors in an essential way.[120] Rather than producing musical compositions, her *Musical Games* seek to teach the guidelines governing the behaviors of clefs, key relationships, and contrapuntal motion. In doing so, they capitalize on the inherent similarity between the procedural nature of music theory, in which the student is generally required to follow strict rules and to think tactically in order to avoid problems in areas such as voice leading, and the universe of moves characteristic of strategy games such as chess or Go.[121]

The *Musical Games* also take advantage of another unusual feature of musical rudiments—their oppositions of flats and sharps, consonance and dissonance, major and minor, duple and triple measure, and high and low registers, binary oppositions that lend themselves particularly well to competitive gameplay. At the same time, the rolls of the dice keep the *Musical Games* from being purely a matter of skill, and give rise, as Anne writes, "to that gentle agitation, arising from the fluctuations of good and bad fortune, which constitute the great charm of other games."[122] Anne's *Musical Games* thus contribute to the discipline of music theory the ludic elements of *agon* and *alea,* competition and luck.[123]

---

[119] Ibid., v.

[120] A comprehensive and engaging study of the history of music and games can be found in Roger Moseley, *Keys to Play: Music as a Ludic Medium from Apollo to Nintendo* (Berkeley and Los Angeles: University of California Press, 2016).

[121] This feature of music theory has been recognized by the many modern-day software start-ups that teach music theory, where the line between game and exercise is often intentionally blurred; see, e.g., Artusi, Solfeg.io, LenMus, EarMaster, and Teoría.

[122] Gunn (née Young), *Introduction*, ix–x.

[123] I thank Braxton Shelley for his observations on this point.

248 HEARING WITH THE MIND

Perhaps uniquely among her many ludomusicological precursors, therefore, Anne seems to have fully recognized and exploited the *ludic potential* of the study of music theory. The fact that she was able to do all this, and to raise innovative music-theoretical issues at a time when women faced so many challenges in the pursuit of an intellectual life, secures her importance in the history of music theory.

Afterword

The initial circulation of Holden's treatise reflects the global reach of the British Empire. By early 1773, the *Essay* was being sold by the bookdealers Purdie & Dixon in Colonial Williamsburg and by Nicholas Langford in South Carolina.[1] A decade later it could be found in the personal libraries of prominent Virginians such as the slave planter (turned opponent of slavery) Robert Carter III and the future president of the United States, Thomas Jefferson.[2] The *Essay* also made its way to India, and a second edition was printed in Calcutta by Aaron Upjohn in 1799.[3] Yet by the early nineteenth century the reception of Holden's ideas seems to have stalled. This can be attributed in part to the small size of the initial print run: as early as 1781, the catalogue of Thomas Philipe, bookseller in Edinburgh, described Holden's *Essay* as "very scarce."[4] In the forward to his edition a few years later, Upjohn regretted that the work was "nearly out of Print, and that few, if any, Copies were to be procured even in Europe."[5] By 1803, according to Anne Young, the

---

[1] See John Edgar Molnar, "Publication and Retail Book Advertisements in the *Virginia Gazette* 1736–1780" (PhD diss., University of Michigan, 1978), 571; *The South Carolina Gazette*, no. 374, January 26, 1773.

[2] A lawyer and planter, Carter eventually attempted to free nearly five hundred enslaved people in his employ, the largest manumission prior to the American Civil War. On Carter's library, see John R. Barden, "'Innocent and Necessary': Music and Dancing in the Life of Robert Carter of Nomony Hall, 1728—1804" (MA thesis, The College of William and Mary, 1983), 43. On Jefferson's library, see *The Catalogue of the Library of Thomas Jefferson* 4 compiled by Millicent Sowerby (Washington, DC: The Library of Congress, 1959), 400.

[3] Upjohn, an English bassoonist, cartographer, and printer, is remembered today for his "Map of Calcutta and its Environs," published in 1794. On Upjohn's infamous character see Peter Robb, "Mr Upjohn's Debts: Money and Friendship in Early Colonial Calcutta," *Modern Asian Studies* 47, no. 4 (2013): 1185–1217; on Upjohn's background see also Kathleen Blechynden, *Calcutta, Past and Present* (London: W. Thacker, 1905), 184.

[4] Thomas Philipe, *A Catalogue of Curious, Rare, and Useful Books in Most Branches of Literature* (Edinburgh, 1781), 29. The same phrase appears in Callcott's description of Holden's treatise in 1799, in conjunction with a wish that the *Essay* would be reprinted. See John Wall Callcott, "Essay on Musical Literature," *The British Miscellany* 2 (1799): 112.

[5] John Holden, *Essay towards a Rational System of Music* (Calcutta: Aaron Upjohn, 1799), i. The content of Upjohn's edition is identical to that of the 1770 Glasgow edition, and it contains in addition a three-page publisher's preface and a new plate featuring: "An approved Method of Tuning the Harpsichord, Piano-forte, etc. somewhat altered from *Pasquali*, by which any Lady or Gentleman, possessing a good Ear, may tune their own Instruments: it is comprized in the last Plate, which I have numbered XIII; but may be cut out, if necessary, without any Injury to the Book, and fixed on Pasteboard, or on the Inside of the Cover of a Music-Book, at Discretion." Ibid., iii.

Hearing with the Mind. Carmel Raz, Oxford University Press. © Oxford University Press 2025.
DOI: 10.1093/9780197786208.003.0008

250   HEARING WITH THE MIND

Essay "ha[d] for many years ceased to be in the view of the public."[6] The practical factors contributing to the work's limited reception would have likely been exacerbated by Holden's death just over a year after he published the treatise, in that he would have been unable to promote or reissue it.

A number of other reasons may have also contributed to the *Essay's* relative obscurity in the nineteenth century and beyond. Given the work's failure to make a significant initial impact, it would not have become known to later generations of musicians. Any nineteenth-century encounter with the work would therefore have had to overcome its use of (by then) old-fashioned mathematical convention, as well as its almost exclusive dependence on Scottish psalmody for its musical examples.[7] Finally, it is possible that the very nature of the task that Holden sets himself—that of speculatively theorizing how our *minds* hear music—may have seemed absurd to late nineteenth-century readers, who would have sought to answer such questions with methods drawn from empirical science rather than introspection.

Anne Young's *Musical Games* likewise did not make much of an impression on the music theory pedagogy of her day. Gunn's fervent advocacy notwithstanding, the *Games* were a commercial failure. The patent, which cost Anne "upwards of £1,000" in 1801, was later, Dalyell reports, "assigned to a musicseller [*sic*] in Edinburgh for £300."[8] Upon his death, moreover, Gunn had thirty copies of his wife's treatise in his possession, suggesting that it, too, did not find an audience. Gunn refers to his wife's productions only once in his correspondence with Margaret Clephane, noting tersely that "I . . . like the unfortunate Musical games may go a begging."[9]

Almost no reference can be found to the *Musical Games* in the nineteenth century, with the exception of a query as to where one might find the instructions issued and answered in an antiquarian magazine in 1857, and a few short entries on Anne in musical dictionaries.[10] The *Musical Games* resurface

---

[6] Gunn (née Young), *Introduction*, xviii. A third edition of Holden's treatise was published in Edinburgh in 1807 (Anne discusses this reprint in her *Introduction*, at ibid.)

[7] The only exceptions are as follows: in the first part of the treatise Holden refers to the opening of an aria by Thomas Arne as exemplifying the programmatic use of a deceptive cadence (part 1, §255, 292). He also comments on the use of accidentals rather than key signatures in editions by Corelli and Rameau (part 1, §262, 239). In the second part of the treatise he notes that the Scottish composer Thomas Erskine, Earl of Kelly, employed quintuplets (part 2, §13, 290).

[8] Dalyell, *Musical Memoirs*, 236.

[9] Gunn, *Clephane Letters*, August 9, 1813. In Kennaway, *John Gunn*, 183.

[10] Anon., "Musical Game," *Notes and Queries . . . for Literary Men, Artists, Antiquaries, Genealogists, etc* 2, no. 4 (July–December 1857): 259, responded to by M. H. on page 421. For entries on Anne Young, see e.g., Fétis, "Gunn (Madame Anne)," in *Biographie universelle des musiciens* (Brussels: Meline, Cans et Compagne, 1837), 4:470–471; "Gunn, John," in *Musical Scotland*, ed. Baptie, 72; "Gunn, Anne," in *British Musical Biography: A Dictionary of Musical Artists, Authors,*

AFTERWORD   251

in print in 1907 in an essay by the prominent anti-suffragist, Bertha Harrison (1851–1916). Harrison remarks that Anne "possessed an uncommon order of mind and must have been a very original teacher," but reports that "her system of teaching did not long survive her. Her book and game are now only to be found in collections."[11] A mere handful of the games survived to this day: there are at least two in the United States, one in Frankfurt, one in Dublin, and a few more in the United Kingdom and in private collections elsewhere.

*

Most of the music-theoretical texts that we still study today exercised a direct influence on later thought: they were widely read according to the standards of their day, and profoundly shaped the ways in which music theorists and even composers subsequently understood music (here we might think of the well-known innovations proposed by figures such as Guido of Arezzo, Jean-Philipp Rameau, or Hugo Riemann). Yet in our investigation of their works, we tend to focus on the intrinsic qualities of theories themselves: their originality, as well as the opportunities they offer for a broader application than the author might have intended (here we might think of the emergence of neo-Riemannian theory or New *Formenlehre* within the Anglo-American sphere over the past few decades). Viewed within this disciplinary context, Scottish Enlightenment music theory holds a relatively unusual position within the field of music theory, as it had little impact in general on subsequent mainstream musical thought beyond Scotland, and a short afterlife within Scotland as well.[12]

Yet we also study theories of music that had a relatively small impact upon the music or music-theoretical thought of their day or later: the *Figurenlehre* of Joachim Burmeister, the microtones of Vicentino, or the mathematical music theory of Leonhard Euler. We do so, generally speaking, because some aspect of these theories seems to resonate with our contemporary concerns: the increasing interest in the *musica poetica* tradition

---

and Composers Born in Britain and its Colonies, ed. James Duff Brown and Stephen Samuel Stratton (Birmingham: Stratton, 1897), 176.

[11] Bertha Harrison, "Games of Music," *The Musical Times* 48, no. 775 (1907): 589–590.

[12] In this respect, Scottish Enlightenment music theory stands in sharp contrast to the musical cultures of late eighteenth-century Scotland, which have continued to occupy researchers to the present. See, e.g., Matthew Gelbart, *The Invention of "Folk Music" and "Art Music": Emerging Categories from Ossian to Wagner* (2007); Semi, *Music as a Science of Mankind* (2012); McAulay, *Our Ancient National Airs*; Brown, *Artful Virtue*; Leon Chisholm, "William McGibbon and the Vernacularization of Corelli's Music," *Eighteenth-Century Music* 15, no. 2 (2018): 143–176.

252   HEARING WITH THE MIND

from the 1960s onward coincided with a broader upsurge of interest in deconstruction and rhetoric in the humanities more broadly; Vicentino's tuning theories seem newly relevant because microtones are now an established part of the vocabulary of Western concert music, while Euler's formula for calculating the sweetness (*suavitas*) of tone combinations indirectly prefigures modern-day methods in computational music psychology that aim to categorize interval combinations according to metrics ranging from harmonicity to roughness.[13] We might think of such theorists as noncausal precursors, whose relevance to us today lies not only in elucidating the historical conditions and environments that made such ideas possible, but also in illuminating the ways in which modern-day research sometimes recapitulates intellectual developments that previously arose in very different circumstances.

John Holden and Walter Young command our attention today as noncausal precursors. Their understanding of musical cognition as the product of a collaboration among various mental faculties—memory, attention, perception, and expectation—in their interaction with our embodied experience and affordances anticipates how we conceptualize the perception of music in the twenty-first century. And it is no coincidence that those four mental faculties, which figure so prominently in Holden's theory, are precisely the ones mentioned by the music psychologist Henkjan Honing as essential to the experience of listening to music.[14] I am aware of no other Enlightenment thinker who possessed so comprehensive a theory of musical cognition.

Like Holden's, Walter Young's ideas are ahead of their time. By locating the origins of our musical preferences in various aspects of our physical and mental experience, his "Essay" anticipates several key tenets underlying psychological and music-theoretical understandings of rhythm of our own day. Indeed, experimental work throughout the long twentieth century has established many of the principles Walter identifies on the basis of introspection alone, ranging from grouping tendencies and subjective rhythmicization to preferred tempi and temporal limits on entrainment.

Along related lines, Anne Young's *Musical Games* are of interest to us today not only because of contemporary curiosity about their author as an eighteenth-century female music theorist and inventor, or with regard to

---

[13] For a review, see Peter M. C. Harrison and Marcus T. Pearce, "Simultaneous Consonance in Music Perception and Composition," *Psychological Review* 127, no. 2 (2020): 216–244.

[14] Henkjan Honing, *Musical Cognition: A Science of Listening,* trans. Sherry Marx and Susan van der Werff-Woolhouse (New Brunswick: Transaction Publishers, 2011), 23.

the simplifications of complex concepts presented in her *Introduction to Music,* but also because gamification is now a ubiquitous feature of human-computer interactions, and hence omnipresent in our daily lives.[15] Indeed, over the past few decades, games and play have emerged as an exciting and prolific field of scholarly inquiry, while gamification has become one of the dominant strategies for the acquisition of learning for children and adults alike.[16]

There are additional ways in which these writers' ideas will seem familiar to a twenty-first century music theorist. Holden and the Youngs make claims about musical cognition that are grounded in experience and introspection. Written in clear and simple English, their writings eschew mystical or aesthetic explanations in favor of cross-modal comparisons and generalizations drawn from every-day behavior. Steeped as they are in Scottish common sense realism, their theories bear the mark of the same intellectual milieu that strongly influenced many disciplines of the modern Anglophone academy, including cognitive psychology, social studies, history, and economics.

A last way in which the ideas of Holden and Walter may seem familiar to contemporary music theorists is the aspiration, shared by both non-causal precursor theorists, to provide a universal proto-cognitive explanation for our experience of music.[17] This aim harmonized with the dominant

---

[15] See inter alia Juho Hamari, "Gamification," in George Ritzer and Chris Rojek, eds., *The Blackwell Encyclopedia of Sociology* (Hoboken, NJ: John Wiley & Sons, 2007); https://doi.org/10.1002/978140 5165518.wbeos1321.

[16] On the study of games in society see the foundational writings of Johan Huizinga's *Homo Ludens* (1938) and Roger Caillois's *Les jeux et les hommes* (1958). The past decade has seen notable books exploring the relationship between music and games including Michiel Kamp, Tim Summers, and Mark Sweeney, eds., *Ludomusicology: Approaches to Video Game Music* (Sheffield: Equinox, 2016); Roger Moseley, *Keys to Play* (Berkeley and Los Angeles: University of California Press, 2016); William Gibbons, *Unlimited Replays: Video Games and Classical Music* (New York: Oxford University Press, 2018); Melanie Fritsch and Tim Summers, eds., *The Cambridge Companion to Video Game Music* (Cambridge, UK: Cambridge University Press, 2021); and Vincent E. Rone, Can Aksoy, and Sarah Pozderac-Chenevey, eds., *Nostalgia and Videogame Music: A Primer of Case Studies for the Player-Academic* (Bristol: Intellect, 2022). On the study of games in education Karl M. Kapp, *The Gamification of Learning and Instruction: Game-Based Methods and Strategies for Training and Education* (Hoboken, NJ: John Wiley & Sons, 2012).

[17] Thus, for example, as we saw in Chapter 2, Holden asserts that the major scale "has always been, and will always be the same in all ages and countries," and he goes on to describe it as "one of those laws which the great Author of Nature prescribed to himself, in the formation of the human mind, that such certain degrees of sound should constitute music." Holden, *Essay,* part 1, §38, 37. Similarly, as we saw in Chapter 5, Walter, defending against historical counterevidence his claim that the ancient Greeks against historical counter-evidence preferred isochronous meter, invokes "the rhythmical constitution of man, which, being a part of his nature, must be fundamentally the same, in all ages and amongst all nations." Young, "Essay," 99. Holden and Walter are here responding to Hume's famous assertion that "there is a great uniformity among the actions of men, in all nations and ages, and that human nature remains still the same, in its principles and operations;" see note 25 in Chapter 5 for further discussion. Hume, *An Enquiry Concerning Human Understanding,* §8, 133.

254 HEARING WITH THE MIND

intellectual currents of their day and, it turns out, with ours.[18] It is also evident in the solution they both propose: an innate and universal mental preference for isochrony, which is to say that they both take an essentially calculative and deterministic approach to the structure of human perception.[19]

Ironically, however, although Holden and Walter set out to write "universal" theories, they do so from a shared faulty premise.[20] Holden claims to generate a comprehensive account of music perception, but only addresses the sacred music of his adopted homeland. In this, his attempt foreshadows subsequent justifications of particular aesthetic values, or later, psychological "preference rules," based solely (and conveniently) on the features of certain familiar repertoires. Similarly, Walter's approach starts out from basic embodied human capacities as well as hypothesized features of our mental abilities but ends up taking a Procrustean approach to accounts of Ancient Greek music in order to somehow fit them into his theory, which is based on his knowledge of the music of his day. In essence, therefore, both thinkers articulate biased ideas about musical perception by taking only the repertoire of their own day into consideration. Here, and elsewhere, they anticipate the tendency in music theory (as practiced in the West) to disregard aspects of musical practice that do not conform to theoretical assertions. This flaw foreshadows the deeply ethnocentric prejudices that music theorists in the twenty-first century have inherited from the canonical music theory of the long nineteenth century.

It is tempting simply to shift the blame away from Holden and Walter to the European ethnocentrism of their day and age. Yet both authors were aware that other cultures possessed different forms of music.[21] And there were also

---

[18] Recent studies focusing on musical "universals" can be found in Patrick E. Savage, Steven Brown, Emi Sakai, and Thomas E. Currie, "Statistical universals reveal the structures and functions of human music," *Proceedings of the National Academy of Sciences* 112, no. 29 (2015): 8987–8992; Samuel A. Mehr et al., "Universality and diversity in human song," *Science* 366, no. 6468 (2019): eaax0868; Nori Jacoby et al. "Commonality and Variation in Mental Representations of Music Revealed By a Cross-Cultural Comparison of Rhythm Priors in 15 Countries," *Nature Human Behaviour* 8, no. 5 (2024): 846–877.

[19] This approach differs from that of Rousseau, whose hypothesis regarding the shared origins of music and language had a comparably long afterlife. On this point, see Jacqueline Waeber, "A Corruption of Rousseau? The Quest for the Origins of Music and Language in Recent Scientific Discourse" in "Colloquy on "Rousseau in 2013: Afterthoughts on a Tercentenary," *Journal of the American Musicological Society* 66, no. 1 (2013): 284–289.

[20] An interesting discussion of Rousseau's approach to the notion of universals and human nature can be found in Matthew Gelbart, "Rousseau and the Quest for Universals," in "Colloquy on Rousseau in 2013: Afterthoughts on a Tercentenary," *Journal of the American Musicological Society* 66, no. 1 (2013): 280–284.

[21] For example, Holden and Walter address the Scottish pentatonic scale in their writings; see Holden, *Essay*, part 2, §42, 313–314; Young, "Preface" in MacDonald, *A Collection of Highland Vocal Airs*, 5–6. Both writers regarded this as an archaic style no longer in use; the stadial historiography

intellectual currents in the Scottish Enlightenment that emphasized the cultural contingency of certain customs and standards, as is evident in the writings of Scottish historian Adam Ferguson.[22]

Both Holden and Walter were, I think, blinded to the cultural contingency of their own musical assumptions by the proto-cognitive nature of their projects. That is to say, in order to justify their universalist approach, which sought explanations for musical preferences in shared aspects of the human constitution, they had to regard the specific musical practices they discussed as comparably universal, otherwise the whole endeavor would collapse. We should keep in mind, therefore, that their treatises seem prescient and familiar to us at least in part because they represent an early attempt to link universal principles of cognition to the form and reception of cultural productions—one which foreshadows the conflicts between universalist ambitions and relativist admonitions that still loom so large on the cultural and intellectual landscape today.

common in Scotland at the time may have implicitly allowed them, by analogy, to disregard the music of other cultures as also developmentally "behind" Europe, as we saw in Chapter 5. Note, however, that Thomas Robertson included a chapter in his *Inquiry* analyzing reports of the music of the South Sea Islanders; see Chapter 4.

[22] Ferguson observed in his *Principles of Moral and Political Science* (1792), that "To the ignorant or to the proud, who consider their own customs as a standard for mankind, every deviation from that standard is considered, not as the use of a different language and form of expression, but as a defect of reason, and a deviation from propriety and correctness of manners." Adam Ferguson, *Principles of Moral and Political Science* (London and Edinburgh: Strahan, Cadell and Creech, 1792), 2:142. Ferguson here provides a clear articulation of the dangers of ethnocentrism, which historian of philosophy Craig Smith has termed the "the methodological problem that would come back to haunt the discipline [of social studies] of which [Ferguson] was a founding figure." Craig Smith, "Adam Ferguson and Ethnocentrism in the Science of Man," *History of the Human Sciences* 26, no. 1 (2013): 64. Smith notes that Gilbert Stuart's *A View of Society in Europe in its Progress from Rudeness to Refinement* (1778) and James Dunbar's *Essays on the History of Mankind in Rude and Cultivated Ages* (1780) also evidence awareness of the problem of ethnocentrism. Ibid., 53.

Appendix

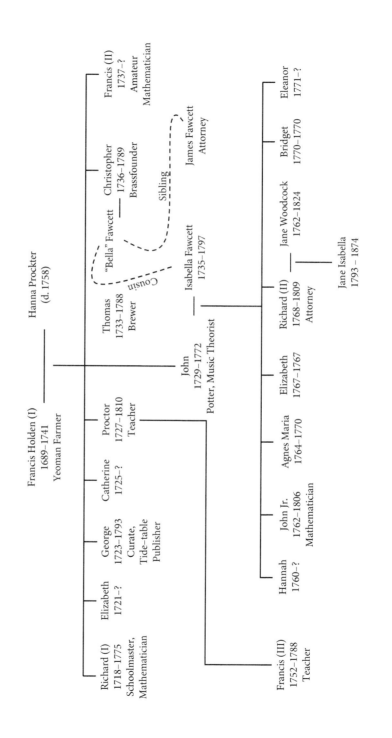

Bibliography

Primary Sources

Adamson, Robert. "Roger Bacon." In *Encyclopedia Britannica*. Vol. 3, 9th ed., edited by Thomas Spencer Baynes, 218–222. Edinburgh: Adam and Charles Black, 1875.

Advertisement [Stockwell Lodgings]. *Glasgow Journal*, no. 1049. September 3–10, 1761.

Advertisement [Holden's Wares]. *Glasgow Journal*, no. 1127. March 3–10, 1763.

Advertisement [Nicholas Langford bookdealers]. *The South Carolina Gazette*, no. 374. January 26, 1773. In *The Performing Arts in Colonial American Newspapers, 1690–1783, Text Database and Index*, compiled and edited by Mary Jane Corry, Kate Van Winkle Keller, and Robert M. Keller. New York, 1997. https://www.cdss.org/elibrary/PacanNew/CITATION/C0218/C0218459.htm [accessed Sep. 12, 2023]

Advertisement [Muir, Wood, & Co.]. *The Caledonian Mercury*. January 2, 1802.

Advertisement [Miss Ocheltrie]. *The Edinburgh Evening Courant*, no. 15593. May 25, 1811.

Advertisement [Miss Ocheltrie]. *Caledonian Mercury*, no. 14694. February 5, 1816.

"Affairs in Scotland (Oct. 1764)." *The Scots Magazine, and Edinburgh Literary Miscellany* 26, edited by William Smellie, 574. Edinburgh: W. Sands, 1764.

Aikin, John. *A Description of the Country from Thirty to Forty Miles Round Manchester*. London: John Stockdale, 1795.

"An Account of a Late Improvement of Church-Music (April 1755)." *The Scots Magazine, and Edinburgh Literary Miscellany* 17, 190. Edinburgh: W. Sands. 1755.

Anderson, James R. ed. *The Burgess & Guild Brethren of Glasgow, 1751–1846*. Edinburgh: Skinner, 1935.

Anderson, John. "Will of John Anderson (so far as it relates to the institution) dated 7th May 1795." Reprinted in *Local and Personal Acts of Parliament*, 165–175. London: H.M. Stationery Office, 1877.

Ashworth, Caleb. *A Collection of Tunes Suited to the Several Metres Commonly Used in Public Worship, with an Introduction to the Art of Singing and Plain Composition*. London: J. Buckland, 1762.

Austen, Jane. *Emma: An Authoritative Text, Backgrounds, Reviews and Criticism*, edited by Stephen M. Parrish. New York: W. Norton, 2000.

Austen, Jane. *Mansfield Park: Authoritative Text, Contexts, Criticism*, edited by Claudia L. Johnson. New York and London: Norton, 1998.

Bacon, Roger. *Epistola de secretis operibus artis et naturae, et de nullitate magiae*, in John S. Brewer, *Fr. Rogeri Bacon, Opera quaedam hactenus inedita*. Vol. 1. Oxford: Longman, Green, Longman, and Roberts, 1859.

Baptie, David. *Musical Scotland, Past and Present: Being a Dictionary of Scottish Musicians from about 1400 Till the Present Time, to Which is Added a Bibliography of Musical Publications Connected with Scotland from 1611*. Paisley: J. and R. Parlane, 1894.

Barker, Andrew. *Greek Musical Writings 2: Harmonic and Acoustic Theory*. Cambridge: Cambridge University Press, 1989.

Bayley, Anselm. *The Alliance of Music, Poetry, and Oratory*. London: J. Stockdale, 1789.

Beattie, James. *Essays On Poetry and Music, as They Affect the Mind, on Laughter, and Judicious Composition, on the Utility of Classical Learning*. Edinburgh: W. Creech, 1776.

Berlioz, Hector. "Strauss. son orchestre, ses valses - De l'avenir du rythme." *Journal des Debates* (November 10, 1837): 1.

260 BIBLIOGRAPHY

Bibliothèque royale de Belgique. *Catalogue de la bibliothèque de FJ Fétis acquise par l'état belge.* Bruxelles: Librairie Européenne C. Muquardt, 1877.

Blechynden, Kathleen. *Calcutta, Past and Present.* London: W. Thacker, 1905.

Bluett, Thomas. *Some Memoirs of the Life of Job: The Son of Solomon the High Priest of Boonda in Africa; Who was a Slave about Two Years in Maryland; and Afterwards Being Brought to England, was Set Free, and Sent to His Native Land in the Year 1734.* London: Richard Ford, 1734.

Bolton, Thaddeus L. "Rhythm." *The American Journal of Psychology* 6, no. 2 (1894): 145–238.

Bond (of Fortrose), Elizabeth. *Letter of a Village Governess; Descriptive of Rural Scenery and Manners; with Anecdotes of Highland Children.* Vol. 1. London: 1814.

Bremner, Robert. *The Rudiments of Music,* 1st ed. Edinburgh: Bremner, 1756.

Bremner, Robert. *The Rudiments of Music,* 2nd ed. Edinburgh: Bremner, 1762.

Brereton, William. *Travels in Holland, the United Provinces, England, Scotland, and Ireland, M. DC. XXXIV-M. DC. XXXV 1.* Manchester: Chetham Society, 1844.

Briseux, Charles-Étienne. *Traité du beau essentiel dans les arts.* Paris: 1753.

Brown, James Duff, and Stephen Samuel Stratton, eds. *British Musical Biography: A Dictionary of Musical Artists, Authors, and Composers Born in Britain and its Colonies.* Birmingham: Stratton, 1897.

Bryan, Margaret. *A Compendious System of Astronomy...in which the Principles of that Science are clearly elucidated, so as to be intelligible to those who have not studied the Mathematics.* London: J. Wallis, 1799.

Bryan, Margaret. *Lectures on Natural Philosophy: The Result of Many Years' Practical Experience of the Facts Elucidated.* London: Thomas Davison, 1806.

Burney, Charles. *The Present State of Music in Germany, the Netherlands, and United Provinces,* vol. 2. London: T. Becket, Strand, 1775.

Burney, Charles. "Account of an Infant Musician." *Philosophical Transactions of the Royal Society of London* 69 (1779): 183–206.

Burney, Charles. *A General History of Music: From the Earliest Ages to the Present Period.* Vol. 4. London: The Author, 1789.

Burney, Charles. "Holden, John" in *The Cyclopaedia: Or, Universal Dictionary of Arts, Sciences, and Literature.* Vol. 18, edited by Abraham Rees, unpaginated. London: Longman, Hurst, Rees, Orme, & Brown, 1811.

Burney, Charles. "Music" in *The Cyclopaedia: Or, Universal Dictionary of Arts, Sciences, and Literature.* Vol. 24, edited by Abraham Rees, unpaginated. London: Longman, Hurst, Rees, Orme, & Brown, 1819.

Burnett, James (Lord Monboddo). *Of the Origin and Progress of Language,* 6 vols. Edinburgh: J. Balfour, 1773–1792.

[Callcott, John Wall]. "Essay on Musical Literature." *The British Miscellany* 2 (1799), 26–32, 109–117.

[Callcott, John Wall]. *A Musical Grammar: In Four Parts.* London: McMillan, 1809.

[Callcott, John Wall]. "Review of *Instructions for Playing the Musical Games.*" *The British Critic* 21 (January 1803): 40–55.

[Callcott, John Wall]. "Review of A. Gunn's *Introduction to Music.*" *The British Critic* 25 (January 1805): 64–72.

[Callcott, John Wall]. "Review of A. Gunn's *Introduction to Music.*" *The British Critic* 25 (February 1805): 163–171.

Campbell, Duncan. "Notes on Church Music in Aberdeen." *Transactions of the Aberdeen Ecclesiological Society* 2 (1888): 15–23.

Catalogue of the Extensive and Valuable Library of the Late John Sidney Hawkins, Esq. F.S.A. ... sold by Auction by Mr. Fletcher ... on Monday, May 8th, 1843 and Eight Following Days. London: Fletcher, 1843.

Chambers, Robert. *A Biographical Dictionary of Eminent Scotsmen.* Vol. 2. Glasgow, Edinburgh and London: Blackie & Son, 1853.

BIBLIOGRAPHY 261

"Charming Sally." In *Voyages: The Transatlantic Slave Trade Database*. Edited by David Eltis, et al. https://www.slavevoyages.org/voyage/database [accessed Sep. 13, 2023].

Cleland, James. *Annals of Glasgow, Comprising an Account of the Public Buildings*, vol. 2. Glasgow: James Hederwick, 1816.

Cook, James. *A Voyage to the Pacific Ocean: Undertaken by the Command of His Majesty, for Making Discoveries in the Northern Hemisphere*. Vol. 1. London: Strahan, 1784.

Cotes, Roger. *Harmonia mensurarum, sive analysis et synthesis per rationum et angulorum mensuras promota*. Cambridge, 1722.

Crawfurd, George. *A Sketch of the Rise and Progress of the Trades' House of Glasgow*. Glasgow: Bell & Bain, 1858.

Dalyell, John Graham. *Musical Memoirs of Scotland with Historical Annotations and Numerous Illustrative Plates*. Edinburgh: T. G. Stevenson, 1849.

Descartes, René. *Compendium of Music*. Translated by Walter Robert. Bloomington, IN: American Institute of Musicology, 1961.

Descartes, René. *Musicae Compendium*. Utrecht: Trajectum ad Rhenum, 1650.

Dibdin, James C. *The Annals of the Edinburgh Stage with an Account of the Rise and Progress of Dramatic Writing in Scotland*. Edinburgh: R. Cameron, 1888.

Dougall, John. *The Modern Preceptor; Or, a General Course of Education*. Vol. 2. London: Vernon, Hood, and Sharp, 1806.

Dunbar, James. *Essays on the History of Mankind in Rude and Cultivated Ages*. London: W. Strahan, 1780.

Duncan, Gideon. *True Presbyterian: Or, a Brief Account of the New Singing, its Author and Progress in General. With an Account of its Procedure, and Way of Carrying On, in the Old Town of Aberdeen in Particular. Together with the Authors Advice to the Ring-Leaders*. Glasgow: 1755.

Ellis, Alexander. "On the History of Musical Pitch." *Journal of the Royal Society of Arts* 28, no. 1,424 (1880): 293–336.

Emerson, William. *The Doctrine of Fluxions*. London: J. Bettenham, 1743.

Equiano, Olaudah. *The Interesting Narrative of the Life of Olaudah Equiano*. London: T. Wilkins, 1789.

Euler, Leonhard. "*Conjecture sur la raison de quelques dissonances généralment reçues dans la musique.*" *Histoire de l'academie royale des sciences et des belles-lettres de Berlin année* 1764. Berlin: Haude et Spener, 1766.

Euler, Leonhard. *Tentamen novae theoriae musicae ex certissimus harmoniae principiis dilucide expositae*. St. Petersburg: Academiae scientiarum, 1739.

Eyre-Todd, George. *The Book of Glasgow Cathedral: A History and Description*. Edinburgh: M'Farlane & Erskine, 1898.

Faujas de Saint-Fond, Barthélemy. *Travels in England, Scotland, and the Hebrides: Undertaken for the Purpose of Examining the State of the Arts, the Sciences, Natural History and Manners, in Great Britain*. Vol. 2. London: James Ridgeway, 1799.

Ferguson, Adam. *An Essay on the History of Civil Society*. Edinburgh: A. Kincaid and J. Bell, 1767.

Ferguson, Adam. *Principles of Moral and Political Science*. Vol. 2. London and Edinburgh: Strahan, Cadell and Creech, 1792.

Fétis, François-Joseph. *Biographie universelle des musiciens* 4. Brussels: Meline, Canset Compagne, 1837.

Fétis, François-Joseph. *Biographie universelle des musiciens* 4, deuxième edition. Paris: Didot, 1862.

Fétis, François-Joseph. *Biographie universelle des musiciens* 5. Brussels: Meline, Cans et Companie, 1839.

Forkel, Nikolaus. *Allgemeine Litteratur der Musik, oder Anleitung zur Kenntniss musikalischer Bücher*. Leipzig: Schwickert, 1792.

Galliard, John Ernest. *Observations on the Florid Song; or Sentiments on the Ancient and Modern Singer*. London: J. Wilcox, 1743.

262 BIBLIOGRAPHY

Gaultier, abbé [Aloisius Edouard Camille]. *Leçons de grammaire*. Paris: 1787.

Gerard, Alexander. *An Essay on Taste*. London: A. Millar, 1759.

Gibson, John. *History of Glasgow: From the Earliest Accounts to the Present Time*. Glasgow: Chapman, 1777.

Grant, Archibald. *Selections from the Monymusk Papers (1713–1755)*, edited by Henry Hamilton. Edinburgh: Scottish History Society, 1945.

Greenwood, James. *An Essay Towards a Practical English Grammar: Describing the Genius and Nature of the English Tongue*. London: R. Tookey, 1711.

Guest, John. *Yorkshire. Historic Notices of Rotherham: Ecclesiastical, Collegiate, and Civil*. Rotherham: White, 1879.

Gunn (née Young), Anne. *An Introduction to Music . . . as Illustrated by the Musical Games and Apparatus*. Edinburgh: C. Stewart & Co, 1803.

Gunn, John. *An Essay, Theoretical and Practical, with Copious and Easy Examples on the Application of the Principles of Harmony, Thorough Bass, and Modulation, to the Violoncello*. London: Preston, 1802.

Gunn, John. *An Essay . . . Towards a More Easy and Scientific Method of Commencing & Pursuing the Study of the Piano Forte*. London: Preston, 1812.

"Gunn, John." In *A Dictionary of Music and Musicians*, edited by Sir George Grove, 641 (Boston: Ditson & Co., 1879).

Hallen, Cornelius. *The Scottish Antiquary: Or, Northern Notes & Queries*. Edinburgh: Constable, 1891.

Harrison, Bertha. "Games of Music." *The Musical Times* 48.775 (1907): 589–92.

Hartley, David. *Observations on Man, his Frame, his Duty, and his Expectations* Part 1 (London: S. Richardson, 1749.

Hauptmann, Moritz. *Die Natur der Harmonik und Metrik*. Leipzig: Breitkopf und Härtel, 1853.

Hayes, Deborah and Jean-Philippe Rameau, "Rameau's Theory of Harmonic Generation: An Annotated Translation and Commentary of *Génération harmonique, ou Traité de musique théorique et pratique*." *PhD Dissertation*. Stanford University, 1968.

Herder, Johann Gottfried. *Die Abhandlung über den Ursprung der Sprache*. Berlin: Voß, 1772.

History of the Speculative Society of Edinburgh. Edinburgh: Speculative Society of Edinburgh, 1845.

Holden, John. *A Collection of Church-Music Consisting of New Setts of the Common Psalm-Tunes . . . for the Use of the University of Glasgow*. Glasgow: Watt and McEwan, 1766.

Holden, John. *Essay towards a Rational System of Music*, 1st ed. Glasgow: Urie, 1770.

Holden, John. *Essay towards a Rational System of Music*, 2nd ed. Calcutta: Upjohn, 1799.

Holden, John. *Essay towards a Rational System of Music*, 3rd ed. Edinburgh: Blackwood, 1807.

Holden, Richard [Sr.]. "Advertisement." *The General Evening Post (London)*. June 12–14, 1760.

Holden, Richard [Jr.]. "Letter from Mr. Holden to his Grace the Duke of Leeds, Read November 26, 1795." *Archaeologia: or, Miscellaneous Tracts Relating to Antiquity* 12 (1796): 207–208.

Holden, Richard [Jr.]. "A Northern Tour in 1808: I—Manchester's 'Lower Orders.' " *The Manchester Guardian*. Thursday, September 10, 1953.

Hume, David. *An Enquiry Concerning Human Understanding*. London: A. Millar, 1748.

Hume, David. "Of National Characters." In *Essays, Moral and Political*, 3rd ed., corrected, 267–288. Edinburgh and London: A. Millar, 1748.

Hume, David. *The Natural History of Religion*. London: A. Millar, 1757.

Hume, William. "John Holden." *Grove's Dictionary of Music and Musicians*. Vol. 2, edited by Sir George Grove, 419. New York: Macmillan, 1906.

Hume, William. "John Holden." *Grove's Dictionary of Music and Musicians*. Vol. 4, edited by Sir George Grove, 678. Philadelphia: Presser, 1895.

Jackson, William. *A Scheme Demonstrating the Perfection and Harmony of Sounds*. Westminster: J. Cluer, 1726.

Kirnberger, Johann Philipp. *Die Kunst des reinen Satzes in der Musik*. Vol. 1. Berlin: Voß, 1771.

BIBLIOGRAPHY   263

Kirnberger, Johann Philipp. *Die Kunst des reinen Satzes in der Musik*. Vol. 2. Berlin and Königsberg: Decker und Hartung, 1776.

Kollmann, Augustus Frederic Christopher. *An Essay on Practical Musical Composition, According to the Nature of that Science*. London: 1799.

Kollmann, Augustus Frederic Christopher. "A List of the Periodical Musical Works of the Historico-Critical Kind, which have Existed Before the Present Musical Register." *Quarterly Musical Register* 1 (1812): 3–6.

Leibniz, Gottfried Wilhelm. "Letter no. 6, Leibniz to Christian Goldbach." In "La correspondance de Leibniz avec Goldbach," edited by Adolf P. Juschkewitsch and Juri C. Kopelewitsch, *Studia Leibnitiana* 20 (1988): 175–89.

Leitch, George. "Notice of a Mahogany Pitchpipe Formerly Used in Cults Parish Church, Fife." *Proceedings of the Society of Antiquaries of Scotland* 40 (December 11, 1905), 43–46.

Lerdahl, Fred and Ray Jackendoff. *A Generative Theory of Tonal Music*. Cambridge, MA: MIT Press, 1983.

Locke, John. *An Essay Concerning Human Understanding*. London: Holt, 1690.

Locke, John. *Some Thoughts Concerning Education, Critical Edition*, edited by John W. Yolton and Jean S. Yolton. Oxford: Clarendon Press, 2000.

Lomas, Sophia Crawford. "The Manuscripts in the Possession of Sir John James Graham of Fintry, K.C.M.G." In *Report on Manuscripts in Various Collections* 5, 185–275. Hereford: His Majesty's Stationary Office, 1909.

Love, James. *Scottish Church Music: Its Composers and Sources*. Edinburgh: W. Blackwood, 1891.

MacDonald, Patrick. *A Collection of Highland Vocal Airs Never Hitherto Published. To Which Are Added a Few of the Most Lively Country Dances or Reels of the North Highland and Western Isles: And Some Specimens of Bagpipe Music*. Edinburgh: 1784.

Mant, Alicia Catherine. *The Study of the Heavens at Midnight During the Winter Solstice, Arranged as Game of Astronomy for the Use of Young Students in that Science*. London: J. Harris, 1814.

Malcolm, Alexander. *A Treatise on Musick, Speculative, Practical & Historical*. Edinburgh, 1721.

Marcet, Jane. *Conversations on Chemistry*. London: Longman & Co., 1805.

Mason, John. *Essays on Poetical and Prosaic Numbers and Elocution*, 2nd ed. London: J. Buckland, 1761.

Maurice, Mary A. *Mothers and Governesses*. London: John W. Parker, 1847.

McUre, John. *Glasghu Facies: A View of the City of Glasgow*. Glasgow: Tweed, 1872.

Meumann, Ernst Friedrich Wilhelm. "Untersuchungen zur Psychologie und Aesthetik des Rhythmus." Habilitationsschrift, Universität Leipzig, 1894.

Meyer, Leonard B. *Emotion and Meaning in Music*. Chicago: University of Chicago Press, 1956.

Miller, Edward and Frederic Palmer Wells. *History of Ryegate, Vermont: From its Settlement by the Scotch-American Company of Farmers to Present Time: With Genealogical Records of Many Families*. St Johnsbury, CT: The Caledonian Company, 1913.

Miller, James. *The Lamp of Lothian, Or, The History of Haddington: In Connection with the Public Affairs of East Lothian and of Scotland: from the Earliest Records to the Present Period*, Haddington: James Allan, 1844.

Mitchell, John Oswald. "Four Old Glasgow Bells." *Publications of the Regality Club, Glasgow* 2 (1893): 33–48.

Mitford, William. *An Inquiry into the Principles of Harmony in Language*, 2nd ed. London: T. Cadell, 1804.

Moore, Thomas. *The Psalm Singer's Pocket Companion*. Glasgow, 1756.

Murray, David. *Memories of the Old College of Glasgow: Some Chapters in the History of the University*. Vol. 2. Glasgow: Jackson, Wylie & Co., 1927.

"[Obituary of] John Runcorn." In *The Gentleman's Magazine and Historical Chronicle* 57, no. 2, edited by Edward Cave, 1195. London: John Nichols, 1787.

"[Obituary of] John Holden [junior]." In *The Annual Register, Or, A View of the History, Politics, and Literature for the Year 1806*, edited by Thomas Morgan, 546. London: Otridge and Son, 1806.

264 BIBLIOGRAPHY

"[Obituary of] Proctor Holden." In *The Monthly Magazine, or British Register* 29, no. 1, edited by Richard Phillips, 502. London: Richard Phillips, 1810.

"Obituary of the Rev. John Holden." In *The Gentleman's Magazine and Historical Chronicle* 76, no. 2 edited by Edward Cave, 878–879. London: John Nichols, 1806.

Philipe, Thomas. *A Catalogue of Curious, Rare, and Useful Books in Most Branches of Literature.* Edinburgh, 1781.

Preston, Thomas. "Correct List of New Publications." *The Literary Magazine or Monthly Epitome of British Literature* (June 1805): 335.

Preston, Thomas. "A March for the Piano Forte, composed by Anne Gunn. 1s. 6d." *The Literary Magazine; or, Monthly Epitome of British Literature* (Oct 1805): 555.

Rameau, Jean-Philippe. *Génération harmonique, ou Traité de musique théorique et pratique.* Paris: Prault fils, 1737.

Rameau, Jean-Philippe. *Traité de l'harmonie reduite à ses principes naturels.* Paris: Ballard, 1722.

Rameau, Jean-Philippe. *A Treatise of Musick: Containing the Principles of Composition.* London: Robert Brown, 1737.

Rameau, Jean-Philippe. *Treatise on Harmony.* Translated by Philip Gossett. New York: Dover Publications, 1971.

Ramsay, John. "Letter No. XXXII. From Mr. J. Ramsay, to the Reverend W. Young, at Erskine, Ochtertyre, October 2, 1737." In *The Works of Robert Burns: With an Account of His Life, and Criticism on His Writing; to which are Prefixed Some Observations on the Character and Condition of the Scottish Peasantry.* Vol. 2, edited by James Currie, 115–119. Liverpool: J. M. M'Creery, 1800.

Ramsay, John. *Scotland and Scotsmen of the Eighteenth Century from the mss. of John Ramsay, esq. of Ochtertyre.* Vol. 1. Edited by Alexander Allardyce. Edinburgh, W. Blackwood and Sons, 1888.

Ramsay, Philip A. *The Works of Robert Tannahill. With Life of the Author, and Memoir of Robert A. Smith, the Musical Composer.* London and Edinburgh: A. Fullarton & Co., 1838.

Ravenscroft, Thomas. *The Whole Booke of Psalmes: With The Humnes Evangelicall, and Songs Spiritual.* London, The Company of Stationers, 1621.

Reid, Robert. *Glasgow, Past and Present: Illustrated in Dean of Guild Court Reports* 3. Glasgow: David Robertson, 1884.

Reid, Thomas. *The Correspondence of Thomas Reid.* Vol. 4, edited by Mark Wood. University Park: Penn State Press, 2002.

Reid, Thomas. *Essays on the Intellectual Powers of Man.* Edinburgh: J. Bell, 1785.

Reid, Thomas. *Essays on the Active Powers of Man.* Edinburgh: J. Bell, 1788.

Reid, Thomas. *An Inquiry Into the Human Mind.* 2nd ed. Edinburgh: A Millar, 1765.

Reid, Thomas. *Practical Ethics: Being Lectures and Papers on Natural Religion, Self-government, Natural Jurisprudence, and the Law of Nations,* edited by Knud Haakonssen. Princeton: Princeton University Press, 1990.

Reid, Thomas. *Thomas Reid's Lectures on the Fine Arts, Transcribed from the Original Manuscript, with an Introduction and Notes,* edited by Peter Kivy. The Hague: Martinus Nijhoff, 1973.

"[Review of] Holden's Essay towards a Rational System of Music." *The Critical Review* 33 (June 1772): 323–325.

"[Review of] Robertson's Inquiry into the Fine Arts." *The English Review* 5 (June 1785): 401–410.

"[Review of] Robertson's Inquiry into the Fine Arts." *The New Review* (May, 1785): 339–349; (June 1785): 389–406.

"[Review of] Robertson's Inquiry into the Fine Arts." *The Monthly Review, or, Literary Journal* 74 (March 1786): 191–199; (April 1786): 245–249.

"[Review of Young's Essay]." *The Critical Review or Annals of Literature* 1 (February 1791): 126–127.

"[Review of Young's Essay]." *The Analytical Review, or History of Literature, Domestic and Foreign* 8 (September–December 1790): 263.

"[Review of Young's Essay]." *The Monthly Review or Literary Journal* 5 (June 1791): 193–197.

"[Review of Young's Essay]." *Göttingische Anzeigen von gelehrten Sachen* 90 (June 4 1791): 903.

"Review of [Margaret Bryan's] *A Compendious System of Astronomy*." *The British Critic and Quarterly Theological Review* 11 (May 1798), 535–539.

"Review of [King's] *A General Treatise on Music*." *The Monthly Review, or, Literary Journal* 32 (August 1800): 382–387.

"[Review of *A Compendious System of Astronomy*] Mrs. Margaret Bryan." *The Lady's Monthly Museum* 7 (August 1801): 73–74.

"Review of New Musical Publications: [Young's] *Musical Game-Tables and Apparatus*." *The Monthly Magazine Or, British Register* 12, part 2, no. 80 (Dec 1, 1801): 428.

"[Review of Young's Games]." In *Englische Miscellen*. Vol. 6, edited by Johann Christian Hüttner, 82. Tübingen: J. Cotta, 1802.

"Review of [Marcet's] *Conversations on Chemistry*." *The Annual Review and History of Literature* 4 (1806): 883.

"Review of [Marcet's] *Conversations on Chemistry*." *The Monthly Review* 50 (July 1806): 330–331.

"Review of Mrs. Bryan's *Lectures on Natural Philosophy*." *The Monthly Review, or, Literary Journal* 51 (December 1806): 380–382.

"Review of [Marcet's] *Conversations on Chemistry*." *The British Critic* 28 (December 1806): 635–64.

Riemann, Hugo. "Ideen zu einer 'Lehre von den Tonvorstellungen.'" *Jahrbuch der Musikbibliothek Peters* 21/22 (1914): 1–26.

Riemann, Hugo. *Musikalische Logik: Hauptzüge der physiologischen und psychologischen Begründung unseres Musiksystems*. Leipzig: C. F. Kahnt, 1874.

Riemann, Hugo. *Musikalische Syntaxis: Grundriß einer harmonischen Satzbildungslehre*. Leipzig: Breitkopf & Härtel 1877.

Riemann, Hugo. *Musik-Lexikon*. Leipzig: Verlag des Bibliographischen Instituts, 1882.

Riemann, Hugo. "Die Natur der Harmonik." In *Sammlung musikalischer Vorträge*. Vol. 4, edited by Paul Graf Waldersee. Leipzig: Breitkopf und Härtel, 1882.

Ritchie, Thomas E. *An Account of the Life and Writings of David Hume, esq.* London: T. Cadell and W. Davies, 1807.

Robertson, Thomas. "Essay on the Character of Hamlet." *Transactions of The Royal Society of Edinburgh* 2, no. 2 (1790): 251–267.

Robertson, Thomas. *The History of Mary Queen of Scots*. Edinburgh: Bell & Bradfute, 1793.

Robertson, Thomas. *Inquiry into the Fine Arts*. Edinburgh: T. Cadell, 1784.

Robison, John. "A Narrative of Mr Watt's Invention of the Improved Engine by Professor John Robison." In *The Life of James Watt: With Selections from his Correspondence*, edited by James Patrick Muirhead, 74–91. London: John Murray, 1858.

Roscoe, Henry. *The Life of William Roscoe* 1. London: T. Cadell, 1833.

Rousseau, Jean-Jacques. "Essai sur l'origine des langues." In *Collection complète des œuvres de Jean-Jacques Rousseau*. Geneva: Volland, 1780.

Rousseau, Jean-Jacques. *Les Confessions* in *œuvres completes I: Les confessions; Autres textes autobiographiques*, edited by Bernard Gagnebin, Marcel Raymond, and Robert Osmont. Paris: Gallimard, 1959.

Rousseau, Jean-Jacques. *Confessions*. Translated by Angela Scholar. Oxford: Oxford University Press, 2000.

Royal Society of Edinburgh. *Former Fellows of the Royal Society of Edinburgh: Biographical Index*. Vol. 2. Edinburgh: The Royal Society of Edinburgh, 2006.

Sancho, Ignatius and Joseph Jekyll. *Letters of the Late Ignatius Sancho, an African: To which are Prefixed, Memoirs of His Life*. London: J Nichols, 1782.

Scott, Hew. *Fasti Ecclesiæ Scoticanæ* 2, no. 1. Edinburgh: William Paterson, 1894.

Scott, Sir Walter. *Old Mortality*. Edited by Jane Stevenson and Peter Davidson. Oxford: Oxford University Press, 2009.

266   BIBLIOGRAPHY

Scott, Sir Walter. *Waverley; Or, 'Tis Sixty Years Since.* Vol. 1. Edinburgh: James Ballantyne & Co., 1814.

Sedgewick, Adam. *A Memorial by the Trustees of Cowgill Chapel with a Preface and Appendix, on the Climate, History and Dialects of Dent.* Cambridge: University Press, 1868.

Serre, Jean-Adam. *Essais sur les principes de l'harmonie.* Paris: Prault fils, 1753.

Smith, Adam. *The Glasgow Edition of the Works and Correspondence of Adam Smith VI: Correspondence.* Edited by Ernest Campbell Mossner and Ian Simpson Ross. Oxford: Oxford University Press, 1987.

Smith, Adam. *Lectures on Justice, Police, Revenue and Arms; Delivered in the University of Glasgow by Adam Smith; reported by a Student in 1763; and Edited, with an Introduction and Notes, by Edwin Cannan.* Oxford: Clarendon Press, 1896.

Smith, Adam. *Lectures on Jurisprudence* (1763). Edited by R. L. Meek, D. D. Raphael, and Peter Stein. Oxford: Clarendon Press, 1978.

Smith, Adam. *Of the Nature of that Imitation which Takes Place in What Are Called the Imitative Arts.* Dublin: Wogan and Byrne, 1795.

Smith, Adam. *The Theory of Moral Sentiments,* 2nd ed. London: A. Milar, 1761.

Smith, Adam. *The Theory of Moral Sentiments. To Which is Added a Dissertation on the Origin of Languages,* 3rd ed. London: A. Millar, 1767.

Smith, Robert. *Harmonics; or, The Philosophy of Musical Sounds.* Cambridge: J. Bentham, 1749.

Smith, Robert Archibald. *Sacred Music, Consisting of the Tunes, Sanctusses, Doxologies, Thanksgivings, &c., Sung in St. George's Church, Edinburgh.* Edinburgh: R. Purdie, 1825.

Somerville, Thomas. *My Own Life and Times, 1741-1814.* Edinburgh: Edmonston & Douglas, 1861.

Southern, John. "On the Vibrations of Musical Strings; with a Mode of Ascertaining the Sound Producible by any Given Number of Vibrations." *The Philosophical Magazine* 40, no. 175 (1812): 333–337.

Speculative Society of Edinburgh. *History of the Speculative Society of Edinburgh, From its Institution in M.DCC.LXIV.* Edinburgh, 1845.

Stanhope, Charles. "The Science of Tuning Instruments with Fixed Tone." *The Repertory of Patent Inventions, and Other Discoveries and Improvements* 9, no. 2 (1806): Appendix, 1–24.

Steele, Joshua. *Prosodia Rationalis: Or, an Essay Towards Establishing the Melody and Measure of Speech, to be Expressed and Perpetuated by Peculiar Symbols.* London: J. Nichols, 1779.

Stephens, Alexander. *Public Characters, 1801-1802.* London: Richard Philipps, 1807.

Stewart, Dugald. *Biographical Memoirs, of Adam Smith, LL. D., of William Robertson, D. D. and of Thomas Reid, D. D.* Edinburgh: G. Ramsay, 1811.

Stewart, Dugald. *Elements of the Philosophy of the Human Mind* 1. London: A. Strahan and T. Cadell, 1792.

Stewart, Robert Walter. "Parish of Erskine." In *The New Statistical Account of Scotland: pt. 1-2 Renfrew, Argyle* 8, edited by John Gordon, 510–526. Edinburgh: W. Blackwood and Sons, 1845.

Stuart, Gilbert. *A View of Society in Europe in its Progress from Rudeness to Refinement.* Edinburgh: J. Bell, 1778.

Sulzer, Johann Georg. *Allgemeine Theorie der Schönen Künste* 3, 2nd ed. Leipzig: Weidemann, 1793.

Taas, William. *Elements of Music.* Aberdeen, 1787.

Tans'ur, William. *A New Musical Grammar: Or, the Harmonical Spectator, Containing All the Technical Parts of Musick, etc.* London: Robinson, 1746.

The Confession of Faith, and the Larger and Shorter Catechisms... now appointed by the General Assembly of the Kirk of Scotland, to be a part of uniformity in religion, etc., 5th ed. London: S. Cruttenden & T. Cox, 1717.

The Gentleman's Diary or Mathematical Repository. Vol. 13, edited by Thomas Simpson. London: Company of Stationers, 1753.

BIBLIOGRAPHY 267

The Gentleman's Diary or Mathematical Repository. Vol. 14, edited by Thomas Simpson. London: Company of Stationers, 1754.

Thom, William. *The Motives which Have Determined the University of Glasgow to Desert the Blackfriar Church, and Betake Themselves to a Chapel In a Letter from Prof. — to H— M—, Esq; Airshire*. Glasgow, 1764.

Tosi, Pier Francesco. *Opinioni de' cantori antichi, e moderni o sieno osservazioni sopra il canto figurato*. Bologna: Lelio dalla Volpe, 1723.

Twelve Tunes for the Church of Scotland, composed in four Parts, according to the Method used by the Master of the Musick School of Aberdeen in 1714. Aberdeen, 1714.

Vitruvius, Pollio. *De architectura*. 30-20 BCE.

Wackenroder, Wilhelm. *Phantasien über die Kunst von einem kunstliebenden Klosterbruder*. Berlin: Johann F. Unger, 1797.

Walker, John. *The Melody of Speaking*. London: The Author, 1787.

Webb, Daniel. *Observations on the Correspondence between Poetry and Music*. Dublin: James Williams, 1769.

Weeton, Ellen. *Journal of a Governess 1811–1825*. Vol. 2, edited by Edward Hall. London: Oxford University Press, 1939.

Wesley, Samuel. "Review of 'Fine Arts. Art. I. *A Musical Grammar in Four Parts.*'" *Annual Review* 5 (1807): 701–703.

Wilberforce, Robert I., and Samuel Wilberforce/ *The Life of Wm. Wilberforce, By His Sons*. Vol. 2. London: John Murray, 1838.

Willis, Arthur J. *Winchester Ordinations, 1660–1829, from Records in the Diocesan Registry, Winchester*. Vol. 1. Hambleden, Lyminge, Folkestone, Kent: A. J. Willis, 1964.

Wise, John. *A System of Aeronautics Comprehending its Earliest Investigations, and Modern Practice and Art*. Philadelphia: Joseph A. Steel, 1850.

Young, Anne. *Elements of Music and of Fingering the Harpsichord*. Edinburgh: [ca. 1790].

Young, Anne. "Specification of the Patent granted to ANN YOUNG, of St. James's Square, in the City of Edinburgh, for an Apparatus consisting of an oblong square Box . . . ." *The Repertory of Arts, Manufactures, and Agriculture: Consisting of Original Communications, Specifications of Patent Inventions, Practical and Interesting Papers, Selected from the Philosophical Transactions and Scientific Journals of All Nations* 16 (London, G. and T. Wilkie, 1802): 9–44.

Young, Thomas. "An Essay on Music." *British Magazine* (October 1800); reprinted in Thomas Young, *Miscellaneous Works of the Late Thomas Young*. Vol. 1. London: J. Murray, 1855.

Young, Thomas. "Letter to Andrew Dalzel of July 8, 1798." In *Memoir of Andrew Dalzel, Professor of Greek in the University of Edinburgh*. Edited by Cosmo Innes, 159–162. Edinburgh: 1861.

Young, Walter. "An Essay on Rythmical Measures." *Transactions of The Royal Society of Edinburgh* 2, no. 2 (1790): 55–110.

Young, Walter. "Statistical Account No. 5: Parish of Erskine." In *The Statistical Account of Scotland Drawn Up from the Communications of the Ministers of the Different Parishes*. Vol. 9, edited by John Sinclair, 58–76. Edinburgh: William Creech, 1793.

Archival Materials

Cambridgeshire Archives, Ely, ref 17/18: The family tree of the Holdens of Westhouse & Clapham.

Cumbria Archive Centre, Kendal:

Will of James Fawcett of Kirkby Stephen, gent., 2 January 1777, in WD U/Box 59/3/T 22–23.

Edinburgh Dean of Guild Court Records, Box 1799/8.

Glasgow City Archives, TD423/2/1: Register of Baptisms at Musselburgh Dalkeith & Glasgow by the Rev. John Falconer 1754–1793.

268 BIBLIOGRAPHY

Glasgow University Archive:

Faculty Minutes, November 9, 1775, vol. 75, 335.
Faculty Minutes, June 10, 1782, vol. 77, 191–192.
Faculty Minutes, April 11, 1783, vol. 77, 233–234.
University of Glasgow Senate Minutes, May 15, 1765, vol. 26643, 41.
University of Glasgow Senate Minutes, January 23, 1766, vol. 26643, 81.
University of Glasgow Senate Minutes, December, 1767, vol. 26643, 129.
University of Glasgow Senate Minutes, June 1, 1769, vol. 26644, 93.
University of Glasgow Senate Minutes, May 18, 1770, vol. 26644, 179.
University of Glasgow Senate Minutes, March 17, 1772, vol. 26690, 59–60.
University of Glasgow Senate Minutes, June 10, 1782, vol. 26692, 191–192.

Glasgow University Library, Special Collections:

MS Gen 4 Minute Book of the College Literary Society, 1790–1799.
Lancashire Record Office: Marriage Bonds and Allegations, 1746–1799.
Lancashire Wills & Probate: Holden, Francis. "Will and Testament." March 13, 1741, Clapham, Yorkshire. Registered probate vol. 87 f.557 MF 1007.

National Records of Scotland:

Window tax rolls (1764), vol. 175/108.
Window tax rolls (1769), vol. 176/113.
Old Parish Registers: Marriages 685/2 180, St Cuthbert's, 51.

The National Archives, Kew:

Articles of Clerkship for Richard Holden (II), articled to James Fawcett, 1782 CP 5/ 121/1.
Holden, Richard (II). "Letter to Francis Ferrand Foljambe of Hertford Street, London." of June 1, 1794, HO 42/31/15 Fol. 59–60.

Secondary Sources

Adelson, Edward H. "Layered Representations for Vision and Video." *Proceedings of the IEEE Workshop on Representation of Visual Scenes.* Cambridge, MA: 1995, 3–9.
Agawu, Kofi. "The Invention of African Rhythm." *Journal of the American Musicological Society* 48, no. 3 (1995): 380–395.
Andersen, Holly K., and Rick Grush, "A Brief History of Time-Consciousness: Historical Precursors to James and Husserl." *Journal of the History of Philosophy* 47, no. 2 (2009): 277–307.
Barden, John R. "'Innocent and Necessary': Music and Dancing in the Life of Robert Carter of Nomony Hall, 1728–1804." *MA Thesis*, The College of William and Mary, 1983.
Baxter, James Reid. "Ecclesiastical Music." In *The Oxford Companion to Scottish History*, edited by Michael Lynch, 431–433. Oxford: Oxford University Press, 2001.
Berry, Christopher. *Social Theory of the Scottish Enlightenment.* Edinburgh: Edinburgh University Press, 1997.
Blackburn, Bonnie J. "Leonardo and Gaffurio on Harmony and the Pulse of Music." In *Essays on Music and Culture in Honor of Herbert Kellman,* edited by Barbara Haggh, 128–149. Paris: Minerve, 2001.
Blackmore, Callum. "Berton's Ludic Pedagogy and the Subdominant Otherwise: Tension and Compromise in the Early Paris Conservatoire Curriculum." *Current Musicology* 104 (Spring 2019): https://doi.org/10.7916/cm.v0i104.5395.
Bonds, Mark Evan. *Absolute Music: The History of an Idea.* Oxford: Oxford University Press, 2014.

BIBLIOGRAPHY 269

Bonge, Dale. "Gaffurius on Pulse and Tempo: A Reinterpretation." *Musica disciplina* 36 (1982): 167–174.

Bow, Charles Bradford, ed. *Common Sense in the Scottish Enlightenment.* Oxford: Oxford University Press, 2018.

Bradford, Richard. *Silence and Sound: Theories of Poetics from the Eighteenth Century.* Rutherford: Fairleigh Dickinson Press, 1992.

Brealey, Peter. "The Charitable Corporation for the Relief of Industrious Poor: Philanthropy, Profit and Sleaze in London, 1707–1733." *History* 98, no. 333 (2013): 708–729.

Brittan, Francesca, and Carmel Raz, eds., "Colloquy: Attention, Anxiety, and Audition's Histories." *Journal of the American Musicological Society* 72, no. 2 (2019): 541–580.

Broadie, Alexander, and Craig Smith, eds. *The Cambridge Companion to the Scottish Enlightenment.* Cambridge: Cambridge University Press, 2019.

Brooks, Jeanice. "Staging the Home: Music in Aristocratic Family Life." In *A Passion for Opera: The Duchess and the Georgian Stage,* edited by Paul W. Boucher, Jeanice Brooks, Katrina Faulds, Catherine Garry, and Wiebke Thormählen, 30–45. Boughton: Buccleuch Living Heritage Trust, 2019.

Brown, Leslie E. *Artful Virtue: The Interplay of the Beautiful and the Good in the Scottish Enlightenment.* New York: Routledge, 2016.

Brown, Leslie E. "The Common Sense School and the Science of Music in Eighteenth-Century Scotland." In *Essays in Honor of John F. Ohl: A Compendium of American Musicology,* edited by Enrique Alberto Arias, 122–132. Evanston: Northwestern University Press, 2001.

Brown, Leslie E. "Thomas Reid and the Perception of Music: Sense vs. Reason." *International Review of the Aesthetics and Sociology of Music* 20, no. 2 (1989): 121–140.

Burgess, Miranda. "Transport: Mobility, Anxiety, and the Romantic Poetics of Feeling." *Studies in Romanticism* 49, no. 2 (2010): 229–60.

Cajori, Florian. *A History of the Logarithmic Slide Rule and Allied Instruments.* New York: Engineering News Publishing Company, 1909.

Caillois, Roger. *Les jeux et les hommes.* Paris: Gallimard, 1958.

Camic, Charles. "Experience and Ideas: Education for Universalism in Eighteenth-Century Scotland." *Comparative Studies in Society and History* 25, no.1 (1983): 50–82.

Carr, Rosalind. *Gender and Enlightenment Culture in Eighteenth-Century Scotland.* Edinburgh: Edinburgh University Press, 2014.

Chenette, Louis F. "Music Theory in the British Isles during the Enlightenment." PhD diss., Ohio State University, 1967.

Christensen, Thomas. *Rameau and Musical Thought in the Enlightenment.* Cambridge: Cambridge University Press, 1993.

Chua, Daniel. *Absolute Music and the Construction of Meaning.* Cambridge: Cambridge University Press, 1999.

Clark, Peter. *British Clubs and Societies, 1580–1800. The Origins of an Associational World.* Oxford: Oxford University Press, 2000.

Clegg, Brian. *The First Scientist: A Life of Roger Bacon.* London: Constable, 2003.

Cohen, David E. "Boethius and the Enchiriadis Theory: The Metaphysics of Consonance and the Concept of Organum." PhD diss., Brandeis University, 1993.

Cohen, David E. "The 'Gift of Nature': Musical 'Instinct' and Musical Cognition in Rameau." In *Music Theory and Natural Order from the Renaissance to the Early Twentieth Century,* edited by Suzannah Clark and Alexander Rehding, 86–127. Cambridge: Cambridge University Press, 2001.

Cohen, David E. "'The Imperfect Seeks its Perfection': Harmonic Progression, Directed Motion, and Aristotelian Physics." *Music Theory Spectrum* 23, no. 2 (2001): 139–169.

"Rousseau as Music Theorist: Harmony, Mode, and (*L'Unité de*) *Mélodie.*" In "Colloquy on "Rousseau in 2013: Afterthoughts on a Tercentenary," edited by Jacqueline Waeber. *Journal of the American Musicological Society* 66, no. 1 (2013): 275–280.

270 BIBLIOGRAPHY

Cohn, Richard L. "Why We Don't Teach Meter, and Why We Should." *Journal of Music Theory Pedagogy* 29 (2015): 5–23.

Copenhaver, Rebecca. "Reid on Memory and Personal Identity." In *The Stanford Encyclopedia of Philosophy*, edited by Edward N. Zalta. Winter 2018 Edition. https://plato.stanford.edu/archives/win2018/entries/reid-memory-identity.

Costa, Shelley. "The 'Ladies' Diary': Gender, Mathematics, and Civil Society in Early-Eighteenth-Century England." *Osiris* 17 (2002): 49–73.

Cranmer, John. "Concert Life and Music Trade in Edinburgh c. 1780–1830." PhD diss., University of Edinburgh, 1991.

Cuneo, Terence, and René van Woudenberg, eds. *The Cambridge Companion to Thomas Reid.* Cambridge: Cambridge University Press, 2004.

Curtius, Ernst Robert. *Europäische Literatur und lateinisches Mittelalter.* Vol. 2. Bern und München: Francke, 1948.

Dahlhaus, Carl. *Die Idee der absoluten Musik.* Kassel: Barenreiter: 1978.

Daiches, David. *The Scottish Enlightenment: An Introduction.* Edinburgh: Saltire Society, 1986.

Damschroder, David. *Thinking about Harmony: Historical Perspectives on Analysis.* Cambridge: Cambridge University Press, 2008.

DeFord, Ruth I. *Tactus, Mensuration, and Rhythm in Renaissance Music.* Cambridge: Cambridge University Press, 2015.

Deutsch, Diana. "The Processing of Structured and Unstructured Tonal Sequences." *Perception & Psychophysics* 28, no. 5 (1980): 381–389.

Donnelly, Michael. "John Holden's Gorbals Pottery, 1762–1786." *Scottish Pottery: The Journal of the Glasgow Branch of the Scottish Pottery Society* 1 (1978): 2–3.

Douthwaite, Julia V. *The Wild Girl, Natural Man, and the Monster: Dangerous Experiments in the Age of Enlightenment.* Chicago: University of Chicago Press, 2002.

Dowling, W. Jay "Rhythmic Groups and Subjective Chunks in Memory for Melodies." *Perception & Psychophysics* 14 (1973): 37–40.

Drake, Carolyn, Mari Riess Jones, and Clarisse Baruch. "The Development of Rhythmic Attending in Auditory Sequences: Attunement, Referent Period, Focal Attending." *Cognition* 77, no. 3 (2000): 251–288.

Duguid, Timothy. "Early Modern Scottish Metrical Psalmody: Origins and Practice." *Yale Journal of Music and Religion* 7, no. 1 (2021): 1–23.

Duguid, Timothy. *Metrical Psalmody in Print and Practice: English 'Singing Psalms' and Scottish 'Psalm Buiks,' c. 1547–1640.* Farnham, Surrey: Ashgate, 2016.

Emerson, Roger L. "Conjectural History and Scottish Philosophers." *Historical Papers / Communications historiques* 19, no. 1 (1984), 63–90.

Emerson, Roger L. *Neglected Scots: Eighteenth Century Glasgwegians and Women.* Edinburgh: Humming Earth, 2015.

Farmer, Henry G. *A History of Music in Scotland.* London: Hinrichsen, 1947.

Fleming, John A. *Scottish Pottery.* Glasgow: EP Publishing, 1973.

Folescu, Marina. "Remembering Events: A Reidean Account of (Episodic) Memory." *Philosophy and Phenomenological Research* 97, no. 2 (2018): 304–321.

Folescu, Marina. "Thomas Reid's View of Memorial Conception." *The Journal of Scottish Philosophy* 16, no. 3 (2018): 211–26.

Fremont-Barnes, Gregory. *The Jacobite Rebellion 1745–46.* London: Bloomsbury Publishing, 2014.

Friday, Jonathan, ed. *Art and Enlightenment: Scottish Aesthetics in the Eighteenth Century.* Exeter: Imprint Academic, 2004.

Fritsch, Melanie, and Tim Summers, eds. *The Cambridge Companion to Video Game Music.* Cambridge: Cambridge University Press, 2021.

Fussell, Paul. *Theory of Prosody in Eighteenth-Century England.* Hamden: Archon Books, 1966.

Gelbart, Matthew. *The Invention of 'Folk Music' and 'Art Music': Emerging Categories from Ossian to Wagner.* Cambridge: Cambridge University Press, 2007.

BIBLIOGRAPHY   271

Gelbart, Matthew. "Rousseau and the Quest for Universals." In "Colloquy on "Rousseau in 2013: Afterthoughts on a Tercentenary." Jacqueline Waeber, ed. *Journal of the American Musicological Society* 66, no. 1 (2013): 280–284.

Gerhard, Anselm. *London und der Klassizismus in der Musik: Die Idee der 'absoluten Musik' und Muzio Clementis Klavierwerke*. Stuttgart and Weimar: J.B. Metzler, 2002.

Ghere, David and Fred Amram. "Inventing Music Education Games." *British Journal of Music Education* 24, no.1 (2007): 55–75.

Gibbons, William. *Unlimited Replays: Video Games and Classical Music*. Oxford: Oxford University Press, 2018.

Goehr, Lydia. *The Imaginary Museum of Musical Works: An Essay in the Philosophy of Music: An Essay in the Philosophy of Music*. Oxford: Clarendon Press, 1992.

Goldstone, Lawrence, and Nancy Goldstone. *The Friar and the Cipher: Roger Bacon and the Unsolved Mystery of the Most Unusual Manuscript in the World*. New York: Crown/Archetype, 2005.

Gowing, Ronald. *Roger Cotes: Natural Philosopher*. Cambridge: Cambridge University Press, 2002.

Grant, Roger M. *Beating Time and Measuring Music in the Early Modern Era*. Oxford: Oxford University Press, 2014.

Greenwood, Andrew. "Song and Improvement in the Scottish Enlightenment." *Journal of Musicological Research* 39, no. 1 (2020): 42–68.

Grey, Thomas. "Review of *Absolute Music: The History of an Idea*, by Mark Evan Bonds." *Music Theory Spectrum* 38, no. 1 (2016): 126–131.

Grote, Simon. *The Emergence of Modern Aesthetic Theory: Religion and Morality in Enlightenment Germany and Scotland*. Cambridge: Cambridge University Press, 2017.

Gurton-Wachter, Lily. *Watchwords: Romanticism and the Poetics of Attention*. Stanford: Stanford University Press, 2016.

Hagner, Michael. "Toward a History of Attention in Culture and Science." *MLN* 118, no. 3 (2003): 670–687.

Hanley, Ryan Patrick. "Social Science and Human Flourishing: The Scottish Enlightenment and Today." *Journal of Scottish Philosophy* 7, no. 1 (2009): 29–46.

Harrison, Peter M. C., and Marcus T. Pearce, "Simultaneous Consonance in Music Perception and Composition." *Psychological Review* 127, no. 2 (2020): 216–244.

Hasty, Christopher. *Meter as Rhythm*. Oxford University Press, 1997.

Henrich, Joseph Steven J. Heine, and Ara Norenzayan. "The Weirdest People in the World?" *Behavioral and Brain Sciences* 33, nos. 2–3 (2010): 61–83.

Herodotus, *The Histories*. Translated by Robin Waterfield. New York: Oxford University Press, 2008.

Hills, Richard L. "James Watt and the Delftfield Pottery, Glasgow." *Proceedings of the Society of Antiquaries of Scotland* 131 (2001): 375–420.

Hill, John Walter. *Joseph Riepel's Theory of Metric and Tonal Order, Phrase and Form: A Translation of his* Anfangsgründe zur musicalischen Setzkunst, Chapters 1 and 2 *(1752/54, 1755) with Commentary*. Hillsdale, NY: Pendragon Press, 2015.

Holman, Peter. "A Little Light on Lorzeno Bocchi: An Italian in Edinburgh and Dublin." In *Music in the British Provinces, 1690–1914,"* edited by Rachel Cowgill and Peter Holman, 61–86. Milton Park, Abdington: Routledge, 2007.

Honing, Henkjan. *Musical Cognition: A Science of Listening*. Translated by Sherry Marx and Susan van der Werff-Woolhouse. New Brunswick: Transaction Publishers, 2011.

Houston, Robert A. "Popular Politics in the Reign of George II: The Edinburgh Cordiners." *The Scottish Historical Review* 72, no. 194.2 (1993): 167–189.

Houston, Robert A. *Scottish Literacy and the Scottish Identity: Illiteracy and Society in Scotland and Northern England, 1600–1800*. Cambridge: Cambridge University Press, 2002.

Hughes, Kathryn. *The Victorian Governess*. London: Bloomsbury, 2003.

Hui, Alexandra. *The Psychophysical Ear: Musical Experiments, Experimental Sounds, 1840–1910*. Cambridge, MA: MIT Press, 2012.

272 BIBLIOGRAPHY

Huizinga, Johan. *Homo Ludens: Proeve Ener Bepaling Van Het Spelelement Der Cultuur.* Groningen, Wolters-Noordhoff, 1938.

Huron, David. "Musical Aesthetics: Uncertainty and Surprise Enhance Our Enjoyment of Music." *Current Biology* 29, no. 23 (2019): R1238–R1240.

Huron, David. *Sweet Anticipation: Music and the Psychology of Expectation.* Cambridge, M.A.: MIT Press, 2008.

Inglis, James. "The Scottish Churches and the Organ in the Nineteenth Century." PhD diss., University of Glasgow, 1987.

Jacobi, Erwin R. "Harmonic Theory in England after the Time of Rameau." Translated by David Kraehenbuhl. *Journal of Music Theory* 1, no. 2 (1957): 126–146.

Jacoby, Nori, et al. "Commonality and Variation in Mental Representations of Music Revealed by a Cross-Cultural Comparison of Rhythm Priors in 15 Countries." *Nature Human Behaviour* 8, no. 5 (2024): 846–877.

Johnson, David. "An Eighteenth-Century Scottish Music Library." *RMA Research Chronicle* 9 (1971): 90–95.

Jorgensen, Owen. *Tuning: Containing the Perfection of Eighteenth-Century Temperament, the Lost Art of Nineteenth-Century Temperament, and the Science of Equal Temperament, Complete with Instructions for Aural and Electronic Tuning.* East Lansing, MI: Michigan State University Press, 1991.

Jungnickel, Christa, and Russell McCormmach. *Cavendish: The Experimental Life.* Lewisburg: Bucknell, 1999.

Kamp, Michiel, Tim Summers, and Mark Sweeney, eds. *Ludomusicology: Approaches to Video Game Music.* Bristol: Equinox, 2016.

Kapp, Karl M. *The Gamification of Learning and Instruction: Game-Based Methods and Strategies for Training and Education.* New Jersey: John Wiley & Sons, 2012.

Kassler, Jamie C. "British Writings On Music, 1760–1830: A Systematic Essay Toward a Philosophy of Selected Theoretical Writings." PhD diss., Columbia University, 1971.

Kassler, Jamie C. *The Science of Music in Britain, 1714–1830: A Catalogue of Writings, Lectures, and Inventions.* 2 vols. New York: Garland Publishing, 1979.

Kassler, Michael. *A.F.C. Kollmann's Quarterly Musical Register (1812): An Annotated Edition.* Farnham: Ashgate Publishing, 2008.

Kassler, Michael. "The Tuning of Maxwell's 'Essay.'" *Studies in Music* 11 (1977): 27–36.

Kennaway, George. *John Gunn: Musician Scholar in Enlightenment Britain.* Woodbridge: The Boydell Press, 2021.

Kinghorn, Jonathan, and Gerard Quail. *Delftfield: A Glasgow Pottery 1748–1823.* Glasgow: Glasgow Museums and Art Galleries, 1986.

Kivy, Peter. *The Seventh Sense: Francis Hutcheson and Eighteenth-Century British Aesthetics.* Oxford: Oxford University Press, 2003.

Klose, Birgit. "Die Erste Ästhetik der Absoluten Musik: Adam Smith und sein Essay 'über die sogenannten Imitativen Künste.'" PhD diss., Philipps-Universität Marburg, 1996.

Langner, Gerald D. *The Neural Code of Pitch and Harmony.* Cambridge: Cambridge University Press, 2015.

London, Justin. "Cognitive Constraints on Metric Systems: Some Observations and Hypotheses." *Music Perception* 19, no. 4 (2002): 536–537.

London, Justin. *Hearing in Time: Psychological Aspects of Musical Meter.* Oxford: Oxford University Press, 2004.

Ludwig, Loren. "J. S. Bach, the Viola da Gamba, and Temperament in the Early Eighteenth Century." *BACH: Journal of the Riemenschneider Bach Institute* 53, no. 2 (2022): 260–300.

Macleod, Jennifer. "The Edinburgh Musical Society: Its Membership and Repertoire, 1728–1797." PhD diss., University of Edinburgh, 2001.

MacKenzie John M., and T. M. Devine, eds. *Scotland and the British Empire.* Oxford: Oxford University Press, 2011.

BIBLIOGRAPHY 273

Madison, Guy. "Sensori-motor Synchronisation Variability Decreases as the Number of Metrical Levels in the Stimulus Signal Increases." *Acta Psychologica* 147 (2014): 10–16.

Margulis, Elizabeth H. "Melodic Expectation: A Discussion and Model." PhD diss., Columbia University, 2003.

Martin, Ann Smart. "Scottish Merchants: Sorting out the World of Goods in Early America." In *Transatlantic Craftmanship: Scotland and the Americas in the 18th and 19th Centuries,* edited by Simon Gilmour and Vanessa Habib, 23–44. Edinburgh: Society of Scottish Antiquaries, 2013.

Mathew, Nicholas. "Interesting Haydn: On Attention's Materials." *Journal of the American Musicological Society* 71, no. 3 (2018): 655–701.

McAuley, Karen. *Our Ancient National Airs: Scottish Song Collecting from the Enlightenment to the Romantic Era.* New York: Routledge, 2016.

McElroy, Davis D. "The Literary Clubs and Societies of Eighteenth-Century Scotland." PhD diss., Edinburgh University, 1952.

McKee, Eric. *Decorum of the Minuet, Delirium of the Waltz: A Study of Dance-Music.* Bloomington & Indianapolis: Indiana University Press, 2012.

McKenzie, J. "School and University Drama in Scotland, 1650–1760." *The Scottish Historical Review* 34, no. 118 (1955): 103–121.

Mehr, Samuel A., Manvir Singh, Dean Knox, Daniel M. Ketter, Daniel Pickens-Jones, Stephanie Atwood, Christopher Lucas, Nori Jacoby, Alena A. Egner, Erin J Hopkins, Rhea M Howard, Joshua K. Hartshorne, Mariela V. Jennings, Jan Simson, Constance M Bainbridge, Steven Pinker, Timothy J. O'Donnell, Max M Krasnow, and Luke Glowacki. "Universality and diversity in human song." *Science* 366, no. 6468 (2019): eaax0868.

Meyer, Leonard B. *Emotion and Meaning in Music.* Chicago: University of Chicago Press, 1956.

Michael, Ian. *English Grammatical Categories: And the Tradition to 1800.* Cambridge: Cambridge University Press, 2010.

Miller, George A. "The Magical Number Seven, Plus or Minus Two: Some Limits on our Capacity for Processing Information." *Psychological Review* 63 (1956): 81–97.

Mizuta, Hiroshi. *Adam Smith's Library: A Catalogue.* Oxford: Clarendon Press, 2000.

Moessner, Lilo. "The Syntax of Older Scots." In *The Edinburgh History of the Scots Language,* edited by Charles Jones, 112–155. Edinburgh: Edinburgh University Press, 1997.

Molnar, John Edgar. "Publication and Retail Book Advertisements in the *Virginia Gazette* 1736–1780". PhD diss., University of Michigan, 1978.

Moseley, Roger. *Keys to Play: Music as a Ludic Medium from Apollo to Nintendo.* Berkeley and Los Angeles: University of California Press, 2016.

Munro, Gordon J. *Scottish Church Music and Musicians, 1500–1700.* PhD diss., University of Glasgow, 1999.

Mutch, Caleb. "The Formal Function of Fortspinnung." *Theory & Practice* 43 (2018): 1–32.

Mutch, Caleb. "How the Triad Took (a) Root." *Journal of Music Theory* 66, no. 1 (2022): 43–62.

Mutch, Caleb. "Studies in the History of the Cadence." PhD diss., Columbia University, 2015.

Narmour, Eugene. *The Analysis and Cognition of Melodic Complexity: The Implication-Realization Model.* Chicago: University of Chicago Press, 1992.

Newman, William R. "The Philosophers' Egg: Theory and Practice in the Alchemy of Roger Bacon." *Micrologus* 3 (1995): 75–101.

Nichols, Ryan. *Thomas Reid's Theory of Perception.* Oxford: Oxford University Press, 2007.

Nichols, Ryan and Gideon Yaffe. "Thomas Reid." *The Stanford Encyclopedia of Philosophy.* Winter 2016 Edition, edited by Edward N. Zalta. https://plato.stanford.edu/archives/win2016/entries/reid/.

Nilson, Peter. "Winged Man and Flying Ships: Of Medieval Flying Journeys and Eternal Dreams of Flight." Trans. Steven Hartman. *The Georgia Review* 50, no. 2 (1996): 267–296.

Omond, Thomas Stewart. *English Metrists: Being a Sketch of English Prosodical Criticism from Elizabethan Times to the Present Day.* Oxford: Clarendon Press, 1903.

Palmieri, Robert M. *Piano: An Encyclopedia,* 2nd ed. New York: Routledge, 2003.

274 BIBLIOGRAPHY

Parlett, David. *Penguin History of Card Games*. London: Penguin UK, 2008.

Patrick, Millar. *Four Centuries of Scottish Psalmody*. Oxford: Oxford University Press, 1949.

Pau, Andrew. "The Harmonic Theories of Jean-Adam Serre." *Intégral* 32 (2018): 1–13.

Payzant, Geoffrey B. "The Organ Controversy in Scotland." *The Dalhousie Review* 32, no. 4 (1953): 44–48.

Pedersen, Olaf. "The 'Philomath' of 18th Century England." *Centaurus* 8 (1963): 238–262.

Perkins, David. "How the Romantics Recited Poetry." *Studies in English Literature, 1500–1900* 31, no. 4 (1991): 655–671.

Phillips, Natalie M. *Distraction: Problems of Attention in Eighteenth-Century Literature*. Baltimore: Johns Hopkins University Press, 2016.

Popkin, Richard H. "The Philosophical Basis of Eighteenth-Century Racism." *Studies in Eighteenth-Century Culture* 3 (1974): 245–262.

Potter, Mary C. Brad Wyble, Carl E. Hagmann, and Emily S. McCourt. "Detecting Meaning in RSVP at 13 ms per Picture." *Attention, Perception, & Psychophysics* 76, no. 2 (2014): 270–279.

Pressnitzer, Daniel. "Mid-Level Audition." *Habilitation à Diriger des Recherches*. Université Paris Descartes et Département d'Etudes Cognitives, Ecole Normale Supérieure, 2009.

Quail, Gerard. "The Millroad Street Pottery Calton: The Background – Thomas Wyse and his Stoneware Shop, 1799–1814." *Scottish Industrial History* 18 (1996): 21–44.

Raz, Carmel. "Anne Young's Musical Games (1801): Music Theory, Gender, and Game Design." *SMT-V: Videocast Journal of the Society for Music Theory* 4, no. 2 (2018): goo.gl/ZXR6Cv.

Raz, Carmel. "An Eighteenth-Century Theory of Musical Cognition? John Holden's *Essay towards a Rational System of Music* (1770)." *Journal of Music Theory* 62, no. 2 (2018): 205–248.

Raz, Carmel. "To 'Fill Up, Completely, the Whole Capacity of the Mind': Listening with Attention in Late Eighteenth-century Scotland." *Music Theory Spectrum* 44, no. 1 (2022): 141–154.

Rehding, Alexander. "Three Music-Theory Lessons." *Journal of the Royal Musical Association* 141, no. 2 (2016): 251–282.

Renton, Alex. *Blood Legacy: Reckoning with a Family's Story of Slaver*. London: Canongate, 2021.

Repp, Bruno H. "Sensorimotor Synchronization: A Review of the Tapping Literature." *Psychonomic Bulletin & Review* 12, no. 6 (2005): 969–992.

Riley, Matthew. *Musical Listening in the German Enlightenment Attention, Wonder and Astonishment*. Burlington, VT: Ashgate, 2004.

Robb, Peter. "Mr Upjohn's Debts: Money and Friendship in Early Colonial Calcutta." *Modern Asian Studies* 47, no. 4 (2013): 1185–1217.

Roebuck, Hettie, Claudia Freigang, and Johanna G. Barry. "Continuous Performance Tasks: Not Just About Sustaining Attention." *The Journal of Speech, Language, and Hearing Research* 59, no. 3 (2016): 501–510.

Rone, Vincent E. Can Aksoy, and Sarah Pozderac-Chenevey, eds. *Nostalgia and Videogame Music: A Primer of Case Studies for the Player-Academic*. Bristol: Intellect Press, 2022.

Rookes, Paul, and Jane Willson. *Perception: Theory, Development and Organisation*. London: Routledge, 2000.

Rosenthal, Caitlin. "Numbers for the Innumerate: Everyday Arithmetic and Atlantic Capitalism." *Technology and Culture* 58, no. 2 (2017): 529–544.

Saintsbury, George. *A History of English Prosody from the Twelfth Century to the Present Day* 2. London: MacMillan, 1908.

Sanger, Keith. "A Letter from the Rev. Patrick MacDonald to Mrs. Maclean Clephane, 1808." *Scottish Gaelic Studies* 26 (2010): 23–34.

Savage, Patrick E., Steven Brown, Emi Sakai, and Thomas E. Currie. "Statistical universals reveal the structures and functions of human music." *Proceedings of the National Academy of Sciences* 112, no. 29 (2015): 8987–8992.

Schaarwächter, Jürgen. *Two Centuries of British Symphonism: From the Beginnings to 1945. A Preliminary Survey*. Vol. 2. Hildesheim: George Olms Verlag, 2015.

BIBLIOGRAPHY   275

Seville, Adrian. *The Cultural Legacy of the Royal Game of the Goose: 400 Years of Printed Board Games.* Amsterdam: Amsterdam University Press, 2019.

Shapin, Steven. "Property, Patronage, and the Politics of Science: The Founding of the Royal Society of Edinburgh." *The British Journal for the History of Science* 7 no. 1 (1974): 1–41.

Shapiro, Lawrence. *Embodied Cognition.* New York and London: Taylor and Francis, 2011.

Seidel, Wilhelm. "Der Essay von Adam Smith über die Musik - Eine Einführung" *Musiktheorie* 3 (2000): 195–204.

Seidel, Wilhelm. "Die Sprache der Musik." In *Festschrift Klaus Wolfgang Niemöller zum 60. Geburtstag,* edited by Jobst P. Fricke, 495–511. Regensburg: Bosse Verlag, 1989.

Semi, Maria. *Music as a Science of Mankind in Eighteenth-Century Britain.* Translated by Timothy Keates. Farnham: Ashgate, 2012.

Shefrin, Jill. " 'Make it a Pleasure and Not a Task': Educational Games for Children in Georgian England." *Princeton University Library Chronicle* 60, no. 2 (1999): 251–275.

Shteir, Ann B. "Elegant Recreations? Configuring Science Writing for Women." In *Victorian Science in Context,* edited by Bernard Lightman, 236–255. Chicago: University of Chicago Press, 1997.

Simpson, Ian J. "Sir Archibald Grant and the Charitable Corporation." *The Scottish Historical Review* 44, no. 137 (1965): 52–62.

Siraisi, Nancy C. "The Music of Pulse in the Writings of Italian Academic Physicians (Fourteenth and Fifteenth Centuries)." *Speculum* 50 (1975): 689–710.

Sisman, Elaine R. "Small and Expanded Forms: Koch's Model and Haydn's Music." *The Musical Quarterly* 68, no. 4 (1982): 444–475.

Skaggs, David C. "John Semple and the Development of the Potomac Valley, 1750–1773." *The Virginia Magazine of History and Biography* 92, no. 3 (1984): 282–308.

Slawson, Wayne. "Forked Tongues: Structural Illusions in Music." In *Reflecting Senses: Perception and Appearance in Literature, Culture and the Arts*, edited by Walter Pape and Frederick Burwick, 250–263 (Berlin: de Gruyter, 2011).

Smith, Craig. "Adam Ferguson and Ethnocentrism in the Science of Man." *History of the Human Sciences* 26, no. 1 (2013): 52–67.

Smith, Craig. "The Scottish Enlightenment, Unintended Consequences and the Science of Man." *Journal of Scottish Philosophy* 7, no. 1 (2009): 9–28.

Sowerby, Millicent. *The Catalogue of the Library of Thomas Jefferson.* Vol. 4, compiled by Millicent Sowerby. Washington, DC: The Library of Congress, 1959.

Steel, Karl. *How Not to Make a Human: Pets, Feral Children, Worms, Sky Burial, Oysters.* Minneapolis: University of Minnesota Press, 2019.

Stewart, Michael A., ed. *Studies in the Philosophy of the Scottish Enlightenment.* Oxford: Oxford University Press, 1990.

Sumera, Magdalena. "The Temporal Tradition in the Study of Verse Structure." *Linguistics* 8, no. 62 (1970): 44–65.

Swetz, Frank J. *The Impact and Legacy of The Ladies' Diary (1704–1840): A Women's Declaration.* Providence: MAA Press, an imprint of the American Mathematical Society, 2021.

Telesco, Paula J. "Identifying the Unknown Source of a Pre-Rameau Harmonic Theorist: Who was Alexander Malcolm's Mysterious Ghostwriter?" *Eighteenth-Century Music* 17, no. 1 (2020): 37–52.

Temperley, Nicholas. "The Old Way of Singing: Its Origins and Development." *Journal of the American Musicological Society* 34, no. 3 (1981): 511–544.

Tenney, James. *A History of "Consonance" and "Dissonance."* New York: Excelsior, 1988.

Thorne, Roland G. *The House of Commons 1790–1820* 3. Woodbridge: Boydell & Brewer, 1986.

Tomalin, Marcus. "Pendulums and Prosody in the Long Eighteenth Century." *The Review of English Studies* 68, no. 286 (2017): 734–755.

Troyano, Leonardo Fernández. *Bridge Engineering: A Global Perspective.* London: Thomas Telford, 1997.

276 BIBLIOGRAPHY

Tucker, George Holbert. *Jane Austen the Woman: Some Biographical Insights.* London: Palgrave Macmillan, 1995.

Venn, John A. *Alumni Cantabrigienses; A Biographical List of All Known Students, Graduates and Holders of Office at the University of Cambridge, from the Earliest Times to 1900.* Part 2, vol. 3. Cambridge: The University Press, 1947.

Voigt, Boris. "Musikästhetik für den Homo oeconomicus: Adam Smith über Gefühle, Markt und Musik." *Zeitschrift für Ästhetik und allgemeine Kunstwissenschaft* 58, no. 1 (2013): 97–120.

Waeber, Jacqueline. "A Corruption of Rousseau? The Quest for the Origins of Music and Language in Recent Scientific Discourse." In "Colloquy on "Rousseau in 2013: Afterthoughts on a Tercentenary," edited by Jacqueline Waeber. *Journal of the American Musicological Society* 66, no. 1 (2013): 284–289.

Wallace, Mark C., and Jane Rendall, eds. *Association and Enlightenment: Scottish Clubs and Societies, 1700–1830.* Lewisburg: Bucknell University Press, 2020.

Watts, Ruth. "Scientific Women Their Contribution to Culture in England in the Late Eighteenth and Nineteenth Centuries." In *Women, Education, and Agency, 1600–2000*, edited by Jean Spence, Sarah Aiston, and Maureen M. Meikle, 49–65. New York and Milton Park, Abingdon, Routledge, 2009.

"Weber's Law." In *A Dictionary of Psychology*, 4th ed. Edited by Andrew M. Colman, 818. Oxford: Oxford University Press, 2015.

Whyte, Ian D. *Scotland before the Industrial Revolution: An Economic and Social History c. 1050–c. 1750.* Milton Park: Routledge, 2014.

Wilfing, Alexander. *Re-Reading Hanslick's Aesthetics: Die Rezeption Eduard Hanslicks im englischen Sprachraum und ihre diskursiven Grundlagen.* Vienna: Hollitzer, 2019.

Willan, Thomas Stuart. *An Eighteenth-Century Shopkeeper: Abraham Dent of Kirkby Stephen.* Manchester: Manchester University Press, 1970.

Winstanley, Denys A. *Early Victorian Cambridge.* Cambridge: Cambridge University Press, 1940.

Wolterstorff, Nicholas. *Thomas Reid and the Story of Epistemology.* Cambridge: Cambridge University Press, 2004.

Woodward, William R. *Hermann Lotze: An Intellectual Biography.* Cambridge: Cambridge University Press, 2015.

Woodworth, Philip L. "Some Further Biographical Details of the Holden Tide Table Makers." *Proudman Oceanographic Laboratory Report* 58 (July 2003): 3–20.

Woodworth, Philip L. "Three Georges and One Richard Holden: The Liverpool Tide Table Makers." *Transactions of the Historic Society of Lancashire and Cheshire* 151 (2002): 19–51.

Wright, Michael. "James Watt: Musical Instrument Maker." *The Galpin Society Journal* 55 (2002): 104–129.

Yaffe, Gideon. "Thomas Reid on Consciousness and Attention." *Canadian Journal of Philosophy* 39, no. 2 (2009): 165–194.

Index

For the benefit of digital users, indexed terms that span two pages (e.g., 52–53) may, on occasion, appear on only one of those pages.

Bold page number indicates the definition of heading.

Tables and figures are indicated by an italic *t* and *f* following the page number.

Family members of John Holden are identified by their relation to him, given in parentheses.

Aberdeen, 96, 98, 103, 163. *See also* Duncan, Gideon; Channon, Thomas
absolute music, 125–26
accents, 48, 56, 141, 147–48, 171–72
Alison, Archibald, 5–6
Ancient Greek music, 195–98, 200, 254
Anderson, John, 23–24, 160
Anglican hymns. *See* hymns
Aristotle (and Aristotelianism), 2–3, 16, 131
Ashworth, Caleb, 105, 106
  Collection of Tunes, A, 105, 106, 108*f*, 122
Associationism, 128–30
attention, 10, 43, **69–76**, 90, 126–27, 128–35 passim 149–50
  distinguishing consonance and dissonance, 77–78
  divided, 70–71, 79–80, 81, 88*f*, 89, 137–39, 235–36 (*see also* fundamental bass: double)
  voluntary (active) vs. involuntary (passive), 129–30, 136–38, 140–41
Austen, Jane, 211, 214

Bach, C. P. E., 247
Bacon, Roger, 15, 15n.32, 16
Beattie, James, 5–6, 8
Boag, Andrew, 35
Bond, Elizabeth, 216–19
Bremner, Robert, 6–7, 96–98, 101n.29, 105
British Empire, 5, 201–3
Bryan, Margaret, 214–16, 217*f*, 238–39, 241–42, 243–46, 244*f*
Burmeister, Joachim, 251–52
Burnett, James (Lord Monboddo), 179

Burney, Charles, 3–4, 8, 18, 97n.15, 143–44, 179–81, 227, 229*f*
Burns, Robert, 5–6

Cadell, Thomas, 143
cadence
  in Anne Young, 226, 234, 237–38
  in Holden, 114–17, 117*f*, 136, 138–39
Callcott, John Wall, 3–4, 11n.21, 203, 210–11, 219, 237–38, 239, 241–42
Calvin, John, 94–96
Cambridge University, 25–26, 35–36, 38–41, 242–43
Campbell, Joshua, 19–20
Captain Cook, 144
Carlyle, Alexander (reverend), 160–61
Carter, Robert, III, 249–50
Channon, Thomas, 98, 99–100, 100n.26, 103
Charles I, King, 94–95
Charlotte, Princess, 242, 246–47
"Charming Sally," (ship), 13
chunking. *See* grouping
circle of fifths, 201–3, 221, 222, 227, 228*f*, 229*f*, 230*f*, 232, 233*f*
Clephane, Lady Margaret, 204, 250
clock chimes, 132–34, 141, 166, 195
Collection of Church Music, A (Holden), 7, 10, 23, 103–24, 142–43, 152
  psalm setting in, 106, 107*f*
Collection of Highland Vocal Airs, A (MacDonalds and Young), 13–14, 178n.85, 197–98, 254–55
Collett, John, 21
colonialism, 5, 12, 13–14

278   INDEX

comma (temperament), 58–59, 67, 72–76, 75f, 87, 121. *See also* Pythagorean tuning; tuning
Common Sense Realism, 127–28
concord. *See* consonance
conjectural history, 158, **162, 162n.27,** 178–79, 195, 200. *See also* universalism
consonance, 77–79, 82, 84–85, 121–22, 137
Corri, Domenico, 6–7
Cotes, Roger, 29
Crotch, William, 179–81, 229, 229f

Dalyell, John Graham, 203, 250
dances (musical genre), 94, 147–48n.81, 172–73, 181, 182f, 183–84, 206
Delftfield (pottery company), 31–32
Descartes, René, 2–3, 127–28, 147–48n.81, 170n.56, 176n.78
difference tones. *See* "implied sounds"
discord. *See* dissonance
dissonance, 56, 77–78, 79–81, 84–85, 105, 121–24, 137, 233–34
"double meaning" (*double emploi*), 68f, 68–69, 68f, 81. *See also under* Rameau
Duncan, Gideon, 99, 102–3

Edinburgh, 100–1, 103, 163–64, 205–6. *See also* Musical Society; Royal Society of Edinburgh
embodiment. *See* rhythm: embodied approach to
English Book of Common Prayer, 94–95
enjoyment. *See* musical pleasure
Enlightenment. *See* Scottish Enlightenment
Erskine, Thomas, 191–93, 193f
Essay towards a Rational Theory of Music (Holden), 1–2, 3–4, 9, 13, 18, 42–91, 135–42, 145, 150, 151, 153–54, 225
  Authorship of, 19
  circulation of, 249–50
  mathematics in, 27–29
  plagiarism of, 42
  psalmody in, 8, 93–94, 106–8, 109, 250
  in Walter Young's "Essay", 169, 170
ethnocentrism, 254–55, 255n.22
Euler, Leonhard, 251–52

factorization, 46, 63–65, 66f, 68f, 89–90, 168–69, 168n.49
Fawcett, Isabella. *See* Holden (family): Isabella Fawcett
Fawcett, James, 33–34
Fechner, Gustav, 87
feminist critique, 10–11

feral children, 179–80
Ferguson, Adam, 5–6, 13, 255
Fétis, François-Joseph, 3–4, 18, 19, 203–4
fifth (scale degree)
  "grave," 74
  and seventh relationship, 60, 61f, 62f, 62f
figured bass, 113–14
first principles, 9, 43–44, 56–57, 127–28
Forkel, Johann Nikolaus, 3–4, 156–57
fourth (scale degree)
  "grave", **58–59,** 59n.51, 68f, 70–71, 72–74
  "perfect", 58–59, 68f, 70–71, 72–74
  See also "double meaning" (*double emploi*)
fugue, 193–94
fundamental bass (Holden; Anne Young), 60–66, 66f, 66f, 68f, 69, 88–89, 88f, 136, 226–27
  double, 77–85, 80f, 88f, 89, 90, 137–38, 235–36
  implied, 116–17, 117f
  progressions, 64, 65–66, 69, 114, 118–19, 136, 140
  projected, 62f, 63, 67
  See also under Rameau

games. *See* pedagogical games
Gaultier, Aloïsius (abbé), 6, 219, 220
General Assembly (Scotland), 96, 104
Gentleman's Diary or The Mathematical Repository, 7, 38–40
  Holden family members contributions, 26–27, 30, 31–32
Gestalt, 42
Gilson, Cornforth, 100–1, 103
Glasgow, 100–1, 163
  Cathedral, 92–93, 113–14
  Literary Society of, 160, 167, 170
Glasgow College and University, 6–7, 13, 21–24, 92, 101–2, 103–5, 135, 142–43
  chapel choir of, 103–5, 110
Grant, Archibald (of Monymusk), 98–99, 100n.26
Greek music. *See* Ancient Greek
grouping, 9, 43, 45, 47f, 64–65, 85, 89–90, 176, 225
  nested, 45, 60–63
  of phrases, 49, 110, 112, 112f, 142
  of pitch, 55–57, 61f, 67
  of rhythm, 48, 88–89, 142, 145–47 (*see also* subjective rhythmicization)
  See also rhythm: dividing and compounding
Gunn, Anne. *See* Young, Anne
Gunn, John, 6–7, 203, 208–12, 213, 250
  Essay, 203, 210–11
Gunter's Line, 28

INDEX 279

Hanslick, Eduard, 125–26, 155
harmonic progressions. *See* fundamental bass: progressions
Harrison, John, 40–41
Hartley, David, 128–29, 150–51
Haydn, Joseph, 6, 184n.107, 184, 185–93, 187*f*, 189*f*, 191*f*, 206
Helmholz, Hermann von, 42–43
Herder, Johann Gottfried, 162
Holden (family), 25, 30–31, 33–34, 35–36, 37–40
    Francis (I, father), 30
    Francis (II, brother), 26, 30
    Francis (III, nephew), 30–31, 38–40
    George (brother), 29–30, 38–40
    Isabella (née Fawcett, wife), 24–25, 24*f*, 33–34, 35, 36–37, 37*f*
    John Jr., 20n.10, 21–22, 25–26, 35–36, 37–38
    Richard (brother), 29–30, 33, 38–40
    Richard (son), 25, 34, 38, 39*f*
Holden Almanack and Tide Tables, 29, 31, 38–40
Home, Henry. *See* Kames, Lord
Hume, David, 5–6, 8, 99, 127–29, 160, 161–62
Hutcheson, Francis, 5–6, 8
Hutton, Charles, 38–40, 245
hymns, 9, 94–95

imitative arts, 125
implied sounds (Holden), 82–85, 82*f*, 83*f*, 84*f*, 88*f*, 89, 90
instrumental music, 125, 148, 149–50, 153–55, 183–84
integer ratios, 1–2, 67n.66, 91, 145, 146*f*
interval ratios. *See* proportions
Introduction to Music (Anne Young), 169, 203, 212–13, 222–38 passim 228*f*, 233*f*, 236*f*, 242, 250, 252–53
isochronous parcels, 44–51, **52–53**, 67, 87–88, 110
    compound parcels, 45–46, **47**, 48–49

Jardine, George, 160, 170
Jefferson, Thomas, 249–50
Just-noticeable difference, 43, 87

Kames, Lord, 5–6, 8
key note. *See* tonic
Kirnberger, Johann Philipp, 48, 169, 199, 247
knowledge acquisition, 203–4, 220, 223, 224–25
Knox, John, 94–96
Koch, Heinrich Christoph, 199
Kollmann, Augustus Frederic Christopher, 3–4

Leibniz, Gottfried Wilhelm, 2, 132
linguistics, 16, 114, 122–23, 151
    analogy with music, 151–52, 153–54
    See also poetic meter; prosody; syntax
Locke, John, 127–28, 131, 132, 219
    Love, James, 18–19, 21

MacDonald, Joseph, 13–14, 197
MacDonald, Patrick, 197
McPherson, James, 5–6
major scale
    ascending, 63–64, 65–66, 68–69, 68*f*
    descending, 58–60, 59*t*, 63–65, 66*f*, 66*f*, 68–69, 68*f*
    See also rule of the octave
Malcolm, Alexander, 144–45, 170
Mant, Alicia Catherine, 214–16, 218*f*
Marcet, Jane, 240–41, 245
Maurice, Mary A., 212, 213
Maxwell, John, 6–7
memory, 128–35 passim . *See also under* perception of music
Mercadier de Belesta, Jean-Baptiste, 237–38
meter. *See under* rhythm
Millar, Edward, 95–96
Millar, John, 5–6
minuet. *See* dances (musical genre)
modulation, 52–56, 70, 73*f*, 75*f*, 140, 221, 224, 229–32
module (Holden), **9**, 42–43, **52–53**, 54*f*, 70–71, 88*f*, 139–41, 172
    division of, 58*f*, 85–86
    shared, 70–74, 73*f*, 75*f*
    triple division of, 64–66
monochord, 52, 91
Montagu (family), 13–14, 13–14n.28, 206
Monymusk Revival, 10, 93, 94–103, 105. *See also* Channon, Thomas; Grant, Archibald; part singing: revival of (in Scotland)
Moore, Thomas, 100–1, 103
Moorgate Hall, 38, 39*f*
Mozart, Wolfgang Amadeus, 179–80, 233–34, 234*f*, 247
music cognition, 2–3, 42, 44, 55–56, 57, 85, 86–87, 88*f*, 135–42, 199, 253
Musical Games (Anne Young), 12, 13–14, 15–16, 201, 202*f*, 209–10, 212–22, 226, 233*f*, 237–48, 250–51, 252–53
musical pleasure, 125, 149, 150, 151–52, 163–64, 172–73, 174–75, 183–200, 225
musical ratios. *See* proportions

National Covenant (of Scotland), 96

part singing, 10, 92–93, 103–4, 116–17, 124. See
also *Essay*: psalmody in; psalm singing
revival of (in Scotland), 98, 99–101, 100n.26,
103, 105 (*see also* Channon, Thomas;
Grant, Archibald; Monymusk Revival)
part writing, 8, 95–96, 106
Pasquali, Niccolò, 6–7
pedagogical games, 203, 212–22, 247. See also
Musical Games (Anne Young)
perception, 85–86, 133
levels of, 87–90, 88*f*
top-down and bottom-up modes of, 9, 15–16,
87, 88*f*, 89–90
perception of music, 1–2, 88*f*, 135–42, 149–50,
155, 254
and anticipation, 122, 138–39, 152–53, 252
and attention, 3, 43, 69–76, 88*f*, 90–91, 136,
147–48, 149–50, 154, 167–68, 170 (*see
also* attention)
and expectation, 44, 50–51, 86, 90–91, 103–4,
107–8, 115, 122
and memory, 3, 43, 49, 50, 69–76, 90–91,
138–39, 142
and pitch, 9, 45, 50, 58, 172
and rhythm, 42, 45, 48, 86, 141, 158 (*see also*
subjective rhythmicization)
and tempo and phrase, 110–12, 112*f*, 175–77
and tonality, 60, 69, 70–71
"perfect chord," 77–79, 83, 83*f*, 84–85, 137–38
pitch pipe, 101–3, 113–14
Platonism, 198
poetic meters (and rhythm), 96, 110, 145, 146*f*,
156, 157
dactyl, 179, 180–81
spondee, 179, 180–81, 195
trochee, 195
Presbyterianism, 5, 41, 92–93
part singing in, 10, 41, 43–44
in Scotland, 94–95, 96, 160–61
Preston, Thomas, 206–7
Priestley, Joseph, 128–29
proportions and ratios, 28, 47
musical ratios, 29, 139–40, 170
and pitch, 52–53, 58–59, 65, 66*f*, 72–76, 73*f*,
75*f*, 78–79
and rhythm, 146*f*, 164–65, 166, 175–77, 183,
226
prosody, 157. *See also* linguistics; poetic meters
psalm singing, 10, 96–97
heterophony in, 96–98
"lining out" style, 96–97
"old" vs. "new" way, 97–98, 102–3, 109–10
in Scotland, 93, 95–96, 118, 250

tempo of, 97–98, 109–10
*Twelve Tunes for the Church of Scotland,
The*, 96
Pythagorean tuning, 70–71, 74 (*see also* tuning;
comma)

Rameau, Jean-Philippe, 2–3, 6, 9, 10, 44, 56,
114–17, 124, 251
cadence, 114–15, 237–38
chord of the large sixth, 234
double emploi, 81, 235–36
fundamental bass, 56, 64, 104
Génération harmonique, 68–69, 120
influence on Holden, 118–21
règle de l'octave, 68–69 (*see also* rule of the
octave)
Traité de l'harmonie, 119
Ramsay, Allan, 5–6
ratios. *See* integer ratios; proportions and ratios
Reid, Thomas, 5–6, 8, 9, 10, 12, 21–23, 35–36,
126–35, 139, 152–53, 154–55, 160, 167,
170
on consciousness and perception, **131–32**
Essay on the Intellectual Power of Man, 129,
131–32, 134
Essays on the Active Powers of Man, 132–33
Inquiry Into the Human Mind, An, 129–30,
132
*Observations on Man, His Frame, His Duty,
and His Expectations*, 129–30
Relativism, 10
rhythm, 163–64
analogy of measure and phrase, 173
analogy of rhythm and pitch, 53, 54*f*, 55, 57,
91, 145–47
dividing and compounding (Walter Young),
164–70, 182*f* (*see also* grouping)
embodied approach to, 165, 167, 174, 175,
178, 254
hypermeter, 173–74, 177–78, 185–94, 186*f*,
187*f*, 189*f*, 191*f*, 192*f*
measure deduction (Walter Young), 171–75,
182*f*
meter, 172–73, 177–78, 181–83
and pleasure, 177–200
See also perception of music: and rhythm;
proportions and ratios: rhythm;
subjective rhythmicization
Riemann, Hugo, 42–43, 68n.70, 251
Riepel, Joseph, 199
Robertson, Thomas (reverend), 10, 141, 142–48
Inquiry Into the Fine Arts, 6–7, 171, 198n.139
Robertson, William (reverend), 5–6, 160–61

INDEX 281

Robison, John, 6–7, 40–41, 160
Roscoe, William, 30–31
Rousseau, Jean-Jacques, 40–41, 162, 254n.19, 254n.20
Rowse, Elizabeth, 214–16, 215f, 216f
Royal Society of Edinburgh, 10, 143, 156–57, 160–61
rule of the octave, 226. *See also* major scale; Rameau, Jean-Philippe: *règle de l'octave*
Rule of Three, 27–28

St. Augustine, 1
St. David's (psalm tune), 92–93, 106, 107f, 108f, 111–12, 111f, 111f, 112f
Scott, Henry (Duke of Buccleuch), 142–43
Scott, Sir Walter, 5–6, 102–3, 238–39
Scottish Enlightenment, 5–6, 10, 127–28, 154–55, 156, 158–60, 161–62, 167–68, 178n.87, 251, 254–55
Scottish Psalter, The, *See* psalm singing: in Scotland
Scottish Reformation, 20–21, 95–96, 105, 161–62
Scottish School of Common Sense. *See* Common Sense Realism
Semphill, Lady, 206, 207f, 208f
Semple, Elizabeth, 242
Serre, Jean-Adam, 6, 79, 235–36
Simpson, Thomas, 38–40. See also *Gentleman's Diary or The Mathematical Repository*
sixth (scale degree), 68–69, 68f, 68f, 70–71
Smith, Adam, 5–6, 10, 11n.19, 12, 125–27, 142–43, 149–55, 160, 170
Smith, John, 229, 231f
Stamitz, Johann, 172–73, 173f, 185, 186f
Steele, Joshua, 165–66
Stevens, Charles Isaac, 42–43
Stewart, Douglas, 133, 154n.103, 162
subdominant, 234–37, 236f
subjective rhythmicization, 9, 43, 48, 56, 141, 144, 169, 171, 199, 252. *See also* grouping: of rhythm; perception of music: and rhythm;
Sulzer, Johann Georg, 3–4
syntax, 85–86, 90, 123. *See also* linguistics

tactus, 175–77
Tans'ur, William, 102
Tartini, Giuseppe, 6, 237–38
Thom, William (reverend), 101–2, 113–14
Thomas, E.E., 221
tide-table. See *Holden Almanack and Tide Tables*
Tolbooth (Glasgow), 19, 20f, 20–21
tonality, 49–50, 51, 106–8, 136, 140
tonic, 56, 63, 70–76, 75f, 106–7, 113, 114–16, 136–37, 139–40, 234
"final" and "medial," 68f, 72, **72n.83**
memory of, 50–51, 67–69, 139, 142, 150
tuning, 70–71, 74, 76, 229–32. *See also* comma; Pythagorean tuning

universalism, 10, 156, 158–60, 161–64, 200, 254–55. *See also* conjectural history; relativism

Vicentino, Nicola 251–52

Wallis, John, 221
Wars of the Three Kingdoms, 94–95
watch ticks. *See* clock chimes
Watt, James, 5–6, 40–41
Weber's Law, 87
Wesley, Samuel, 179–80, 239
Wilberforce, William, 12, 38
women writers, 206, 238–48 passim

Yorkshire, 7–8
Young, Anne, 4, 10–11, 201–48, 249–50, 252–53
Elements of Music and Fingering the Harpsichord, 206, 207f, 208f, 209f, 227
Musical Cards, 203, 222
See also *Introduction to Music*; *Musical Games*
Young, David, 158–59, 158n.11, 197, 205
Young, Thomas, 156–57
Young, Walter (reverend), 4, 10, 11, 13–14, 154n.103, 156–200, 205–6, 211, 243–45, 252–54